Springer Water

The book series Springer Water comprises a broad portfolio of multi- and interdisciplinary scientific books, aiming at researchers, students, and everyone interested in water-related science. The series includes peer-reviewed monographs, edited volumes, textbooks, and conference proceedings. Its volumes combine all kinds of water-related research areas, such as: the movement, distribution and quality of freshwater; water resources; the quality and pollution of water and its influence on health; the water industry including drinking water, wastewater, and desalination services and technologies; water history; as well as water management and the governmental, political, developmental, and ethical aspects of water.

G. M. Tarekul Islam · Sonia Binte Murshed ·
Shammi Haque
Editors

Sustainable Water Management and Environmental Resilience

Springer

Editors
G. M. Tarekul Islam
Institute of Water and Flood Management (IWFM)
Bangladesh University of Engineering and Technology (BUET)
Dhaka, Bangladesh

Sonia Binte Murshed
Institute of Water and Flood Management (IWFM)
Bangladesh University of Engineering and Technology (BUET)
Dhaka, Bangladesh

Shammi Haque
Institute of Water and Flood Management (IWFM)
Bangladesh University of Engineering and Technology (BUET)
Dhaka, Bangladesh

ISSN 2364-6934 ISSN 2364-8198 (electronic)
Springer Water
ISBN 978-3-031-93349-3 ISBN 978-3-031-93350-9 (eBook)
https://doi.org/10.1007/978-3-031-93350-9

This Springer imprint is published by the registered company Springer Nature Switzerland AG
The registered company address is: Gewerbestrasse 11, 6330 Cham, Switzerland

Preface

Water resources, environmental sustainability, and community resilience are critical global challenges that require innovative approaches and solutions. This book "*Sustainable Water Management and Environmental Resilience*" presents a collection of selected papers from the 9th International Conference on Water and Flood Management (ICWFM), held from October 14 to 16, 2023, in Dhaka, Bangladesh, highlighting research efforts aimed at tackling key challenges in water management, sanitation, ecosystem conservation, and climate adaptation. The book addresses broader hydrological and environmental concerns, including the hydrologic alteration of major rivers, pollution in coastal waters, and the impact of infrastructural projects on water quality. The studies on erosion monitoring, water security, and the effectiveness of nature-based solutions emphasize the synergy between technology, governance, and community-driven interventions.

We firmly believe that this collection will be beneficial to researchers, policymakers, and practitioners working in water resources management, environmental conservation, and sustainable development. We hope that this book will inspire further research and innovation in addressing critical challenges, particularly in vulnerable regions like Bangladesh and beyond.

We extend our sincere gratitude to the members of the Scientific Committee and the reviewers for dedicating their valuable time to reviewing the papers. Our appreciation also goes to Springer Nature for publishing this volume. This serves as a strong incentive for early-career researchers to participate in the conference and showcase their work.

Dhaka, Bangladesh

Prof. G. M. Tarekul Islam
Dr. Sonia Binte Murshed
Dr. Shammi Haque

Contents

Chapter 1
Assessment of Hydrologic Alteration for the Ganges River in Bangladesh: A Pre and Post Farakka Barrage and Treaty Scenario

Md. Reaz Akter Mullick, Shyamal Acharya, Md. Hazrat Ali, Md. Abu Sayed, and Saima Zaheen

Abstract This study investigates the hydrological alteration of the Ganges River in the context of three distinct periods: pre-Farakka (1960–1975), post-Farakka (1976–2019), and the post-Ganges Water Sharing Treaty (GWT) period (1997–2019) using Indicators of Hydrologic Alteration (IHA) software. The Ganges River, shared by India, China, Nepal, and Bangladesh, is vital for the socio-economic and environmental well-being of the region. The Barrage at Farakka that was constructed in 1975 altered the natural flow of the Ganges, impacting downstream riparian country Bangladesh. Despite a water sharing agreement in 1977 and subsequent treaties, unilateral withdrawals by India persisted, leading to major impact to the downstream riparian as well as river health. With the historical discharge data of the Hardinge Bridge point the IHA software was used to assess the hydrologic alteration for the Ganges River in Bangladesh. Results show that some improvement has been achieved in post treaty period regarding river hydrology. The Hydrologic Alteration (HA) in pre-treaty period was 59 whereas in post-treaty period the HA is 55. However, Environmental flow parameters are suffering significantly in the post Farakka period.

Md. R. A. Mullick (✉) · S. Acharya · Md. H. Ali · Md. A. Sayed
Department of Civil Engineering, Chittagong University of Engineering and Technology (CUET), Chittagong, Bangladesh
e-mail: reazmullick@cuet.ac.bd

S. Acharya
e-mail: acharya_shyamal@cuet.ac.bd

Md. H. Ali
e-mail: contact@enghazrat.com

Md. A. Sayed
e-mail: hello@sayed.page

S. Zaheen
University of Texas at El Paso (UTEP), El Paso, TX, USA
e-mail: saimalaboni003@gmail.com

G. M. Tarekul Islam et al. (eds.), *Sustainable Water Management and Environmental Resilience*, Springer Water, https://doi.org/10.1007/978-3-031-93350-9_1

The findings contribute valuable insights for policymakers seeking to balance water sharing, economic needs, and ecological sustainability in the Ganges basin.

Keywords Ganges river · Farakka barrage · Hydrologic alteration · Ganges water sharing treaty · Environmental flow

1.1 Introduction

The Ganges basin is located in between 70° E to 88° 30′ E and 21° N to 31° N [13] and shared among four countries, with India having the highest percentage at 79%, followed by Nepal at 14%, Bangladesh at 4%, and China at 3% [12]. Both the monsoon-driven precipitation during July–October and snow melting in Himalayas during the dry season (April–June) contribute to the Ganges flow [16]. Flowing down from Nepal, the Ganges is feed from four major tributaries namely, the Mahakali, the Gandak, the Kosi and the Karnali. According to Islam [10], these tributaries account for approximately 41% of the Ganges annual flow and about 71% of its dry season flow. Rising on the southern slopes of the Himalayas at Gangotri, the river flows east through the Indian states of Uttar Pradesh, Bihar and West Bengal to the border of Bangladesh. Approximately 407 million people from China, India, Nepal, and Bangladesh rely on the Ganges for their livelihoods and daily needs [5].

The basin share of Bangladesh is very small compared to India and Nepal; however, both for socio-economically and environmentally the country critically depends on the river. The unceasing supply of water from the upstream transboundary rivers (54 entering from India and 3 from Myanmar) is critically important for agricultural, industrial and domestic purposes; for maintaining river morphology, maintaining forestry and fisheries; and for conserving balance of sea water intrusion from the Bay of Bengal [12]. Dry season flow is more critical for these issues. The Ganges River contributed significantly to the formation of Bangladesh by depositing silt (coarse sediment approximately 190 million tons/year) and formed the values and culture of its people for ages [18]. The river also supports the endangered species such as Ganges River dolphin (Platanista gangetica) along with other important flora and fauna [5]. The Ganges River and its flow is essential for maintaining the environment and ecology in the southwest area of Bangladesh, supporting the livelihoods of approximately 30 million people in that region [18]. Particularly in dry season, the Ganges depended area of Bangladesh namely the districts of Kushtia, Pabna, Jessore, Khulna and Faridpur depend on Ganges flow for irrigation water [14]. Of the total annual flow of Ganges, 80% occurs during the monsoon period and thus huge groundwater subtraction is observed during the dry period [17]. The Ganges flow is also important to retard salinity in the southwest region of the country along with and the Sundarbans region [2].

Until 1975 meaning that the period before the commissioning of the Farakka Barrage, the Ganges flow can be considered was natural and during dry season it was sufficient. In order to sustain adequate navigability at the Kolkata port, the Farakka

Barrage was commissioned on April 21, 1975, located in Murshidabad, West Bengal, India. The Farakka point is 17 km upstream from the India-Bangladesh border [10]. During the dry season, the dam diverts 1133 cumec of water from the Ganges to the Hooghly River, providing salinity-free water supply for Kolkata city and ensuring proper navigation at Kolkata port. However, the hydrology of the Ganges became changed abruptly after commissioning the barrage. At the Hardinge Bridge point in Bangladesh, discharge data from 1934–1995 was analysed, revealing that average dry season (January–May) flow was 2340 cumec for the pre-barrage period whereas it is approximately 1236 cumec for the post barrage time [13]. Ganges water diverted unilaterally by India during the dry season, the government of Bangladesh sought a negotiation. After extensive discussions, the first water-sharing agreement was signed on November 5, 1977, lasting only five years. This agreement allotted 60% of the Ganges flow at Farakka to Bangladesh and included a guarantee clause. After its expiration in 1982, two Memorandum of Understandings (MOUs) were adapted which did not include any guarantee clause and finally lasted in 1988. India continued its hegemony over Ganges water sharing and the lowest diurnal flow from 1989 to 1995 at Hardinge Bridge was noted to be 267 cumec on April 6, 1993 during no-treaty period [18].

The significant reduction in flow in the Ganges and its tributaries led to noteworthy economic and environmental impacts to the southwestern region of Bangladesh. The situation prompted the government to reengage in discussions regarding water sharing with India. Ultimately, on December 6, 1996, a 30-year treaty was signed with India concerning the sharing of the Ganges water during the dry season. According to the treaty, India and Bangladesh each are guaranteed 991 cumec (35,000 cusec) of water in alternating three 10-day periods from March 11 to May 10 (Article II), as detailed in annexure I [8].

The treaty does not contain any dispute settlement mechanism and also there is no stating of guarantee clause like 1977 agreement [13]. The treaty, valid for thirty years, may be a pioneering step toward resolving the long-term dispute over the Ganges flow in the dry season. Analysis shows that while India generally abides by the treaty, it unilaterally withdraws water during critical dry periods, affecting regional livelihoods [20]. In the Post-Farakka (1976–2015) time period the maximum, average, and minimum flow have decreased by about 22%, 48%, and 72%, respectively particularly in the dry season months of January–May and after the Ganges Water Treaty (1997–2015), these reductions are around 23, 43, and 65% [14]. This indicates minimal improvement in Ganges flow conditions in Bangladesh in post-treaty period.

Hydrologic alteration is characterized as any adjustment in the timing or quantity of natural river flow and stream flow patterns, often driven by human activities and climate change [5]. The construction of barrages and dams represents a primary source of such anthropogenic hydrologic changes. According to Ghanbarpour et al. [6], the scale of dam construction and associated water diversion activities is substantial, resulting in numerous environmental impacts at both regional and local levels, as exemplified by the Farakka Barrage effects on the southwestern region of Bangladesh. Ganges river basin flow has been observed as manipulated both in upstream and downstream region due to several human induced diversions and activities. To understand these alterations there are many available literatures which are basically qualitative approaches and give an overall idea on the existing situation of the flow regime. In this context, we attempt to evaluate this manipulation in quantitative terms so that this numerical approach of Indicators of Hydrologic Alteration (IHA) would assist the policymakers to take decisions regarding the allowable extent of hydrologic alteration. In fact, this study is the first to adopt this simple and systematic method of IHA for Ganges River flow alteration. This paper adds to the current literature on the Ganges basin by evaluating the hydrological implication as well as environmental repercussion in Bangladesh caused by India's unilateral water withdrawal at the Farakka barrage. Discharge data (1960–2019) from the Hardinge Bridge station in Bangladesh were analysed using the IHA methodology. The IHA approach, widely used to assess eco-hydrological impacts from flow regime changes due to river management and climate change, employs 33 indicators over five categories: magnitude, duration, timing, frequency, and rate of change. This research evaluates Ganges environmental flow management by assessing five essential flow components critical to river ecosystem health: extreme low flows, low flows, high-flow pulses, low floods, and large floods.

Based on the available historic data of Ganges flow, this research contains two specific objectives; firstly, to assess the hydrologic alteration of post-Farakka (1976–2019) period compared to the pre-Farakka period (1960–1975) on Ganges River hydrology by using the IHA (Indicator of Hydrologic Alteration) tool. Secondly, to identify the rationality of Ganges water sharing treaty by comparing discharges of post-GWT with pre-Farakka and post-Farakka but pre-GWT period.

1.2 Methodology

1.2.1 Study Area

The Ganges is one of the main rivers of the Indian subcontinent originating at a height of about 10,300 feet in Gangotri glacier on the slope of the Himalayan mountains in the Uttar Pradesh, India [18] and flowing about 2510 km on the east direction through a vast terrain to the Bay of Bengal [5]. On its way towards the sea, many tributaries contribute to its flow from Nepal and India.

After leaving Uttar Pradesh, the Ganges flows through Bihar, Rohtas and West Bengal province [13] (Fig. 1.1a). It starts to flow south after entering West Bengal and divides into two arms. The left channel flows eastward and enters Bangladesh as shown in Fig. 1.1b and the right channel flows south continuing as the Bhagirathi in West Bengal. On the bank of Hooghly (name of Bhagirathi flowing south of Kolkata) river the Kolkata port is located. Two tributaries join Ganges at the downstream of Farakka—Mahananda and Baral [18]. In Bangladesh, the Gorai, being the primary distributary, departs the Ganges around 65 km above the confluence of the Ganges

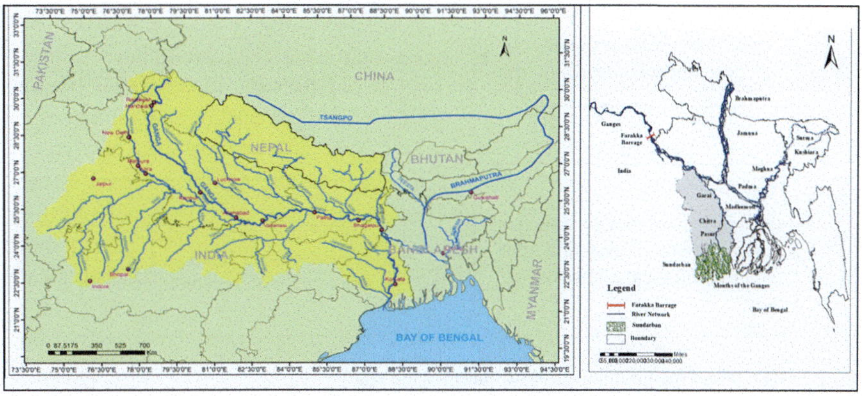

Fig. 1.1 **a** The Ganges River Basin [3]. **b** The flow path of Ganges River in the downstream with the location of Farakka Barrage

and Brahmaputra rivers [12]. The confluence of the Ganges and Brahmaputra rivers, collectively known as the Padma, occurs near Chandpur, where it joins the Meghna. The accumulated flow of the three big rivers continues with the name of the Meghna and finally empties into the Bay of Bengal. According to Chowdhury and Ward [4], the total catchment area of the Ganges Basin is 907,000 km^2, with Bangladesh accounting for 45,548 km^2 (4%) of the total area. Annual rainfall in Uttar Pradesh (upper Gangetic Plain) ranges from 760 to 1020 mm, while in Bihars (Middle Ganges Plain) it ranges from 1020 to 1520 mm, and in the delta region it ranges from 1520 to 2540 mm [21]. The average annual precipitation of the basin varies from 800 to 1200 mm [9] and the average annual discharge from (1949–73) is 12,105 cumec. The river is mainly monsoonal 80% of total annual flow occurs in monsoon [11] and supports the fishing industry which contributes to the economy of the basin area greatly. At Farakka the silt load of the Ganges is 1235 ton per square km per year [1] that makes the soil alluvial and fertile for producing various types of crops. Thus, a large number of people earn their livelihood depending on the river. It also supports important plant species which take an essential part in water conservation and soil erosion control [5]. The non-monsoonal water crisis increased largely after the commissioning of Farakka dam as evident from several studies.

1.2.2 Data and Method

Using April 1–March 31 as a water year, daily discharge data (1960–2019) data for the Ganges River as acquired from the Bangladesh Water Development Board (BWDB) for the point of Hardinge Bridge station was analysed. The quality of BWDB data is high since it is used for preparing various action plans such as Flood Action Plan (FAP), National Water Plan (1986) and many other reports. Owing to

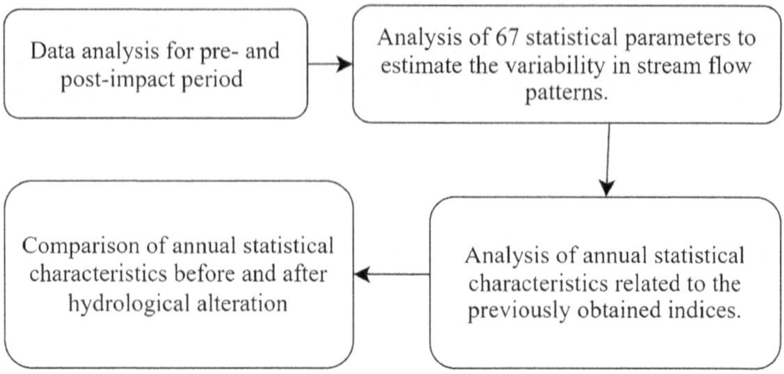

Fig. 1.2 IHA based methodological flow chart for the study

liberation war in Bangladesh, discharge data from April, 1971 to March, 1972 is missing and discarded from the analysis.

The Indicators of Hydrologic Alteration (IHA) software [19] has been employed in this study to assess the extent of hydrologic alteration in the Ganges basin. The software was employed in over 30 hydrological studies conducted in Australia, the United States of America, South Africa, and Canada. The evaluation of IHA can be conducted following the 4 steps as given in Fig. 1.2.

Following these steps from Fig. 1.2, it is possible to quantitatively analyse the effect of hydrological alteration caused by the anthropogenic activities like the construction of a barrage or reservoir. In this research, the alteration in the stream flow pattern was statistically investigated using the IHA recommended by [15].

The flow parameters of IHA are characterized into five regime groups consisting the magnitude, timing, frequency, duration, and rate of change [19]. These characteristics are further classified into 5 IHA statistical groups which are mentioned in Table 1.1.

When evaluating alterations between two-time frames, the IHA software facilitates the use of the Range of Variability Approach (RVA). This approach helps maintain the distribution of annual IHA attribute values as close to pre-development variations as possible [15]. During RVA analysis, pre-impact data are divided into three categories: the lowest category (values $\leq$ 33rd percentile), the middle category (values between the 34th and 67th percentiles), and the highest category (values > 67th percentile) [19].

The IHA computes five different parameters of Environment Flow Components (EFCs): extreme low flows, low flows, high flow pulses, small floods, and large floods as given in Table 1.2. This parametric allocation of EFCs is based on the understanding by ecological research that indicates that river hydrographs can be distributed into a repeating set of hydrographic patterns relevant to ecological health and sustainability [19].

Table 1.1 Summary of hydrologic parameters used in the IHA and their characteristics according to [19]

IHA parameter	Hydrologic parameter
Group 1. Monthly water conditions by magnitude	Mean flow for each calendar month
Group 2. Annual extreme water conditions by magnitude and duration	Annual minima, 1-day, 3-day, 7-day, 30-day and 90-day mean Annual maxima, 1-day, 3-day, 7-day, 30-day and 90-day mean Number of days observing zero-flow Base flow index: 7-day minimum flow/mean flow for a year
Group 3. Annual extreme water conditions by timing	Julian date of each annual 1-day maximum and 1-day minimum
Group 4. High and low pulses (frequency and duration)	Number of high and low pulses within each water year Mean or median duration of high and low pulses (days)
Group 5. Water condition changes (rate and frequency)	Rise and Fall rates: Mean or median of all positive differences between consecutive daily values Number of hydrologic reversals

Table 1.2 Summary of environmental flow component (EFC) parameter

EFC type	Hydrologic parameters
Group 1. Monthly low flows	Mean or median values of low flows during each calendar month
Group 2. Extreme low flows	Frequency of extreme low flows during each water year or season
Group 3. High flow pulses	Frequency of high flow pulses during each water year or season
Group 4. Small floods	Frequency of small floods during each water year or season
Group 5. Large floods	Frequency of large floods during each water year or season

The high flow threshold is 75th percentile of daily flows and all flows equal to or less than this threshold is categorized as low flows. All flows equal to or less than 50th percentile of daily flows which is the low flow threshold are categorized as low flow events [7].

1.3 Results

1.3.1 The Ganges Flow Characteristics

Figure 1.3 shows that the monthly median flows show a rising pattern from May to August and a declining pattern from September to April for the Ganges basin. The monthly median flows are lower during the post-dam period for all months except July, September and October.

Based on this analysis (Table 1.3), the findings suggest that the High-Season flow months are July, August, September, and October. These months have the highest average flow, exceeding 10% of total flow. The intermediate-season flow months are June, November, and December, which have a moderate flow, ranging from 3 to 10% of total flow. The Low-Season flow months are January–May, which have the lowest average flow, falling below 3% of total flow.

The bulk of the Ganges River's flow occurs during the high Season flow months, which are followed by the intermediate and low Season flow months, respectively.

Table 1.4 gives a summary of the changes in discharge at the Hardinge point of the Ganges over the study period.

Table 1.4 and Fig. 1.4 show the mean monthly flow of the Padma River in cubic meters per second (m^3/s) for three periods:

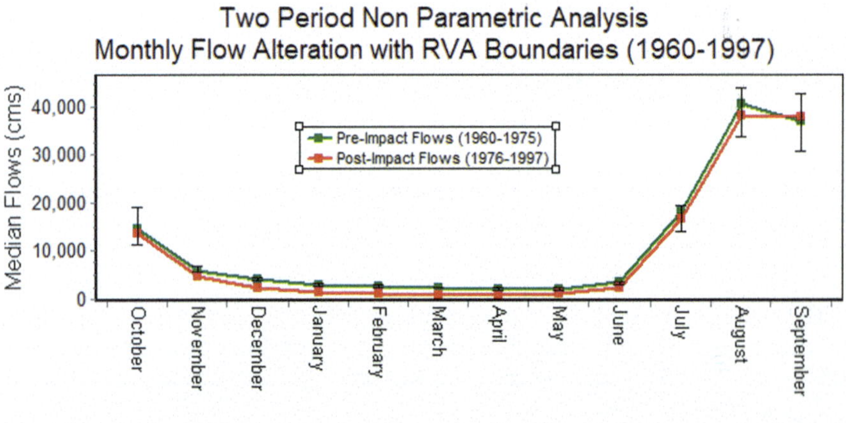

Fig. 1.3 Pre- and post-barrage monthly median flow at Hardinge station point

Table 1.3 Percentage of flow based on season

Flow category	Months	Total flow (%)
High season flow	July–October	80
Intermediate season flow	June, November, December	11
Low season flow	January–May	9

Table 1.4 Changes in Ganges river discharge

Season	Month	Pre-Farakka (1960- 1975)			Post-Farakka (1976–1996)			Post-GWT (1997–2019)		
		Period I			Period II			Period III		
		Mean	Maximum	Minimum	Mean	Maximum	Minimum	Mean	Maximum	Minimum
Low flow	January	3077	4240	2270	1661	2810	1220	2869	8031	1095
	February	2621	3650	1945	1238	1785	486	2071	5508	924.6
	March	2294	3310	1530	978.8	1700	311	1767	5922	540.9
	April	2108	2630	1515	938.7	1755	381	1768	6187	662.4
	May	2212	3000	1610	1349	2160	554	2060	6468	813.3
Intermediate flow	June	4209	6880	2090	3405	6320	702.5	4216	7530	1425
	November	6648	11,350	2297	5193	10,550	2685	7080	15,970	2746
	December	4101	6030	2313	2699	4170	2010	4286	14,760	1732
High flow	July	16,340	26,700	9880	17,420	35,900	5120	19,940	33,460	6403
	August	33,540	49,600	22,000	35,380	57,600	20,900	32,580	46,340	8824
	September	32,830	62,350	17,700	36,010	56,850	16,650	30,470	54,110	11,210
	October	17,250.00	43,600	7440	15,030	38,800	7810	16,670	29,580	4890
Season	Month	Percent (%) changes in period I & II			Percent (%) changes in period I & III			Percent (%) changes in period II & III		
		Mean	Maximum	Minimum	Mean	Maximum	Minimum	Mean	Maximum	Minimum
Low flow	January	46.02	34	46.26	6.76	– 186	51.76	42.11	65	– 11.42
	February	52.77	51	75.01	20.98	– 209	52.46	40.22	68	47.44
	March	57.33	49	79.67	22.97	– 248	64.65	44.61	71	42.5
	April	55.47	33	74.85	16.13	– 253	56.28	46.91	72	42.48
	May	39.01	28	65.59	6.87	– 199	49.48	34.51	67	31.88
Intermediate flow	June	19.10	8	66.39	– 0.17	– 19	31.82	19.24	16	50.7

(continued)

Table 1.4 (continued)

Season	Month	Percent (%) changes in period I & II			Percent (%) changes in period I & III			Percent (%) changes in period II & III		
		Mean	Maximum	Minimum	Mean	Maximum	Minimum	Mean	Maximum	Minimum
	November	21.89	7	− 16.89	− 6.50	− 51	− 19.55	26.65	34	2.22
	December	34.19	31	13.1	− 4.51	− 254	25.12	37.03	72	− 16.05
High flow	July	− 6.61	− 34	48.18	− 22.03	7	35.19	12.64	− 7	20.04
	August	− 5.49	− 16	5	2.86	20	59.89	− 8.59	− 24	− 136.85
	September	− 9.69	9	5.93	7.19	5	36.67	− 18.18	− 5	− 48.53
	October	12.87	11	− 4.97	3.36	24	34.27	9.84	− 31	− 59.71

Note (−) sign indicates flow is increasing, (+) sign means flow is reduced with respect to base flow. Consider Pre Farakka period as base flow for determining % change in (i, ii) and (i, iii) and consider post Treaty period as base flow for determining % change of (ii, iii)

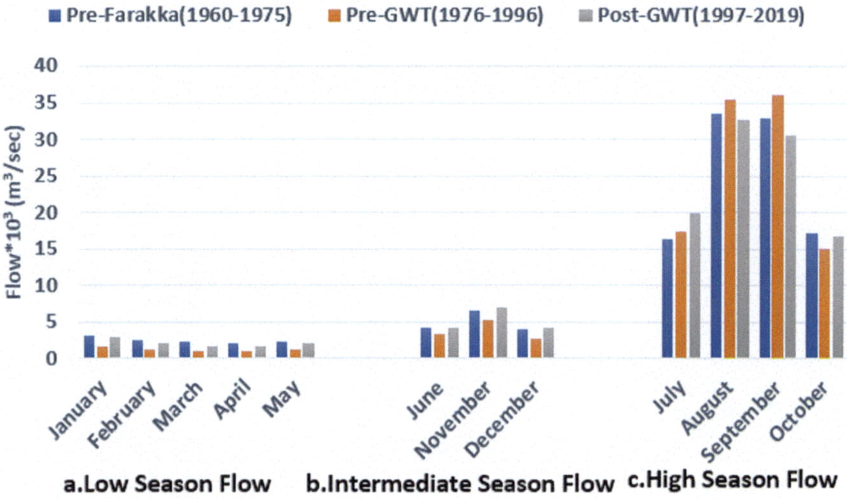

Fig. 1.4 Monthly mean flow analysis

Low Flow Months (Jan–May)

- Pre-Farakka versus Post-Farakka: There was a significant decrease in discharge across all months, with reductions ranging from 39% (May) to 57% (March) in Table 1.4. This reduction confirms the substantial impact of the Farakka Barrage on low-flow months (dry season).
- Pre-Farakka versus Post-Treaty: Similar to the pre-Farakka versus post-Farakka comparison, all months show a decrease in discharge, with percentages ranging from 7% (January) to 23% (March) mentioned in Table 1.4. This reduction highlights the continued challenges of dry-season water availability.

Intermediate Flow Months (Jun, Nov, and Dec)

- Pre-Farakka versus Post-Farakka: Discharge reductions are minor than low-flow months, ranging from a slight decrease of 19% in June to 34% in December, as specified in Table 1.4.
- Pre-Farakka versus Post-Treaty**:** Not similar to the post-Farakka comparison, changes are more pronounced, with increases in June (almost 1%), December (5%), and November (7%) described in Table 1.4.

High Flow Months (July–Oct)

- Pre-Farakka versus Post-Farakka: Discharge increases are observed in all months without October (decrease by 13%), ranging from 6% (August) to 10% (September).
- Pre-Farakka versus Post-Treaty: all months show reductions without July (increased by 22%), reductions ranging from 3% in August to 7% in September. though this reduction is less significant. Almost similar flow as pre-Farakka period.

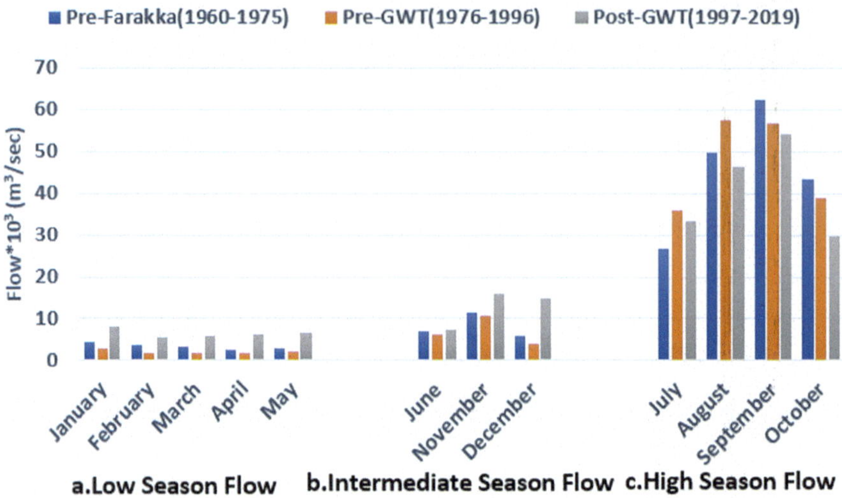

Fig. 1.5 Monthly maximum flow analysis

Overall, the Ganges Treaty has not fully addressed the issue of reduced Padma River flow, particularly during the dry season. While it has shown some positive signs in intermediate and high flow months, significant concerns remain regarding dry season water security and the livelihoods of millions of Bangladeshis.

Pre-Farakka to Post-Farakka: Following the constructing of the Farakka Barrage, the flow of the Padma River has dramatically dropped in all months except the high flow season Fig. 1.5. February has the biggest decline (51%) mentioned in Table 1.4.

Pre-Farakka to Post-Treaty: In comparison to the pre-Farakka era Fig. 1.5, the Padma River's flow has typically risen as a result of the Ganges Treaty during both low and intermediate flow months (January–May, June, November, December). December had the biggest rise, which is recorded at 145%. And flow has been decreased in high flow months without July.

Bangladesh has had positive effects from the Ganges Treaty. Because, it has improved the Padma River's flow in the dry (low flow) and intermediate flow months, which is advantageous for water supply, agriculture, and navigation. Yet, it has decreased (not significant) flow during the months of high flow, which may cause issues with sedimentation and erosion.

The Padma River minimum flow values for the months of January through December are displayed in Fig. 1.6 and Table 1.4 for three different time periods: pre-Farakka (1960–1975), post-Farakka (1976–1987), and post-GWT (1997–2019).

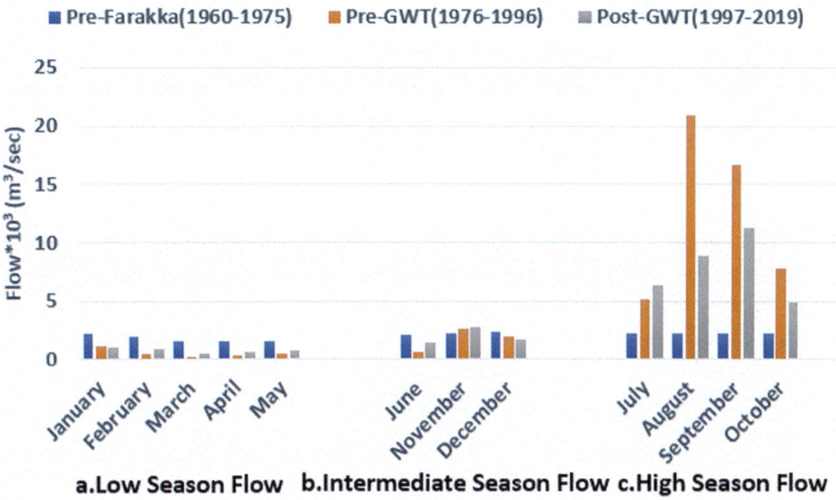

Fig. 1.6 Monthly minimum flow analysis

January, February, March, April, and May Are Low-Flow months

Analysis of pre and post the Farakka: All months had a considerable fall in minimum flows and the months of February (75%) and March (80%) had the biggest declines described in Table 1.4.

Pre treaty versus Post treaty: In general, minimum flows increased after the treaty, except for January (11% decreases) and percentage falls were 47%, 43%, 42%, and 32% in February, March, April, and May respectively (Table 1.4). This reduction is less in amount than post treaty period.

June, November, and December Are the months with Intermediate Flows

Pre-Farakka versus Post-Farakka**:** For the post-Farakka and post-Treaty eras, minimum flows fell in June and December but grew in November specified in Fig. 1.6. However, there is less flow in the post-treaty zone.

July, August, September, and October Are High-Flow months

Except for October, minimum flows fell in every month mentioned in Fig. 1.6 and Table 1.4 prior to and following the Farakka. But the flow always decreases in the post-treaty phase compared to post Farakka and after the treaty described in Fig. 1.6. By analyse Fig. 1.6, The construction of the Farakka Barrage significantly reduced minimum flows in the dry season (February–July and November) and increased them in the wet season (August–October).

However, the treaty has not fully restored flows to pre-Farakka levels, and some months, particularly in the dry season, still show significant reductions. The reduced dry season flows can have a number of negative impacts on Bangladesh, including: reduced agricultural productivity due to water scarcity, loss of biodiversity in riverine ecosystems, reduced hydropower generation, increased wet season flows may also have some negative impacts, such as increased flooding.

1.3.2 Degree of Hydrologic Alteration Analyses

The term "hydraulic alteration" describes how human actions, such as building dams, channelizing waterways, and diverting water, may modify the way water flows through rivers, streams, and other bodies of water.

From the IHA Non-Parametric RVA Scorecard analysis, we obtained the expected and observed values based on 5 parameter groups. We then used simple spreadsheet calculations to determine the degree of hydraulic alteration.

1.4 Discussion

Decreased dry season flow and increased wet season flow result into high flow pulses and floods. The flow over the period ranges from 500 to 40,000 cumec annually. Based on RVA approach (Fig. 1.3) eight months (Jan-June, Nov, Dec) are considered as low flow season where the flow falls under 13,167 m^3/s, which is 33% of highest flow. Only two months are treated as intermediate flow season- July & Oct when the mean flow is within 34–67% and the remaining two months August and September is in high flow season when the flow is more than 67%. It indicates that Ganges flow is not uniformly distributed and prone to monsoonal precipitation.

From the mean flow analysis Table 1.4, Discharge increased in the post-treaty period compared to the post-Farakka period in January (42%), February (40%), March (45%), April (47%), and May (35%) in low flow months and discharge increased in the months of November showing an increase of 26%, followed by June (19%) and December (37%). The hydrologic alteration is 59% for pre- and post-Farakka, and 55% for pre- and post-treaty times (Table 1.5). Hydrologic alteration is also reduced in the post-treaty period. This suggests an improved scenario compared to the immediate post-Farakka period. Though, in high flow months a mixed impact of the treaty was observed where flow has been increased in July (13%) and October (10%) but decreased in August (9%) and September (18%) from Table 1.4.

The study primarily focuses on analysing discharge data from the Hardinge Bridge station to assess the impacts of the Farakka Barrage on the Ganges River's hydrology, which may limit the general findings to other regions or rivers. The research predominantly examines the hydrological changes of the Ganges River in Bangladesh, potentially overlooking the broader transboundary implications of the Farakka Barrage on neighbouring countries sharing the Ganges basin. The paper may lack a detailed discussion on the socio-economic impacts of the Farakka Barrage and water diversion on local communities, fisheries, agriculture, and overall livelihoods in the region. However, these are the crucial aspects to consider in comprehensive

Table 1.5 Degree of hydrologic alteration (HA) analyses

	Comparing post Farakka period pre-treaty (1976–1996) with pre Farakka period (1960–1975) flow			Comparing post Farakka period post-treaty (1997–2019) with pre Farakka period (1960–1975) flow		
	Expected	Observed	Degree of HA in %	Expected	Observed	Degree of HA in %
Parameter Group #1						
January	7.875	1	87	9.471	0	100
February	9.188	0	100	10.82	1	91
March	7.875	0	100	9.471	1	89
April	7.875	0	100	9.471	0	100
May	7.875	1	87	9.471	1	89
June	7.875	1	87	9.471	5	47
July	7.875	7	11	9.471	3	68
August	7.875	7	11	9.471	5	47
September	7.875	9	14	9.471	5	47
October	7.875	9	14	9.471	8	16
November	7.875	2	75	9.471	10	6
December	7.875	1	87	9.471	4	58
Parameter Group #2						
1-day minimum	7.875	0	100	9.471	1	89
3-day minimum	7.875	0	100	9.471	1	89
7-day minimum	7.875	0	100	9.471	1	89
30-day minimum	7.875	0	100	9.471	0	100
90-day minimum	7.875	0	100	9.471	0	100
1-day maximum	7.875	3	62	9.471	5	47
3-day maximum	7.875	4	49	9.471	6	37
7-day maximum	7.875	6	24	9.471	8	16
30-day maximum	7.875	8	2	9.471	9	5
90-day maximum	7.875	10	27	9.471	9	5
Number of zero days	21	21	0	23	23	0
Base flow index	7.875	1	87	9.471	1	89
Parameter Group #3						

(continued)

Table 1.5 (continued)

	Comparing post Farakka period pre-treaty (1976–1996) with pre Farakka period (1960–1975) flow			Comparing post Farakka period post-treaty (1997–2019) with pre Farakka period (1960–1975) flow		
	Expected	Observed	Degree of HA in %	Expected	Observed	Degree of HA in %
Date of minimum	7.875	2	75	9.471	5	47
Date of maximum	7.875	3	62	9.471	3	68
Parameter Group #4						
Low pulse count	17.06	17	0	18.94	13	31
Low pulse duration	7.875	3	62	8.118	5	38
High pulse count	14.44	14	3	14.88	18	21
High pulse duration	6.563	9	37	8.118	6	26
Parameter Group #5						
Rise rate	9.188	1	89	10.82	4	63
Fall rate	9.188	9	2	16.24	14	14
Number of reversals	7.875	0	100	9.471	1	89
	Average alteration		59			55

Note $D(\text{Degree of hydrologic alteration}) = \frac{\text{observed count} - \text{expected count}}{\text{expected count}}$

water resource management studies. The analysis relies on historical data and may not fully capture the current or future impacts of climate change, land-use changes, or other evolving factors that could further influence the hydrology and environmental flow components of the river.

1.5 Conclusion

The study highlights significant changes in hydrology and its components after the construction of the Farakka Barrage, underscoring its profound impact on the river's hydrology and ecology. The Farakka Barrage has disrupted the natural flow regime, affecting downstream regions like Bangladesh and illustrating the environmental consequences of such infrastructure projects. The Ganges Water Sharing Treaty (GWST) was established to address water-sharing disputes between India and Bangladesh, especially during the dry season. Analysis of discharge data from the Hardinge Bridge station showed a reduction in maximum, average, and minimum discharges during the post-GWST period compared to the pre-Farakka period. This suggests that there has been limited improvement in the flow conditions of the Ganges in Bangladesh following the treaty.

The results as obtained from the study highlight the positive impacts from the treaty on the hydrology of the Ganges River particularly in the Bangladesh part benefiting regional water management, agriculture, and socio-economic stability. The tenure of the treaty is for 30 years and that will be over in 2026. Such hydrological analysis of discharge of the Ganges River within Bangladesh shows the current river health as well as improvement of the river health due to treaty. Such analysis will be more informative while discussing the extension of the treaty in coming days. Additionally, we suggest future research areas such as continuous monitoring of the Ganges River flow, assessing ecological effects due to hydrological changes, and analysing socio-economic advantages of enhanced water management strategies.

As the Ganges Water Sharing Treaty approaches its conclusion in 2026, there is a pressing need for its renewal with updated and well-defined regulations. Renewing the treaty with clear guidelines can ensure that Bangladesh receives equitable benefits from the shared water resources. It is essential for both India and Bangladesh to negotiate and implement a renewed treaty that addresses the evolving needs and challenges of the region while fostering cooperation and sustainable management of the Ganges River system.

References

1. Abbas N, Subramanian V (1984) Erosion and sediment transport in the Ganges River Basin (India). J Hydrol 69
2. Bangladesh Water Development Board (2001) Environmental and social impact assessment of Gorai River Restoration Project
3. Chauhan M, Kumar V, Rahul A, Dikshit P, Dwivedi SB (2017) Spatial distribution of the suspended sediments in River Ganga at Varanasi, Uttara Pradesh, India. Int J Earth Sci Eng 10:533–540
4. Chowdhury R, Ward N (2004) Hydro-meteorological variability in the greater Ganges–Brahmaputra–Meghna basins. Int J Climatol 24(12):1495–1508
5. Gain A, Giupponi C (2014) Impact of the Farakka Dam on thresholds of the hydrologic flow regime in the lower Ganges River Basin (Bangladesh). Water 6:2501–2518
6. Ghanbarpour MR, Salimi S, Hipel KW (2013) A comparative evaluation of flood mitigation alternatives using GIS-based river hydraulics modelling and multicriteria decision analysis. J Flood Risk Manag 6(4):319–331
7. Gunawardana SK, Shrestha S, Mohanasundaram S, Salin KR, Piman T (2020) Multiple drivers of hydrological alteration in the transboundary Srepok River Basin of the 1 Lower Mekong Region 2
8. GWST (1996) Treaty between the government of the Republic of India and the government of the People's Republic of Bangladesh on sharing of the Ganga/Ganges Waters at Farakka
9. Haggat K (1994) Rivers of life. University Press Limited
10. Islam MR (1987) The Ganges water dispute: its international legal aspects
11. Kolås Å, Barkved L, Bhattacharjee J, Edelen K, Hoelscher K, Holen S, Jahan F, Jha HB, Miklian J (2013) Water scarcity in Bangladesh. www.prio.org

12. Mirza MMQ (1997) Hydrological changes in the Ganges system in Bangladesh in the post-Farakka period. Hydrol Sci J 42(5):613–631
13. Rahaman M (2009) Integrated Ganges Basin management: conflict and hope for regional development. Water Policy 11:168–190
14. Rahman MM, Rahaman MM (2018) Impacts of Farakka barrage on hydrological flow of Ganges river and environment in Bangladesh. Sustain Water Resourc Manag 4(4):767–780
15. Richter BD, Baumgartner JV, Powell J, Braun DP (1996) A method for assessing hydrologic alteration within ecosystems. Conserv Biol 10(4)
16. Rouillard J, Benson D, Gain A (2014) Evaluating IWRM implementation success: are water policies in Bangladesh enhancing adaptive capacity to climate change impacts? Int J Water Resourc Dev
17. Shamsul Huda ATM (2001) In: Juha AKB, Uitto I (ed) Constraints and opportunities for cooperation towards development of water resources in the Ganges basin. The United Nations University Press
18. Tanzeema S, Faisal IM (2001) Sharing the Ganges: a critical analysis of the water sharing treaties. Water Policy 3(1):13–28
19. The Nature Conservancy (2009) Indicators of hydrologic alteration version 7.1 user's manual
20. Thomas KA (2017) The Ganges water treaty: 20 years of cooperation, on India's terms. Water Policy 19(4):724–740
21. Transboundary River Basin Overview-Ganges-Brahmaputra-Meghna (2011) www.fao.org/

Chapter 2
Isotopic and Hydrogeochemical Evaluation of Ganges and Jamuna Floodplain Aquifers in Bangladesh: Integrated Water Quality Index and Health Risk Appraisal

Md. Moniruzzaman, Hafiz Al-Asad, Hazzaz Bin Hasan, Ratan Kumar Majumder, Shamim Ahmed, Md. Abdul Quaiyum Bhuiyan, and Md. Ariful Ahsan

Abstract Developing countries and the global population face a significant challenge in ensuring an adequate supply of fresh groundwater. Human activities have had a detrimental impact on groundwater, which is a crucial source of fresh water. Currently, there is growing apprehension regarding the presence of nitrates (NO_3^-) and trace metals (As, Cr, Ni, Pb, and Fe) in groundwater due to the extensive use of fertilizers and other human-made sources like sewage or industrial wastewater discharge. Thus, this study assesses hydrochemical characteristics that influence groundwater quality, assessing contamination levels, recharge mechanisms, and potential health risks of the Ganges Floodplain (GF) and Jamuna Floodplain (JF) in Bangladesh. A total of 105 water samples were gathered and analyzed, encompassing chemical parameters, trace metals, and isotopic composition to calculate groundwater quality. The findings indicated that the standard allowable limits WHO are exceeded by the concentrations of NO_3, As, Cr, Ni, Pb, Fe, and Mn in the shallow and intermediate aquifer, while it is lower at the deep Holocene aquifer of GF and JF.

Md. Moniruzzaman (✉) · R. K. Majumder · Md. A. Q. Bhuiyan · Md. A. Ahsan
Isotope Hydrology Division, Institute of Nuclear Science and Technology, Atomic Energy Research Establishment, Bangladesh Atomic Energy Commission, Ganakbari, Savar, Dhaka, Bangladesh
e-mail: monir@korea.ac.kr; monir1.gm@gmail.com

H. Al-Asad
Department of Chemistry, Mawlana Bhashani Science and Technology University, Santosh, Tangail, Bangladesh

H. B. Hasan
Department of Environmental Science, Bangladesh University of Professionals (BUP), Dhaka, Bangladesh

S. Ahmed
Department of Geology and Mining, University of Rajshahi, Rajshahi, Bangladesh

G. M. Tarekul Islam et al. (eds.), *Sustainable Water Management and Environmental Resilience*, Springer Water, https://doi.org/10.1007/978-3-031-93350-9_2

Water types prevalent in the GF and JF are primarily Ca–HCO_3 to Na–HCO_3 types. The results of $\delta^{18}O$ and δ^2H showed how precipitation has recently recharged the Floodplain regions. The PCA was employed to discern pollution sources and controlling factors, revealing that groundwater quality is primarily influenced by geological factors and various anthropogenic activities. The IWQI indicates that groundwater from shallow aquifers is deemed unsuitable for drinking purposes in both the GF and JF. In contrast, deep Holocene aquifers are found to be suitable for drinking. Additionally, the majority of shallow groundwater samples highlight remarkable levels of pollution in JF and GF according to the HPI, HEI, and CD values. It is possible to conclude that the results of a comprehensive investigation will alert decision-makers and the local populace, empowering them to implement efficient management strategies to protect groundwater resources in Bangladesh and mitigate potential health risks for the residents.

Keywords Groundwater contamination · Stable isotope composition · IWQI · Water pollution indices · Health risk assessment

2.1 Introduction

Groundwater stands as a crucial resource, significantly influencing various aspects of life, including residential water supply, agriculture, mining, industry, and urban development, contributing to the economic development of countries similar to Bangladesh [75]. Around the world, groundwater continues to be an essential source of drinking water, meeting approximately 33% of the world's freshwater requirements [96]. Approximately 70% of those living in cities and 95% living in rural areas rely on groundwater for residential and drinking needs in Bangladesh [87]. With the escalating demand for water, more groundwater is being abstracted daily [56].

The quality and quantity of groundwater are generally contingent upon the presence of chemical components, hydrogeological conditions, aquifer properties as well as seasonal variations that play vital roles [91]. However, the issue of groundwater contamination has gained widespread attention globally. In recent years, groundwater contamination has become a serious concern impacting human health and livelihoods in various countries, including Bangladesh. The Ganges, Brahmaputra, and Meghna rivers created enormous floodplains on the surface of the Bengal basin, but these floodplains are complex in terms of geomorphology and structural design. The Jamuna is the Brahmaputra's downstream path after the disastrous flood and earthquake of the 1780s. Large amounts of water are released by the Jamuna at the same time that massive amounts of sediment are brought in annually. The Brahmaputra-Jamuna River moves laterally at a rate of at least tens of metres annually. The most recent Pleistocene border has been buried and the Brahmaputra-Jamuna valley has been filled with Holocene sediments. The hydrogeology and groundwater chemistry near the Jamuna flood plain have been significantly impacted by the sedimentation, stratigraphy, and geomorphology caused by the Jamuna River. This challenge has been exacerbated by rapid population growth, agricultural intensification, industrial activities, and urbanization [26, 41, 43]. Groundwater quality is greatly impacted and usually degrades as a result of point and non-point sources of pollutants, such as anthropogenic activities like agriculture, mining, landfilling, domestic sewage, municipal waste, waste dumping, fast urbanization, and industrialization [3]. Furthermore, geological processes such as rock weathering, dissolution, interaction of rock with water, and soil leaching and environmental changes have a major impact on groundwater quality by adding several harmful trace elements [93].

Several researchers have previously conducted studies on the possible risks to humans health resulting from groundwater contamination in various nations, including Bangladesh [2, 3, 6, 24], India [1, 88], Pakistan [49, 74], China [25, 26], Iran [67], and Ghana [17].

Groundwater quality is critical to human health since it is the most important source of water for human consumption. Groundwater quality and level has been depleted due to excess uptake for human consumption, irrigation, and industrial uses. Most of the people in Bangladesh are associated with agricultural and livestock farming activities. A significant no. of industries like textile, leather tanning, steel, food manufacturing, and battery has been developed in recent time. These releases heavy metal containing effluent to the environment. The fertilizer used in agriculture, manure in farmland activities, and heavy metal from industries can infiltrate into the aquifer during the recharge of the shallow and intermediate floodplain aquifer and can contaminate the groundwater [6]. These undesirable elements play a crucial role in degrading groundwater quality, contributing not only to water scarcity but also posing significant threats to human health due to their toxicity, duration, and carcinogenic properties [39, 51, 94]. For instance, there is a positive correlation between the Mg^{2+} and Ca^{2+} content in drinking water and the incidence of kidney stones [52]. The nitrate content in drinking water is closely associated with infant's blue baby syndrome and various health issues, including vomiting, stomach cancer [70], methemoglobinemia, gastric cancer, congenital disabilities, and gastrointestinal diseases [88, 97]. Furthermore, arsenic contamination in drinking water, especially in the Ganges Plain has led to severe skin diseases [18, 35], significantly compromising the safety of drinking water and hindering social and economic development [42].

Furthermore, prolonged exposure to excess levels of trace elements in drinking water through various pathways, including dermal contact and oral ingestion, may lead to a range of adverse health effects. These effects encompass neurological, cardiovascular, hematological, skin-related issues, kidney problems, bladder cancer, hyperkeratosis, diabetes mellitus, Parkinson's disease, mental and neurological disorders, weight loss, and joint pain [16, 46]. As a result, groundwater quality assessments have drawn a lot of attention from throughout the globe. Various traditional methods, including fuzzy logic, multivariate techniques, hierarchical analyses, and water quality indexes [49, 74], have been employed over the past few decades. Nevertheless, these methods may have certain limitations. The use of the Entropy weight method addresses these limitations by eliminating the influence of subjective factors on water quality parameters and providing rational weights to the parameters. When combined with Water Quality Index (WQI), Entropy weight proves to be an effective tool for quantifying groundwater quality [72, 76, 91]. Numerous indices are used to evaluate pollution, including the Degree of Contamination (Cd), Heavy Metal Pollution Index (HPI), and Heavy Metal Evaluation Index (HEI). Moreover, multivariate statistical techniques like principal component analysis (PCA) and cluster analysis (CA) can be used to quickly address pollution problems in many parts of the world, including Bangladesh. These approaches not only aid in understanding the factors influencing groundwater systems but also serve as effective tools for water quality management. They provide rapid solutions to pollution issues, as exemplified in many regions worldwide, including Bangladesh [15]. The health risk assessment (HRA) model was developed by the US Environmental Protection Agency (USEPA) and is widely used to assess the degree of groundwater contamination and the risk of carcinogenic and non-carcinogenic human health outcomes, respectively [20]. Additionally, the source of recharge and regional flow of groundwater are identified through the measurement of the stable isotopic composition of groundwater [40].

As the most significant source of water for human use, groundwater quality is indispensable to maintaining human health [6]. Some research has been conducted in Bangladesh to determine the quality of water or the health risk. Our study examined the chemistry of groundwater, its source, its quality using an innovative approach called the Integrated Water Quality Index (IWQI), the degree of pollution brought on by human and agricultural activity, and the health risks associated with drinking groundwater simultaneously. Numerous studies have been individually carried out on groundwater quality in the Ganges Floodplain and the Jamuna Floodplain. However, there is a notable absence of comparative analyses between these two regions. Recognizing the critical need for comprehensive water resources management and public health considerations, this study seeks to fill this gap by conducting a comparative evaluation of hydrogeochemical characteristics, groundwater source, recharge mechanism and quality. Identifying contamination sources, mapping contamination distribution, and understanding recharge mechanisms for effective future management and mitigation of health hazards.

2.2 Materials and Methods

2.2.1 Study Area

The study region is divided into two separate areas: the Ganges Floodplain of Chapai Nawabganj and the Jamuna Floodplain of Tangail. Situated in the center region of Bangladesh, close to Dhaka, is the Tangail district, which includes the Jamuna Floodplain (JF). The JF is located between latitudes 24° 01′ and 24° 30′ N and longitudes 89° 44′ and 90° 18′ E (Fig. 2.1). Tangail is located to the west of the Jamuna River. The Lohajang, Elengjani, Bairan, Atai, Kiru, and Jhinai rivers are among the other significant ones. Of the 3.6 million people living in the district,

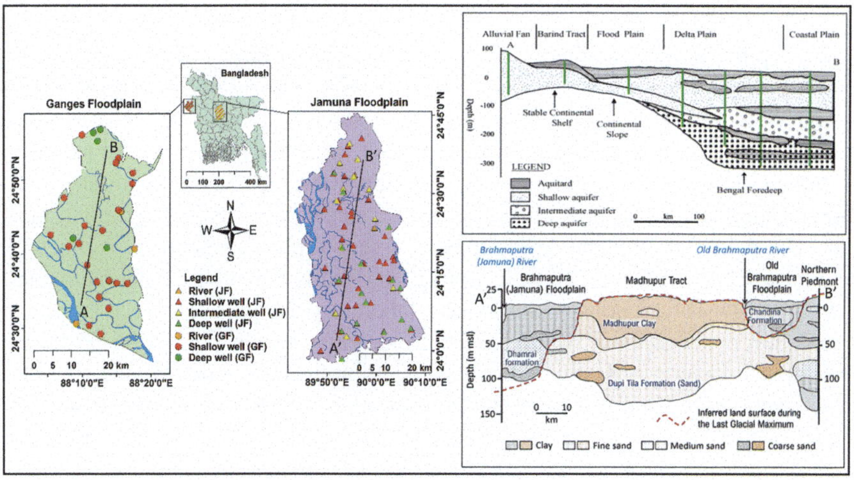

Fig. 2.1 Study area map with sampling location and geology of the study area

49.53% work in agriculture. The district experiences mild, dry winters and hot, humid tropical wet seasons due to its tropical monsoon climate. 1800 mm of rain falls annually, with the monsoon season accounting for about 70% of it. 27.5 °C is the average temperature. The research area's average topographic height is 15 m, with a little slope from north to south, based on data from the USGS Digital height Map (DEM).

Geologically speaking, the research region is made up of the Pleistocene Madhupur tract highland in the east and the low-lying Jamuna floodplain (Holocene) area in the west. Conversely, the Ganges Floodplain (GF) is situated in the northwest of Bangladesh, namely in the Chapai Nawabganj area. The GF is located between latitudes 24 °15′ and 25° 10′ N and longitudes 89° 01′ and 88° 45′ E. The two principal rivers in this region are the Mahananda and the Ganges (Padma). Huge sediments from the upper stream are carried by these rivers. The process of sedimentation is ongoing in this portion of the research area. As a result, during recharging, anthropogenic pollutants can readily combine with the water.

2.2.2 *Sampling and Analytical Procedure/sample Collection, Processing, and Analysis*

A total of 105 samples were systematically collected from the Jamuna Floodplain (n = 76) and Ganges Floodplain (n = 29), representing various aquifer stages, including river water, shallow, intermediate, and deep layers, at diverse locations as illustrated in Fig. 2.1. The specific sampling points were meticulously recorded using a handheld GPS device. The wells owner provided information about the borehole's depths. Before sampling, the wells underwent a brief pumping process to eliminate any stagnant water. Each collected sample was carefully distributed into three distinct HDPE bottles: 500 ml for anion analysis, another 500 ml for cation analysis, and a smaller 50 ml portion dedicated to isotopic analysis. All samples, excluding the isotopic ones, underwent filtration using a 0.45 μm membrane filter. Cation samples underwent acidification with concentrated HNO_3 for further analysis. The field measurements encompassed essential physicochemical parameters, including pH, EC (Electrical Conductivity), Eh (Redox Potential), temperature, and TDS (Total Dissolved Solids). Subsequent laboratory analyses employed advanced techniques: concentrations of Ca^{2+}, Mg^{2+}, and heavy metals (As, Cr, Ni, Pb) were meticulously determined using Atomic Absorption Spectroscopy (Model: Analytic Jena, Zeenit 700). Simultaneously, the concentrations of Na^+ and K^+ were quantified via a flame photometer, while anions like Cl^-, $SO_4{}^{2-}$, and $NO_3{}^-$ underwent measurement through ion chromatography (Model: Ion Chromatograph, Dionex DX-3000, USA). The determination of $HCO_3{}^-$ concentrations employed the titrimetric method. The isotopic composition was precisely evaluated utilizing the Liquid Water Isotope Analyzer (LWIA-24-EP). The American Public Health Association's (APHA-2012) methods were followed for all studies carried out at the Institute of Nuclear Science and Technology (INST), Atomic Energy Research Establishment (AERE), Savar, Dhaka, Bangladesh.

2.2.3 *Water Quality and Pollution Assessment Indices*

2.2.3.1 Integrated Water Quality Index (IWQI)

Compared to the current techniques, the EWQI is thought to be a more practical way to evaluate the quality of the water [6, 7, 66]. The four stages of the EWQI computation technique are as follows:

$$X = \begin{bmatrix} x_{11} & x_{12} & \ldots & x_{1n} \\ x_{21} & x_{22} & \ldots & x_{2n} \\ \vdots & \vdots & & \vdots \\ x_{m1} & x_{m2} & \ldots & x_{mn} \end{bmatrix} \tag{2.1}$$

Using Eq. (2.2) and (2.3), get the standard matrix Y. There are significant dimensional differences in evaluating hydrochemical indices. The data must be standardized before the EWQI can be calculated. is the maximum value and is the minimum value.

$$Y_{ij} = \frac{x_{ij} - \left(x_{ij}\right)_{min}}{\left(x_{ij}\right)_{max} - \left(x_{ij}\right)_{min}} \tag{2.2}$$

$$Y = \begin{bmatrix} y_{11} & y_{12} & \cdots & y_{1n} \\ y_{21} & y_{22} & \cdots & y_{2n} \\ \vdots & \vdots & & \vdots \\ y_{m1} & y_{m2} & \cdots & y_{mn} \end{bmatrix} \tag{2.3}$$

Using Eq. (2.2) and (2.3), get the standard matrix Y. There are significant dimensional differences in evaluating hydrochemical indices. The data must be standardized before the IWQI can be calculated. is the maximum value and is the minimum value.

$$p_{ij} = \frac{(1 + y_{ij})}{\sum_{j=1}^{m}(1 + y_{ij})} \tag{2.4}$$

$$e_j = -\frac{1}{\text{In}(m)} \sum_{i=1}^{m} p_{ij}\,\text{In}(p_{ij}) \tag{2.5}$$

$$w_{ij} = \frac{(1 - e_j)}{\sum_{i=1}^{m}(1 - e_j)} \tag{2.6}$$

Using Eqs. (2.4)–(2.6), the information entropy (e_j) and entropy weight (w_j) have been determined. Sample I's index j value is displayed as P_{ij}.

$$q_j = \frac{C_j}{S_j} \times 100 \tag{2.7}$$

$$IWQI = \sum_{j=1}^{n} w_j \times q_j \tag{2.8}$$

The following Eq. (2.7) and (2.8) are used to calculate the IWQI value. "q_j" is the quantitative grading scale for hydrochemical indices, determined by using hydrochemical concentrations (C_j) and the World Health Organization's drinking standards (S_j). Subsequently, the EWQI value is calculated using the entropy weight "w_j" and the quantitative grading scale "q_j".

2.2.3.2 Heavy Metal Pollution Index

The HPI serves as a robust methodology for evaluating the comprehensive impact of individual heavy metal concentrations in water, as highlighted by Moldovan et al. [45]. The unit weightage (Wi) and the standard specified limit (Si) of each specific heavy metal are the two main pillars upon which this technique is primarily built [71]. The calculation of HPI for groundwater, based on the concentrations of Fe, Mn, As, Ni, Cr, and Pb, was executed using the following equation, as outlined by Mohan et al. [44]:

$$HPI = \frac{\sum_{i=1}^{n} Q_i W_i}{\sum_{i=1}^{n} W_i} \tag{2.9}$$

In this equation, the sub-index and unit weight of the ith parameter are denoted by Q_i, and W_i while n signifies the number of investigated parameters. The determination of Q_i was accomplished through the following equation:

$$Q_i = \sum_{i=1}^{n} \frac{|M_i(-)I_i|}{S_i - I_i} \times 100 \tag{2.10}$$

In this case, M_i, I_i, and S_i stand for the heavy metal's studied value, its highest standard concentration, and its desired value, respectively.

2.2.3.3 Heavy Metal Evaluation Index

The HEI methodology provides a complete understanding of water quality concerning heavy metal content. It was computed using the equation as follows [45, 60, 61]:

$$HEI = \sum_{i=1}^{n} \frac{M_i}{S_i} \tag{2.11}$$

Here, M_i is the monitored value and S_i is the maximum acceptable concentration (MAC) of the ith parameter.

2.2.3.4 Degree of Contamination

The collective effects of multiple parameters of water quality deemed detrimental to household water use are assessed through the Contamination Degree (CD) [4, 8]. It is calculated using the following equation:

$$CD = \sum_{i=1}^{n} CFi \tag{2.12}$$

$$HEI = \sum_{i=1}^{n} \frac{CA_i}{CN_i} - 1 \tag{2.13}$$

where CF_i, CA_i, and CN_i represent the factor of contamination, observed data, and the higher allowable concentration of the ith component, respectively. *N* denotes the normative value, and CN_i is treated the Maximum Allowable Concentration (MAC) adopted from WHO suggested values.

2.2.4 Human Health Risk Evaluation

Assessing health risks is essential to supporting assessments and management of water quality. According to Bortey-Sam et al. [17], the risk assessment for human health is the process of assessing the likelihood of any given degree of unfavorable health impacts occurring within a specified period. Typically, this evaluation is focused on determining the risk level and is expressed in terms of health risks that are carcinogenic and non-carcinogenic [82]. The assessment takes into account both the oral and dermal ways, and the following models were used to compute the elements' chronic daily intake (CDI): [37, 78, 92].

$$CDI_{oral} = \frac{(CW \times IR \times EF \times ED)}{(BW \times AT)} \tag{2.14}$$

$$CDI_{dermal} = \frac{\left(CW \times S_A \times ED \times EF \times ET \times CF \times K_p \times 10^{-3}\right)}{(BW \times AT)} \tag{2.15}$$

In the equations above, CDI_{Oral} and CDI_{Dermal} represent the exposure dose in mg/kg/day unit through oral ingestion and dermal pathways in mg/kg/day unit, respectively, and are calculated using the parameter values from Table 2.1.

Table 2.1 Description and value of calculated parameters for non-carcinogenic risk by oral exposure

Input parameters	Values	Unit	References
CDI-chronic daily intake	–	Mg kg^{-1} day^{-1}	
CW-concentration in water	–		
IR-ingestion rate	M-2.2	L/day	Karim [37], Rahman et al. [55], and Wu et al. [92]
	F-1.6		
	C-1		Karim [37], Rahman et al. [55], and Wu et al. [92]
EF-exposure frequency	365	Days/year	Karim [37] and Rahman et al. [55], USEPA [81], and Wu et al. [92]
ED-exposure duration	M-70 F-67 C-10	Year	Karim [37], USEPA [81], and Wu et al. [92]
ET-exposure time		Min/day	
BW-body weight	M-70 F- 57.3 C-25	kg	Karim [37] and USEPA [81]
AT-average time	M-25550 F-24455 C-3650	Days	Karim [37], USEPA [81], and Wu et al. [92]

To calculate the non-carcinogenic health risk associated with exposure to contaminants, Hazard Quotients (HQ) were also computed for both oral and dermal pathways, following the methodology outlined by USEPA [78]. According to USEPA [80], if the value of HQ is less than 1, there is an unacceptable risk of negative non-carcinogenic impacts on health. If the value of HQ exceeds 1, however, there is an acceptable level of risk.

$$HQ = \frac{CDI}{RfD} \tag{2.16}$$

where RfD represents the reference oral and dermal absorption doses of a specific contaminant (μg/kg/day), obtained from the risk-based concentration table [79].

To evaluate the overall potential for non-carcinogenic effects posed by multiple chemicals, the Hazard Quotients (HQs) calculated for each chemical are summed to create a Hazard Index (HI) [78], as expressed by the formula:

$$HI = HQ_1 + HQ_2 + \cdots + HQ_n \tag{2.17}$$

No chronic risks were considered to exist at the site if HI does not exceed unity (HI < 1), and conversely, if HI > 1, it implies an increased likelihood of non-carcinogenic health risks [78, 95].

Carcinogenic Risk

The carcinogenic risk (CR) represents the added likelihood of developing a specific cancer throughout one's life due to exposure to a carcinogenic substance, and it is a dimensionless measure [62]. The calculation of the carcinogenic risk for each component is determined by Eq. (2.18).

$$CR = ADD \times CSF \tag{2.18}$$

In this equation, CSF represents the oral cancer slope factor $(\text{mg/kg/day})^{-1}$. According to USEPA [83], a CR in the range of 1×10^{-6} to 1×10^{-4} is generally regarded as acceptable, however, a CR value larger than 10^{-4} indicates a higher possibility of cancer risk [23, 31].

2.2.5 *Statistical Analyses*

The collected data underwent statistical analysis using SPSS software (Version-25). For assessing the degree of contamination, identifying potential sources, and exploring correlated relationships among the studied components, multivariate statistical techniques such as principal component analysis and Pearson's correlation matrix were employed [30].

2.3 Results and Discussion

2.3.1 *General Chemistry of Groundwater*

Tables 2.2 and 2.3 compares groundwater hydrochemical data to World Health Organization (WHO) and Bangladesh standards, with minimum (Min), maximum (Max), and mean values with standard deviation for river water, shallow, intermediate, and deep wells samples.

In JF, the pH levels across river water, shallow, intermediate, and deep wells vary between 6.8–7.4, 6.1–7.1, 6.2–7.5, and 6.1–7.1, respectively, with mean values of 7.0, 6.6, 6.7, and 6.7. Notably, the samples range from slightly acidic (88.16%) to alkaline. Conversely, the pH values of the river water, shallow and deep wells in GF range from 6.9–7.4, 6.8–7.2, and 6.8–7.2 with their average values of 7.3, 7.1, and 7.0 indicating that all the samples in these categories lean towards being mildly acidic to alkaline (62.07%). In JF, the electrical conductivity (EC) values for river water, shallow, intermediate, and deep wells span from 237–509 μS/cm, 243–1301 μS/cm, 238–547 μS/cm, and 258–680 μS/cm, respectively. The mean values for these categories are 369 ± 135.64 μS/cm, 483.89 ± 215.30 μS/cm, 385.92 ± 100.07 μS/cm, and 404.13 ± 105.88 μS/cm. Contrastingly, in GF, the EC values for river water, shallow, and deep wells range from 149–235 μS/cm, 505–1850 μS/cm, and 663–1911 μS/cm, with mean values of 178.4 ± 49.04 μS/cm, 1053.24 ± 425.16 μS/cm, and 1190.80 ± 453.66 μS/cm, respectively. Notably, the EC of groundwater exhibits a decreasing trend with depth in JF while the reverse condition is found in the GF. Similar EC values are observed in different regions of Bangladesh [2, 3, 6, 65]. In JF, the Total Dissolved Solids (TDS) concentration in river water, shallow, intermediate, and deep wells ranges from 113.4–246 mg/L, 116.7–643 mg/L, 114–265 mg/L, to 124–331 mg/L, respectively having mean values 177.93 ± 66.37 mg/L, 234.83 ± 106.53 mg/L, 186.25 ± 49.04 mg/L, and 195.16 ± 51.88 mg/L. In contrast, the TDS content in GF varies from 74.3–117.5 mg/L, 252.5–925 mg/L, and 331.5–955.5 mg/L. The mean values for river water, shallow well, and deep well samples are, respectively, 89.2 ± 24.52 mg/L, 526.62 ± 212.58 mg/L, and 595.4 ± 226.83 mg/L. Based on Freeze and Cherry's classification [21], all the analyzed waters fall into the freshwater category, as their TDS levels are below the threshold of 1000 mg/L. The findings indicate that in both JF and GF, Ca^{2+} and Na^{+} are the predominant cations, while HCO_3^- and Cl^- are the leading anions. The order of prevalence is $Ca^{2+} > Na^{+} > Mg^{2+} > K^{+}$ for cations and $HCO_3^- > Cl^- > SO_4^{2-} > NO_3^-$ for anions. This outcome aligns with the conclusions drawn in the studies conducted by Bodrud-Doza et al. [15] and Al-Asad et al. [6]. The average Ca^{2+} concentrations in JF were determined to be 46.23 ± 2.55 mg/L, 52.50 ± 15.35 mg/L, 50.86 ± 12.63 mg/L, and 21.36 ± 12.72 mg/L for river water, shallow, intermediate, and deep samples, respectively. Concurrently, the mean concentrations of Na^{+} were recorded as 18.96 ± 11.91 mg/L, 17.05 ± 8.58 mg/L, 14.48 ± 5.83 mg/L, and 61.71 ± 25.69 mg/L for the corresponding samples. In contrast, the mean Ca^{2+} concentration in GF for river water, shallow, and deep samples was found to be 4.70 ± 0.75 mg/L, 108.71 ± 35.05

Table 2.2 Statistical representation of chemical parameters for Jamuna floodplain

Parameter	Jamuna floodplain (JF)															
	River water				Shallow Well				Intermediate well				Deep Well			
	Min	Max	Avg	Std	Min	Max	Avg	Std	Min	Max	Avg	Std	Min	Max	Avg	Std
pH	6.8	7.4	7	0.4	6.1	7.1	6.6	0.2	6.2	7.5	6.7	0.3	6.1	7.1	6.7	0.3
EC	237	508	369	135.64	243	1301	483.89	215.3	238	547	385.92	100.07	258	680	404.13	105.88
TDS	113.4	246	177.9	66.37	116.7	643	234.8	106.53	114	265	186.25	49.04	124	331	195.16	51.88
Ca	44.2	49.09	46.23	2.55	10.43	88.08	52.5	15.35	37.74	83.55	50.86	12.63	12.09	62.15	21.36	12.72
Mg	2.89	22.7	11.76	10.07	1.84	22.7	10.21	5.59	1.91	12.7	6.88	3.74	1.18	21.91	7.37	5.67
Na	7.89	31.57	18.96	11.91	3.31	47.33	17.05	8.58	1.58	25.1	14.48	5.83	9.47	106.43	61.71	25.69
K	11.88	21.88	18.55	5.77	1.15	32.82	9.28	7.99	2.41	13.09	5.33	2.85	2.19	23.68	9.65	6.52
HCO_3	189.32	242.21	211.22	27.59	95.23	301.56	187.04	47.75	135.45	304.32	193.86	50.27	159.03	302.79	234.79	46.34
Cl	2.07	23.69	11.69	11	1.72	142.03	28.53	28.63	2.5	35.99	13.1	10.32	2.52	55.74	11.47	12.78
NO_3	0.47	11.46	4.64	5.95	1.02	57.72	17	14.36	1.57	13.29	6.7	4.34	0.07	17.23	2.55	4.57
SO_4	12.3	25.87	18.22	6.95	0.84	62.31	16.85	14.33	2.06	33.73	9.71	8.34	0.23	30.59	8.98	9.61
Fe	0.01	0.13	0.06	0.06	0.95	3.15	2.08	0.73	0.02	2.17	0.72	0.64	0	0.21	0.08	0.07
Mn	0.01	0.05	0.02	0.03	0	0.2	0.05	0.06	0	0.23	0.1	0.07	0	0.04	0.01	0.01
As	5.00E−05	5.00E−05	5.00E−05	0	0.08	0.15	0.113	0.019	0.019	0.114	0.074	0.031	5.00E−05	5.00E−05	5.00E−05	1.00E−20
Cr	5.00E−06	5.00E−06	5.00E−06	0	0.025	0.095	0.054	0.018	0.027	0.057	0.044	0.01	5.00E−06	5.00E−06	5.00E−06	0
Ni	1.00E−05	1.00E−05	1.00E−05	0	0.007	0.024	0.013	0.003	0.007	0.018	0.012	0.003	1.00E−05	1.00E−05	1.00E−05	0
Pb	5.00E−06	5.00E−06	5.00E−06	0	0.018	0.175	0.093	0.039	0.017	0.101	0.052	0.029	5.00E−06	5.00E−06	5.00E−06	0
δ^2H ‰	− 58	− 15.6	− 32.2	22.6	− 43.5	− 9.6	− 25.8	7.6	− 38.6	− 17.9	− 24.6	7.3	− 42.9	− 14.7	− 25.6	8.2
$\delta^{18}O$ ‰	− 9.7	− 3.6	− 5.7	3.5	− 6.6	− 2.1	− 4.3	1.1	− 5.9	− 2.7	− 4.1	0.9	− 6.8	− 2.8	− 4.4	1.1

Table 2.3 Statistical representation of chemical parameters for Ganges floodplain

Parameter	Ganges floodplain (GF)											
	River water				Shallow well				Deep well			
	Min	Max	Avg	Std	Min	Max	Avg	Std	Min	Max	Avg	Std
pH	6.9	7.8	7.3	0.5	6.8	7.4	7.1	0.2	6.8	7.2	7	0.2
EC	149	235	178.4	49.04	505	1850	1053.2	425.16	663	1911	1190.8	453.66
TDS	74.3	117.5	89.2	24.52	252.5	925	526.62	212.58	331.5	955.5	595.4	226.83
Ca	3.86	5.3	4.7	0.75	24.9	173	108.71	35.05	38.1	169	75.96	56.24
Mg	1.56	5.3	3	2.01	24.07	50.69	35.73	8.43	26.51	53.61	35.44	10.6
Na	11.6	15.5	13.73	1.98	13.4	54.5	35.12	10.91	26.9	43.4	34.55	6.47
K	4.6	5.21	5	0.34	1.75	28.22	7.98	7.38	1.75	7.28	3.99	2.64
HCO_3	20.8	166.4	74.88	79.69	141.44	312	203.34	48.89	178.88	393.12	259.58	100.05
Cl	5.43	9.34	7.63	2	0.96	115.91	32.96	31.49	9.39	22.81	17.79	5.67
NO_3	1.54	2.19	1.9	0.33	1.24	75.23	14.57	19.8	5.82	14.28	9.65	3.28
SO_4	3.24	12.59	6.41	5.35	0.45	64.38	19.88	20.26	1.14	49.67	19.72	19.56
Fe	0.07	0.23	0.15	0.08	0.01	3.99	1.11	1.47	0.17	2.28	0.72	0.88
Mn	1.27	1.68	1.53	0.23	0.02	0.99	0.51	0.34	0.22	0.32	0.26	0.05
As	0.002	0.003	0.002	6.00E−04	0.014	0.127	0.067	0.027	0.01	0.04	0.02	0.012
Cr	0.005	0.005	0.005	2.00E−04	2.00E−05	0.008	0.0044	0.003	2.00E−04	0.006	0.002	0.003
Ni	–	–	–	–	–	–	–	–	–	–	–	–
Pb	0.049	0.056	0.052	0.004	0.014		0.0768	0.041	0.007		0.0351	0.024
δ^2H ‰	− 6.1	− 5.6	− 5.9	0.3	− 46.4	− 26.6	− 36.5	6.1	− 37.5	− 25.7	− 32.7	5.3
$\delta^{18}O$ ‰	− 1.8	− 1.5	− 1.6	0.2	− 7.8	− 3.6	− 5.4	1.2	− 5.7	− 3.3	− 4.9	1.1

mg/L, and 75.96 ± 56.24 mg/L, respectively. Similarly, the mean Na^+ concentrations were determined as 13.73 ± 1.98 mg/L, 35.12 ± 10.91 mg/L, and 34.55 ± 6.47 mg/L for the respective samples. Because of the ion exchange process, it was discovered that the intermediate aquifer had a greater calcium concentration, nevertheless, the deep aquifer has a larger sodium value because of the reverse cation exchange process silicate weathering in the study area [85]. In the JF area, the concentration of Mg^{2+} ranged from 2.89 to 27.70 mg/L, 1.84 to 22.70 mg/L, 1.91 to 12.70 mg/L, and 1.18 to 21.91 mg/L for river water, shallow, intermediate, and deep samples, respectively. The mean concentrations of this component were determined as 11.76 ± 10.07 mg/L, 10.21 ± 5.59 mg/L, 6.89 ± 3.74 mg/L, and 7.37 ± 5.67 mg/L for the corresponding samples. Moreover, in the GF area, the mean concentrations of Mg^{2+} were found to be 3.00 ± 2.01 mg/L, 35.73 ± 8.43 mg/L, and 35.44 ± 10.60 mg/L for river water, shallow, and deep wells. It's noteworthy that Mg^{2+} values in all samples fall within the WHO-2011 prescribed limits. Further, the K^+ values for river water, shallow, intermediate, and deep well samples in GF areas are found to be 5.00 ± 0.34 mg/L, 7.98 ± 7.381 mg/L, and 3.99 ± 2.64 mg/L, respectively, while the average values for JF are 18.55 ± 5.77 mg/L, 9.29 ± 7.99 mg/L, 5.34 ± 2.85 mg/L, and 9.65 ± 6.52 mg/L in respective areas. 11.84% and 6.90% samples exceed the WHO [86] standard limit in JF and GF areas respectively. This finding indicates that the excessive use of fertilizer during irrigation has led to greater concentrations of potassium in surface and shallow groundwater in both JF and GF areas 6 (Figs. 2.2 and 2.3).

The bicarbonate anion concentration ranged from 189.32–242.21 mg/L, 95.23–301.56 mg/L, 135.45–304.32 mg/L, and 159.03–302.79 mg/L while the average values for river water, shallow, intermediate, and deep well samples in the JF areas are 211.22 ± 27.59 mg/L, 187.04 ± 47.75 mg/L, 193.86 ± 50.27 mg/L, and 234.79 ± 46.34 mg/L. In addition, it is found that the concentrations of HCO_3^- in GF ranged from 20.80–166.40 mg/L for river water, 141.44–312.00 mg/L for shallow wells, and 178.88–393.12 mg/L for deep wells with the determined mean concentration to be 74.88 ± 79.69 mg/L, 203.35 ± 48.89 mg/L, and 259.58 ± 100.05 mg/L respectively. The second most prevalent anion Cl^- is in the JF area, with an average concentration of 11.69 ± 11.00 mg/L, 28.53 ± 28.63 mg/L, 13.11 ± 10.32 mg/L, and 11.47 ± 12.78 mg/L for river water, shallow, intermediate and deep well samples. However, in GF areas, the mean concentrations of chloride are found as 7.63 ± 2.00 mg/L, 32.96 ± 31.49 mg/L, and 17.79 ± 5.67 mg/L for river water, shallow and deep wells sequentially. All the samples are within the WHO-recommended limits. The concentrations of NO_3^- range between 0.47–11.46 mg/L, 1.02–57.72 mg/L, 1.57–13.29 mg/L, 0.07–17.23 mg/L, with mean values of 4.64 ± 5.96 mg/L, 17.00 ± 14.36 mg/L, 6.70 ± 4.34 mg/L, and 2.55 ± 4.57 mg/L for shallow, intermediate, and deep well samples, respectively. Additionally, for river water, shallow wells, and deep wells, the ranges are from 1.54–2.19, 1.24–75.23, and 5.82–14.28, with mean concentrations of NO_3^- in GF determined to be 1.90 ± 0.33 mg/L, 14.57 ± 19.80 mg/L, and 9.65 ± 3.28 mg/L surface water, shallow well and deep well respectively. 43.42% samples of the JF and 34.48% of the GF exceeds the standard permissible for drinking. The nitrate concentrations are found higher in the surface and shallow wells cause due to anthropogenic activities such as nitrogen-based fertilizers application,

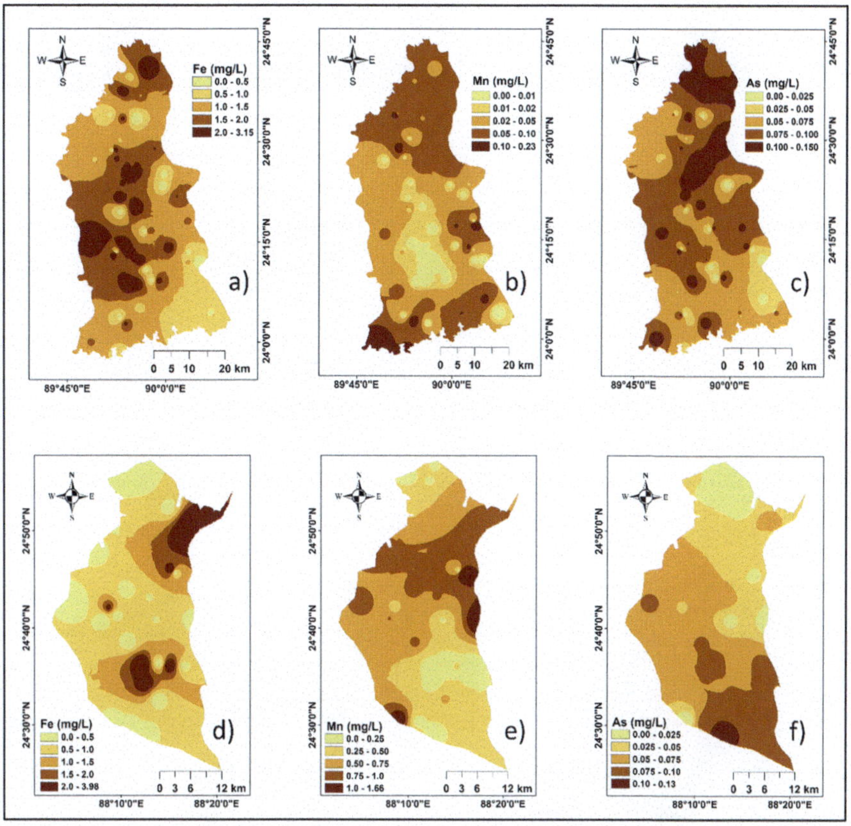

Fig. 2.2 Spatial distribution of Fe, Mn, and As in JF and GF areas

livestock farming, inadequate sanitation practices, and sewage leakage. This decline in nitrate levels can be linked to the processes of denitrification and adsorption as water percolates through the clay layer above the deep aquifer [6, 47]. For river water, shallow, middle, and deep well samples in the JF region, the mean SO_4^{2-} concentrations are 18.22 ± 6.95 mg/L, 16.85 ± 14.33 mg/L, 9.71 ± 8.345 mg/L, and 8.98 ± 9.61 mg/L, respectively. It is found that the average SO_4^{2-} concentrations in the GF are 6.41 ± 5.36 mg/L, 19.88 ± 20.26 mg/L, and 19.72 ± 19.56 mg/L in respective areas. The average iron (Fe) and manganese (Mn) values in JF are 0.06 ± 0.06 mg/L, 2.08 ± 0.73 mg/L, 0.72 ± 0.64 mg/L, and 0.08 ± 0.07 mg/L as well as 0.02 ± 0.03 mg/L, 0.05 ± 0.06 mg/L, 0.10 ± 0.08 mg/L, and 0.01 ± 0.01 mg/L for river water, shallow, intermediate, and deep well samples respectively. Correspondingly, mean values of these ions in GF areas are found 0.15 ± 0.08 mg/L, 1.11 ± 1.47 mg/L, and 0.72 ± 0.88 mg/L as well as 1.53 ± 0.23 mg/L, 0.51 ± 0.34 mg/L, and 0.26 ± 0.05 mg/L for river water, shallow and deep wells chronologically (Fig. 2.4a, b). In JF, samples containing 68.42% Fe and 17.11% Mn exceeded the WHO-mandated limit, whereas in GF, samples containing 44.83% Fe and 93.10% Mn did the same.

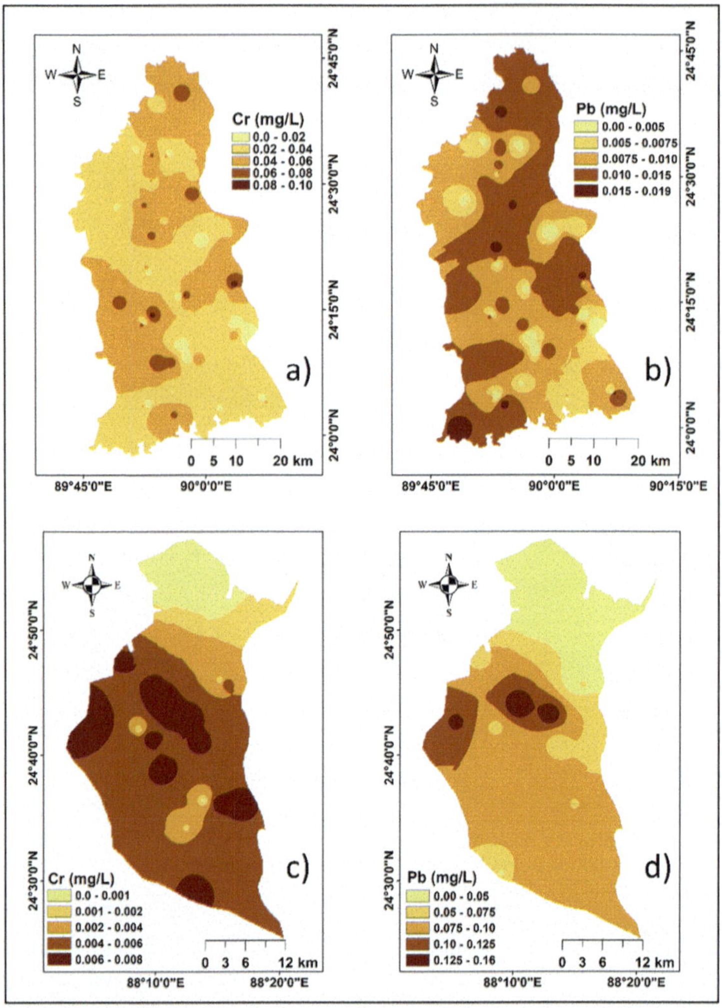

Fig. 2.3 Spatial distribution of Cr, and Pb in JF and GF areas

The elevated levels of Fe and Mn in groundwater can be attributed by the dissolution of iron and manganese-containing minerals in geological formations, such as iron oxides and manganese-bearing rocks as well as agricultural activities, including the use of fertilizers and pesticides, urban wastewater, etc. The redox conditions prevailing in certain aquifer zones can also influence the mobilization and release of iron and manganese into the groundwater [63]. In JF regions, the values of As, Cr, Ni, and Pd range from 0.08 to 0.15 mg/L, 0.025 to 0.095 mg/L, 0.018 to 0.175 mg/L, and 0.07 to 0.024 mg/L with the mean values of 0.113 ± 0.019 mg/L, 0.055 ± 0.018 mg/L, 0.093 ± 0.039 mg/L, and 0.013 ± 0.003 mg/L for shallow wells. For intermediate wells, the values ranged from 0.019 to 0.114 mg/L, 0.027 to 0.057 mg/L, 0.017 to

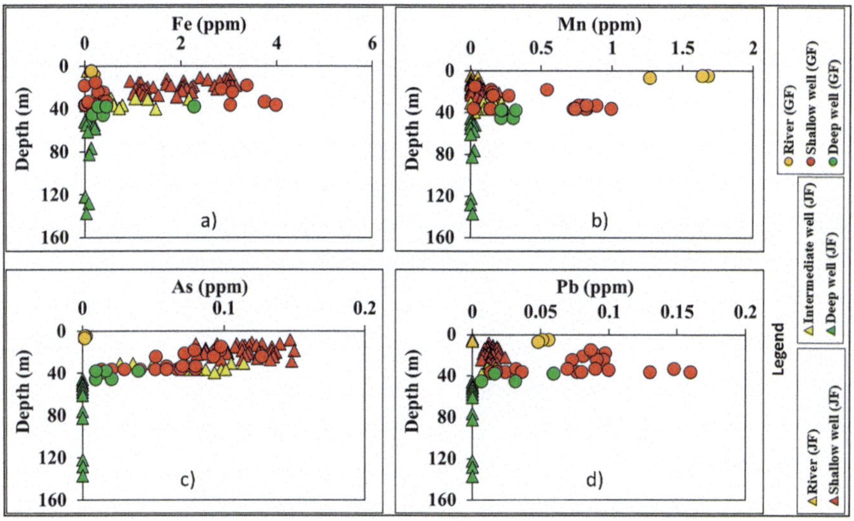

Fig. 2.4 Vertical distribution of Fe, Mn, As, and Pb in JF and GF areas

0.101 mg/L, and 0.007 to 0.018 mg/L, with corresponding mean values of 0.074 ± 0.031 mg/L, 0.044 ± 0.010 mg/L, 0.052 ± 0.029 mg/L, and 0.012 ± 0.003 mg/L. In contrast, the average concentration of As in the GF area is 0.002 ± 0.001 mg/L, 0.067 ± 0.027 mg/L, and 0.020 ± 0.012 mg/L, Cr is 0.005 ± 0.0002 mg/L, 0.004 ± 0.0003 mg/L, and 0.002 ± 0.003 mg/L, and Pb is 0.052 ± 0.004 mg/L, 0.077 ± 0.041 mg/L, and 0.035 ± 0.024 mg/L in the surface water, shallow and deep aquifers respectively. The acceptable limit is exceeded by 71.05% As, 39.47% Cr, and 55.26% Ni in JF samples and by 586.2% As and 65.52% Pb in GF samples. These results show that a comparatively lower concentration of these elements is found in the GF than in the JF (Figs. 2.2a–f, 2.3a–d). It is caused by the industrial activities raised in the JF 6. The concentration of trace metals, including Fe, Mn, As, Cr, Ni, and Pb, in the Holocene deep aquifer and surface water samples is reported to be below the detection limit. Interestingly, these trace metals exhibit a discernible declining trend with depth (Fig. 2.4a–d). This phenomenon can be attributed to the presence of a clay layer that acts as a barrier, separating the upper shallow and intermediate aquifers from the deep aquifer [2, 3].

2.3.2 *Hydrochemical Facies*

The hydrochemistry of water is regulated by major ions. Water types are determined using the Piper diagram [53]. As illustrated in Fig. 2.5, all the shallow and intermediate samples in the JF area exhibit Ca–HCO_3^- types, while most of the deep aquifers (Holocene) display a transition from Ca–HCO_3^- to Na–HCO_3^- types. Conversely,

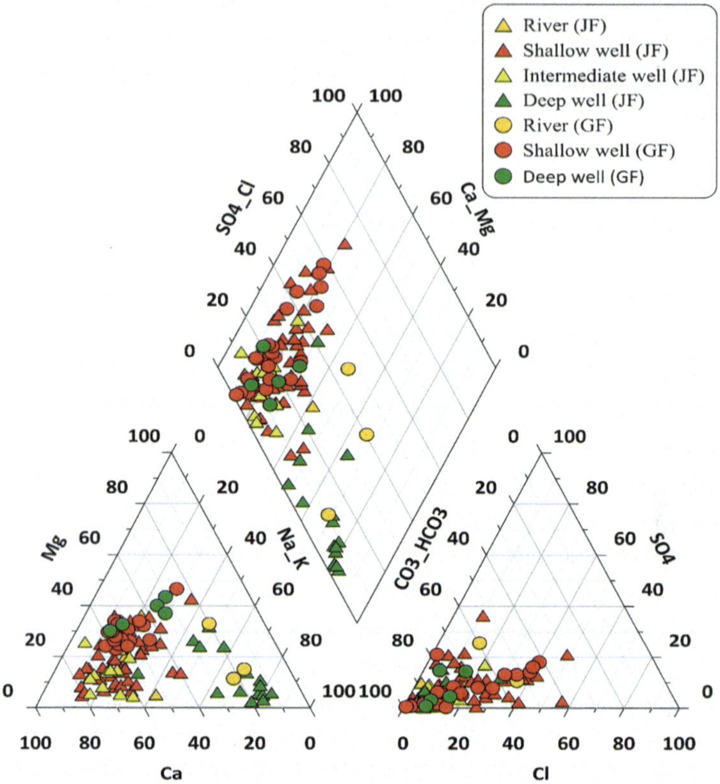

Fig. 2.5 Piper trilinear diagram for groundwater

deep well samples are mostly of the Ca–HCO_3^- type, whereas the majority of shallow well samples in the GF area are of the Ca–HCO_3^- to Ca–Cl^- type.

2.3.3 Isotopic Analysis

The isotopic approach plays a crucial role in pinpointing groundwater sources, as well as ascertaining their age, rate of flow into aquifers, and residence duration. Additionally, it facilitates the exploration of connections between different aquifers and the interconnections between surface water and groundwater. Figure 2.6a displays the isotopic composition (δ^2H and $\delta^{18}O$) plots of groundwater. The δ^2H values exhibit a range from –58.0 ‰ to – 15.6‰, – 43.5‰ to – 9.6‰, – 38.6‰ to – 17.9‰ and – 42.9‰ to – 14.7‰ for river water, shallow, intermediate, and deep samples of the Holocene aquifer, respectively, with mean values of – 32.2‰, – 25.8‰, – 24.6‰ and – 25.6‰ in JF regions. Furthermore, the values of δ18O range from – 9.71‰ to – 3.59‰, – 6.6‰ to – 2.1‰, – 5.9‰ to – 2.7‰ and – 6.8‰ to

− 2.8‰. The corresponding mean values are − 5.7‰, − 4.3‰, − 4.1‰ and − 4.4‰ for same locations. The δ^2H for river water, shallow, and deep samples of the Holocene aquifer in GF regions ranged from − 6.1‰ to − 5.6‰, − 46.4‰ to − 26.6‰, and − 37.5‰ to − 25.7‰, with mean values of − 5.9‰, − 36.5‰, and − 32.7‰, respectively. Furthermore, the corresponding δ18O values in the same areas range from − 1.8‰ to − 1.5‰, − 7.8‰ to − 3.6‰, and − 5.7‰ to − 3.3‰, with mean values of − 1.7‰, − 5.4‰, and − 4.9‰ for the samples of river water, shallow, and deep aquifer, respectively. Figure 2.6a illustrates the small difference in isotopic composition between the GF and JF regions. The majority of samples from these areas closely align with the Global Meteoric Water Line (GMWL) and Local Meteoric Water Line (LMWL), with only a few exceptions noted in deep well samples of GF. Notably, surface water, shallow, middle, and deep from the Holocene region have very high variance in isotopic composition, very close to GMWL and LMWL. These aquifers' water comes from recent rainfall-induced recharge. Due to potential evaporation, some shallow and intermediate water is below the GMWL and LMWL, indicating that these two aquifers are semi-confined to unconfined. However, this information enhances our understanding of the recharge mechanisms inferred from the isotopic composition. The d excess line supports this recharge mechanism both in the JF and GF areas (Fig. 2.6b).

2.3.4 *Geostatistical Analyses*

Multivariate statistical techniques, including correlation analysis, hierarchical cluster analysis (HCA), and principal components analysis (PCA), serve as valuable tools to detect and classify the possible pollutant sources in various sample locations [69]. In this study, Principal Component Analysis (PCA) was conducted on groundwater quality data, employing Varimax rotation with Kaiser Normalization. PCA enables the assessment of relationships between component levels and important variables that have an impact on the hydrogeochemical processes of the aquifer and groundwater quality by defining a new set of variables [34]. Varimax rotation, utilized in this analysis, maximizes the sum of the variance of the factor coefficients, providing a clearer understanding of potential sources influencing water systems [13, 22, 29, 55].

2.3.4.1 Correlation Analysis

The results of the correlation analysis (Tables 2.4 and 2.5) of the data from groundwater samples show strong correlations, which provide insight into possible influencing factors in the hydrochemical composition that contribute to the understanding of groundwater quality and its underlying sources.

In both JF and GF areas, compelling strong positive correlations are observed among key parameters such as EC, TDS, HCO_3^-, Cl^-, SO_4^{2-}, NO_3^-, Ca^{2+}, Mg^{2+}, Na^+, and K^+. This interconnectedness implies that the concentrations of these ions play a

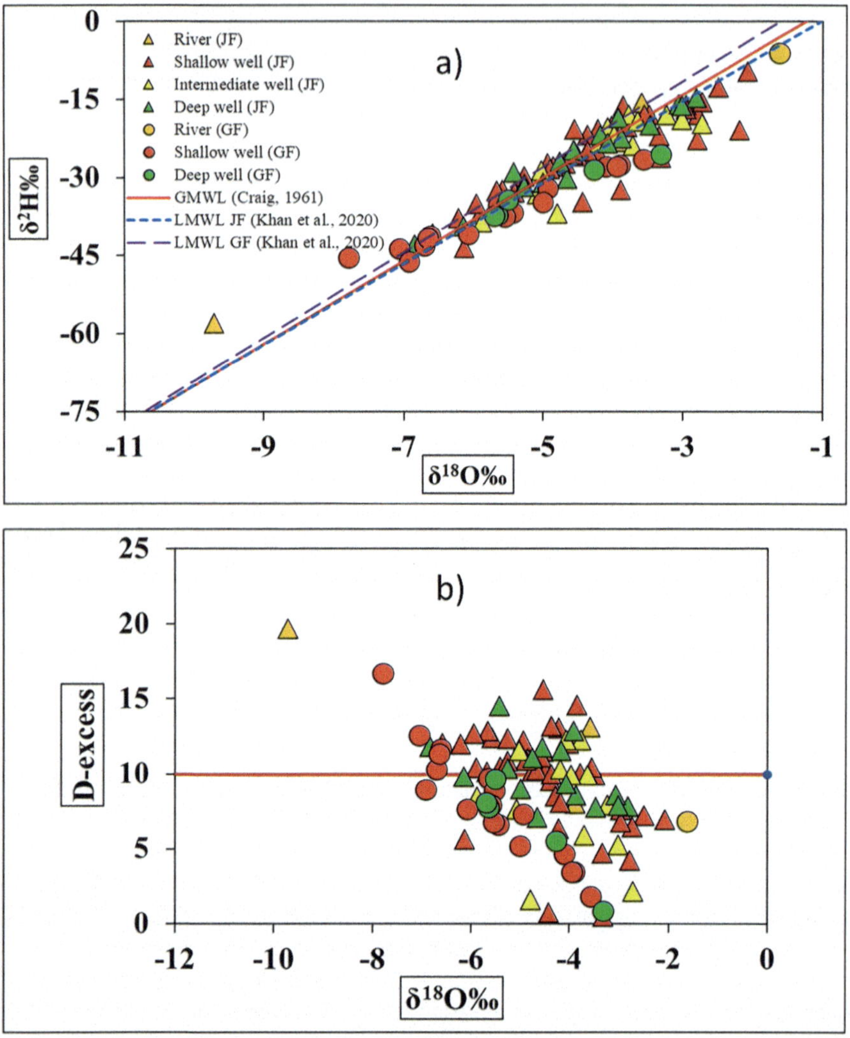

Fig. 2.6 Isotopic analysis of groundwater

pivotal role in influencing groundwater chemistry. The origin of these associations could stem from various factors, including anthropogenic activities like domestic wastewater infiltration, pesticide usage, as well as natural geological processes like ion exchange, rock-water interaction, and the weathering of rocks, calcite, and dolomite. This interdependence is denoted as the pollution factor, highlighting the intricate relationship between these parameters and the impact of human activities on groundwater quality [27, 59]. Moreover, a robust positive correlation is identified among trace elements (Fe, As, Cr, Ni, and Pb) in both areas, suggesting a commonality in their sources, potentially linked to anthropogenic activities or mineral dissolution.

Table 2.4 Pearson correlation of Jamuna floodplain (JF)

	Depth	EC	TDS	HCO_3	Cl	NO_3	SO_4	Ca	Mg	Na	K	Fe	Mn	As	Cr	Ni	Pb
Depth	1																
EC	−0.16	1															
TDS	−0.16	0.99	1														
HCO_3	0.34	−0.15	−0.15	1													
Cl	−0.26	0.75	0.75	−0.35	1												
NO_3	−0.42	0.31	0.31	−0.50	0.47	1											
SO_4	−0.36	0.31	0.31	−0.46	0.34	0.44	1										
Ca	−0.62	0.46	0.46	−0.22	0.64	0.44	0.28	1									
Mg	−0.30	0.14	0.13	0.26	0.12	−0.01	−0.04	0.12	1								
Na	0.74	−0.07	−0.07	0.54	−0.22	−0.38	−0.26	−0.67	−0.20	1							
K	0.10	0.23	0.23	0.05	0.20	0.13	0.12	0.06	0.09	0.02	1						
Fe	−0.58	0.33	0.33	−0.31	0.45	0.58	0.33	0.57	0.24	−0.50	0.03	1					
Mn	−0.15	−0.15	−0.15	0.02	−0.04	−0.11	−0.06	0.21	0.02	−0.24	−0.05	−0.04	1				
As	−0.64	0.19	0.19	−0.41	0.30	0.45	0.27	0.56	0.13	−0.65	−0.08	0.70	0.23	1			
Cr	−0.59	0.16	0.17	−0.40	0.24	0.41	0.14	0.47	0.14	−0.66	−0.10	0.71	0.15	0.77	1		
Ni	−0.56	0.15	0.16	−0.23	0.12	0.40	0.26	0.37	0.26	−0.51	−0.12	0.63	0.20	0.70	0.75	1	
Pb	−0.60	0.15	0.15	−0.37	0.25	0.37	0.24	0.55	0.10	−0.65	−0.14	0.67	0.35	0.80	0.80	0.74	1

Table 2.5 Pearson correlation of Ganges floodplain (GF)

	Depth	EC	TDS	HCO3	Cl	NO3	SO4	Ca	Mg	Na	K	Fe	Mn	As	Cr	Pb
Depth	1															
EC	0.52	1														
TDS	0.52	1.00	1													
HCO3	0.58	0.60	0.60	1												
Cl	−0.27	0.26	0.26	−0.01	1											
NO3	−0.16	−0.05	−0.05	0.14	0.66	1										
SO4	−0.19	0.44	0.44	0.24	0.67	0.50	1									
Ca	0.20	0.57	0.57	0.43	0.50	0.42	0.54	1								
Mg	0.53	0.72	0.72	0.61	0.39	0.25	0.48	0.84	1							
Na	0.30	0.63	0.63	0.40	0.60	0.37	0.50	0.53	0.61	1						
K	−0.42	0.18	0.18	−0.10	0.83	0.41	0.64	0.52	0.33	0.37	1					
Fe	0.02	0.10	0.10	−0.11	0.23	0.10	0.01	0.33	0.21	0.36	0.35	1				
Mn	−0.42	−0.33	−0.33	−0.36	−0.48	−0.44	−0.40	−0.43	−0.58	−0.52	−0.36	−0.16	1			
As	−0.04	0.02	0.02	0.10	0.32	0.26	0.23	0.54	0.39	0.18	0.41	0.15	−0.40	1		
Cr	−0.40	0.06	0.06	0.04	0.02	−0.20	0.24	0.28	0.10	−0.17	0.35	−0.25	0.28	0.44	1	
Pb	−0.17	−0.01	−0.01	0.12	0.09	0.06	0.33	0.38	0.31	0.00	0.26	−0.16	0.06	0.65	0.77	1

This correlation aligns with previous studies emphasizing the influence of human-related factors on trace metal concentrations in groundwater [10, 28, 73, 90]. Interestingly, a negative correlation is observed between depth and trace elements in both JF and GF areas, indicating a decrease in the rate of contamination with increasing depth. This depth-related trend underscores the importance of considering vertical variations when assessing groundwater quality and contamination risks.

2.3.4.2 Principal Component Analysis (PAC)

The results of the principal component analysis, including factor loadings, eigenvalues, percentages of variation, total cumulative variance, and community values are shown in Table 2.6. From the dataset, the principal component analysis identified four distinct factors in the JF (Fig. 2.7a, c), while in the Ganges Floodplain (GF), five factors were identified based on the groundwater quality dataset, each having eigenvalues (Fig. 2.7b,d) greater than 1. To further understand the underlying parameter structure, a scree plot of the retained PCA number has been created (Table 2.6). Four factors were retrieved from the principal components analysis in JF, accounting for 68.83% of the total variance, while five factors were extracted from GF, accounting for 83.86% of the total variance. PC1, which represented 37.10% of the variance in the overall physicochemical characteristics of groundwater in JF, was influenced by significant positive loadings of Ca, Fe, As, Cr, Pb, and Ni concentrations while being negatively loaded by Na. This suggests that the presence of heavy metals in groundwater is primarily attributed to the untreated effluent discharge from industries, vehicular emissions along highways, and corrosion of brakes, tires, and engine parts [6, 36, 54, 68]. Furthermore, the significant loading of arsenic, iron, and manganese concentrations is thought to be caused by the redox state of the groundwater, chemical weathering, and microbial decomposition of As, Mn, and Fe-bearing rocks in the aquifer [12, 32, 33]. The second extracted factor in PCA2 revealed high positive factor loadings for EC, TDS, Cl, and Ca, contributing to approximately 15.95% of the total variance. This implies that these parameters are influenced by factors such as increased evaporation due to extreme aridity, rock-water interaction with ion exchange, and the presence of irrigation return flow [15, 19]. With the positive load of HCO_3 and Mg concentrations, PC3 explained 9.15% of the overall variance, indicating that silicate mineral weathering is thought to be the main source of HCO_3 in groundwater. HCO_3, however, can also be produced by bacteria breaking down organic pollutants in aquifers [9, 77]. PC4, accounting for 6.63% of the total variance, was loaded positively by NO_3 concentration and negatively by pH and Mn. This implies that NO_3 may originate from domestic and agricultural wastes in the surrounding areas [11, 16, 19, 46].

However, in the GF areas, PC1 accounted for 37.77% of the variance in the groundwater's physicochemical characteristics overall. This variance was driven by the high quantities of EC, TDS, HCO_3, Ca, Mg, and Na, which positively loaded the groundwater, and the pH, which negatively loaded the groundwater. The elevated concentrations of dissolved ions in groundwater, contributing to the EC loadings,

Table 2.6 Varimax rotated principal component analysis and commonality in JF and GF

Rotated component matrix											
Jamuna floodplain (JF)						Ganges floodplain (GF)					
Parameters	PCA 1	PCA 2	PCA 3	PCA 4	Communality	PCA1	PCA2	PCA3	PCA4	PCA5	Communality
pH	– 0.26	– 0.02	0.12	**– 0.53**	0.37	**– 0.73**	0.04	0.16	0.09	– 0.05	0.57
EC	0.10	**0.94**	– 0.01	0.04	0.90	**0.96**	0.15	0.03	0.04	0.02	0.94
TDS	0.10	**0.94**	– 0.01	0.04	0.90	**0.96**	0.15	0.03	0.04	0.02	0.94
HCO_3	– 0.37	– 0.09	**0.77**	– 0.17	0.76	**0.66**	– 0.16	0.12	0.48	– 0.28	0.79
Cl	0.24	**0.84**	– 0.19	0.09	0.80	0.11	**0.91**	0.03	0.22	0.16	0.92
NO_3	0.45	0.31	– 0.37	**0.50**	0.65	– 0.13	**0.64**	– 0.08	**0.59**	– 0.08	0.78
SO_4	0.22	0.33	– 0.49	0.28	0.47	0.36	**0.78**	0.21	0.12	– 0.18	0.83
Ca	**0.61**	**0.54**	– 0.13	– 0.20	0.72	**0.51**	0.36	0.47	0.38	0.28	0.83
Mg	0.28	0.18	**0.74**	0.12	0.67	**0.70**	0.19	0.32	0.47	0.15	0.88
Na	**– 0.78**	– 0.05	0.23	0.13	0.68	**0.60**	0.45	– 0.10	0.32	0.26	0.74
K	– 0.15	0.39	0.09	0.14	0.20	0.02	**0.84**	0.32	0.00	0.30	0.90
Fe	**0.76**	0.28	0.02	0.35	0.79	0.08	0.12	– 0.09	0.07	**0.93**	0.90
Mn	0.34	– 0.13	– 0.04	**– 0.72**	0.65	– 0.21	– 0.31	0.05	**– 0.79**	– 0.11	0.77
As	**0.88**	0.09	– 0.11	0.07	0.80	– 0.06	0.14	**0.74**	0.44	0.26	0.83
Cr	**0.88**	0.03	– 0.03	0.09	0.79	0.03	0.11	**0.87**	– 0.35	– 0.18	0.94
Pb	**0.90**	0.04	– 0.11	– 0.09	0.84	– 0.01	0.09	**0.92**	0.03	– 0.13	0.86
Ni	**0.84**	0.00	0.11	0.11	0.72	–	–	–	–	–	–
Eigenvalue	6.31	2.71	1.56	1.13	–	6.04	2.84	2.24	1.17	1.13	–
Variance (%)	37.10	15.95	9.15	6.63	–	37.77	17.72	13.98	7.33	7.06	–
Cumulative (%)	37.10	53.05	62.20	68.83	–	37.77	55.49	69.48	76.81	83.87	–

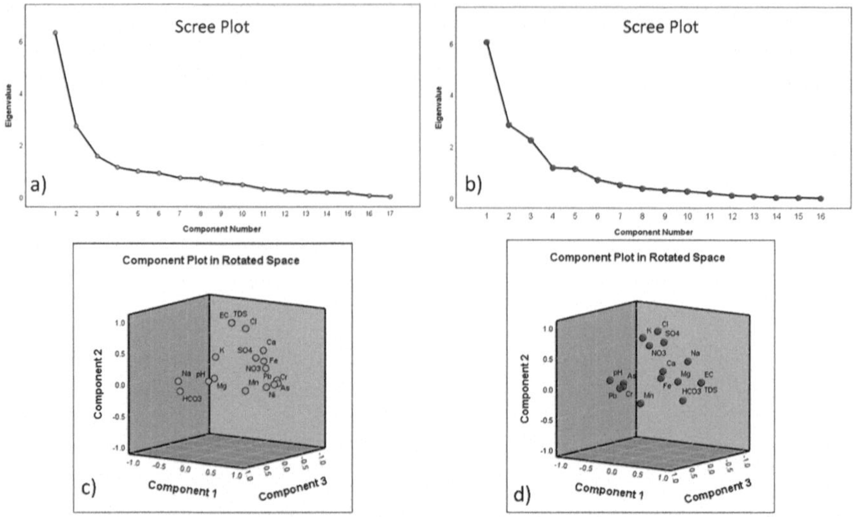

Fig. 2.7 Principal component analysis (scree plot and component plot in rotated space) in the Jamuna floodplain **a, c** and Ganges floodplain **b, d**

were attributed to rock-water interaction with ion exchange and agricultural effluent [15, 19]. The PC2 represented 17.72% of the total accounted variance, with positive loadings for Cl, NO_3, SO_4, and K concentrations. This suggests the influence of domestic wastes, septic tank leakage, and agricultural applications of fertilizers and agrochemicals as potential sources [14, 15, 19, 47]. The concentrations of As, Cr, and Pb positively loaded PC3, which explained 13.98% of the variance overall. The enrichment of arsenic in the groundwater may be attributed to natural processes [12, 57]. The NO_3 concentrations loaded PC4, which accounted for 7.33% of the overall variation, with a negative load of Mn. Higher NO_3 concentrations are likely the result of agricultural activities, urban outflow discharge, and agricultural effluent [19, 38, 47]. PC5 accounted for 7.06% of the total variance, loaded by Fe concentration. The presence of iron, a metal abundant in the Earth's crust, suggests that this factor may be associated with the weathering of iron-bearing rock [2, 3, 33].

2.3.5 Integrated Water Quality Index (IWQI) of Groundwater of the Study Area

According to the WHO-2011 drinking water quality standard, the IWQI values for the samples are displayed in Table 2.7. The IWQI values are ranked into five distinct classes according to the water quality: Excellent (< 50), Good (50–100), Medium (100–150), Poor (150–200), and Extremely Poor (> 200) [1]. In the JF area, IWQI values range from 131.38 to 230.30, 71.22 to 111.23, and 33.18 to 62.55, respectively in the shallow, intermediate, and deep samples. The corresponding mean concentrations are 169.92 ± 24.86, 92.37 ± 12.28, and 46.64 ± 7.66. Additionally, IWQI values range from 36.86 to 166.01, and 40.33 to 77.37 in the GF area for shallow wells, and deep wells, with mean concentrations of 105.07 ± 34.86, and 54.08 ± 14.99, respectively. The findings of the study indicate that seven shallow samples in JF fall into the extremely poor category, while the remaining samples range from medium to poor. Intermediate samples are classified as good to medium, and deep Holocene groundwater samples are rated as excellent to good, making them suitable for drinking. All of the deep wells' water is excellent to good for drinking quality, while the GF's shallow well samples are found to be of good to medium quality (Fig. 2.8c). The spatial distribution of IWQI highlights that water quality is relatively fresher in deep aquifers compared to shallow aquifers. The consumption of contaminated water from shallow aquifers the residents may be in danger of health problems (Fig. 2.8a, b).

2.3.6 Groundwater Pollution Index Assessment

To comprehensively assess the collective impact of determined heavy metals on groundwater quality and distinguish the pollution levels in the studied area, distinct pollution indices, including the Heavy Metal Pollution Index (HPI), Heavy Metal Evaluation Index (HEI), and Degree of Contamination (Cd), are constructed for each sample (Table 2.7).

In line with the classification proposed by Vetrimurugan et al. [84], HPI is categorized into five ranks: excellent (0–25), good (26–50), poor (51–75), very poor (76–100), and unsuitable (>100). The HPI values for the shallow, intermediate, and deep samples in the JF area are 397.78 to 691.46, 143.82 to 527.12, and 0.27 to 2.10, respectively. The respective average concentrations are 1.00 ± 0.58, 365.48 ± 117.58, and 536.98 ± 81.41. For shallow and deep wells in GF zones, the HPI values ranged from 60.35–256.13 and 31.6–98.52, respectively, with mean concentrations of 161.94 ± 50.21 and 64.62 ± 31.31. These classifications mean that deep well water is safe to drink, but all of the samples from shallow and intermediate wells in JF and the majority of the shallow water samples in GF are unsuitable for drinking (Fig. 2.9a). The shallow, intermediate, and deep samples in the JF area have HEI values ranging from 17.53 to 30.02, 10.38 to 16.67, and 0.03 to 0.94, respectively. The respective average concentrations are 0.36 ± 0.26, 13.57 ± 2.45, and 22.48 ± 3.18, respectively. Nonetheless, the groundwater sample HEI values in the designated GF regions range from 2.48 to 12.74 and 4.56 to 23.83, with corresponding mean concentrations of 12.77 ± 5.91 and 6.02 ± 4.11, respectively. In both JF and GF locations, shallow groundwater samples had the highest HEI value, which progressively decreased with depth (Fig. 2.9b). Furthermore, the Degree of Contamination (CD) was utilized to support the determination of heavy metal pollution levels [5]. The CD values for the shallow, intermediate, and deep samples in the JF area are, respectively, 11.53 to 24.02, 4.38 to 10.67, and − 5.97 to − 5.06. The respective

Table 2.7 Water quality, pollution index, and human health risk assessment

Parameters				Jamuna floodplain (JF)				Ganges floodplain (JF)			
				Min	Max	Avg	Std	Min	Max	Avg	Std
EWQI			River water	47.84	115.18	70.93	38.33	47.29	54.00	50.25	3.42
			Shallow	131.38	230.30	169.92	24.86	36.86	166.01	105.07	34.86
			Intermediate	71.22	111.23	92.37	12.28	–	–	–	–
			Deep	33.18	62.55	46.64	7.66	40.33	77.37	54.08	14.99
HPI			River water	0.59	2.82	1.47	1.18	19.77	23.52	21.47	1.90
			Shallow	397.78	691.46	536.98	81.41	60.35	256.13	161.94	50.21
			Intermediate	143.82	527.12	365.48	117.58	–	–	–	–
			Deep	0.27	2.10	1.00	0.26	31.65	98.52	64.62	31.31
HEI			River water	0.17	0.89	0.51	0.36	2.81	2.95	2.87	0.07
			Shallow	17.53	30.02	22.48	3.18	4.56	23.83	12.77	5.91
			Intermediate	10.38	16.67	13.57	2.45	–	–	–	–
			Deep	0.03	0.94	0.36	0.26	2.48	12.74	6.02	4.11
CD			River water	− 5.83	− 5.11	− 5.49	0.36	− 2.19	− 2.05	− 2.13	0.07
			Shallow	11.53	24.02	16.48	3.18	− 0.44	18.83	7.77	5.91
			Intermediate	4.38	10.67	7.57	2.45	–	–	–	–
			Deep	− 5.97	− 5.06	− 5.64	0.26	− 2.52	7.74	1.02	4.11
Health risk assessment	NCR	Children	River water	0.034	1.177	0.426	0.651	0.888	1.092	0.981	0.103
			Shallow	13.091	23.313	17.209	2.644	2.870	17.557	10.769	3.616
			Intermediate	3.941	16.708	11.420	4.084	–	–	–	–
			Deep	0.012	0.449	0.114	0.000	1.593	5.805	3.170	1.713
		Adult	River water	0.027	0.925	0.335	0.511	0.697	0.858	0.770	0.081

(continued)

Table 2.7 (continued)

Parameters				Jamuna floodplain (JF)				Ganges floodplain (JF)			
				Min	Max	Avg	Std	Min	Max	Avg	Std
			Shallow	10.286	18.317	13.521	2.078	2.255	13.794	8.461	2.841
			Intermediate	3.096	13.128	8.973	3.209	–	–	–	–
			Deep	0.010	0.353	0.090	0.122	1.252	4.561	2.490	1.346
	CR	Children	River water	0.000	0.000	0.000	0.000	0.000	0.000	0.000	0.000
			Shallow	0.008	0.022	0.014	0.003	0.001	0.008	0.004	0.002
			Intermediate	0.005	0.012	0.009	0.002	–	–	–	–
			Deep	0.000	0.000	0.000	0.580	0.001	0.002	0.001	0.001
		Adult	River water	0.000	0.000	0.000	0.000	0.000	0.000	0.000	0.000
			Shallow	0.006	0.017	0.011	0.002	0.001	0.006	0.003	0.001
			Intermediate	0.004	0.010	0.007	0.002	–	–	–	–
			Deep	0.000	0.000	0.000	0.000	0.000	0.002	0.001	0.001

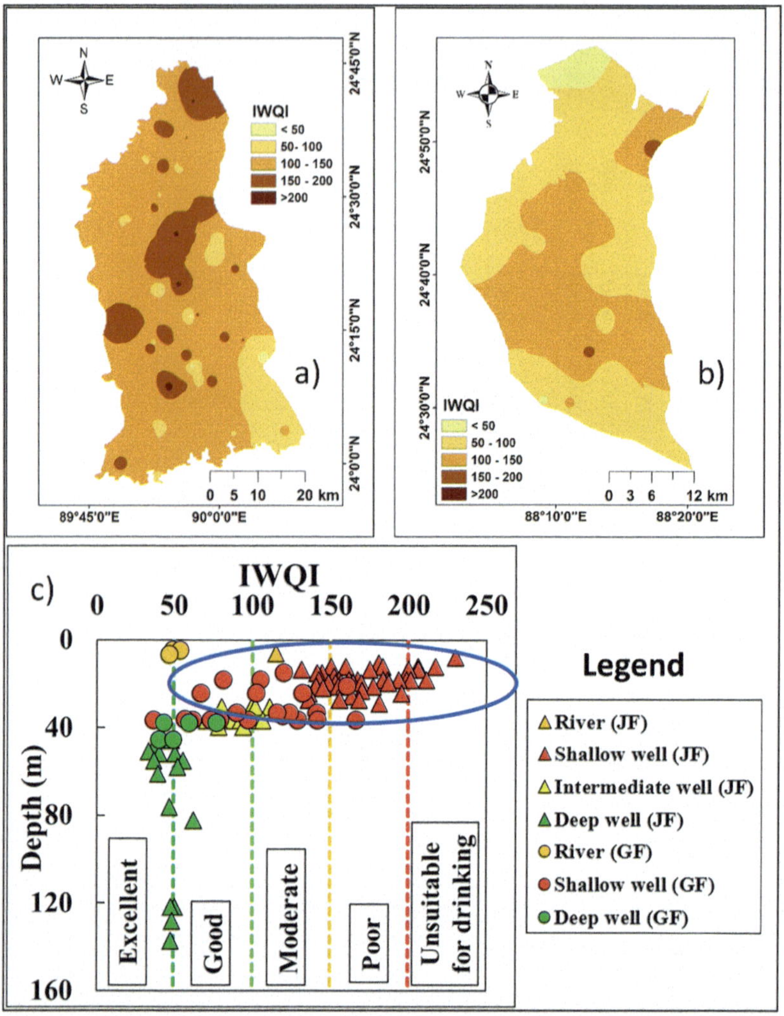

Fig. 2.8 Integrated weighted water quality index **a** and **b** spatial distribution of IWQI and **c** Depth versus IWQI

mean values are 16.48 ± 3.18, 7.57 ± 2.45, and – 5.64 ± 0.26. The average concentrations of 7.77 ± 5.91 and 1.02 ± 4.11 were found in the groundwater samples of shallow and deep wells, with CD values ranging from – 0.44 to 18.83 and – 2.52 to 7.74 in GF, respectively. These findings also reveal that the highest and lowest CD values of groundwater samples are found in JF and GF in shallow and deep wells, respectively (Fig. 2.9c).

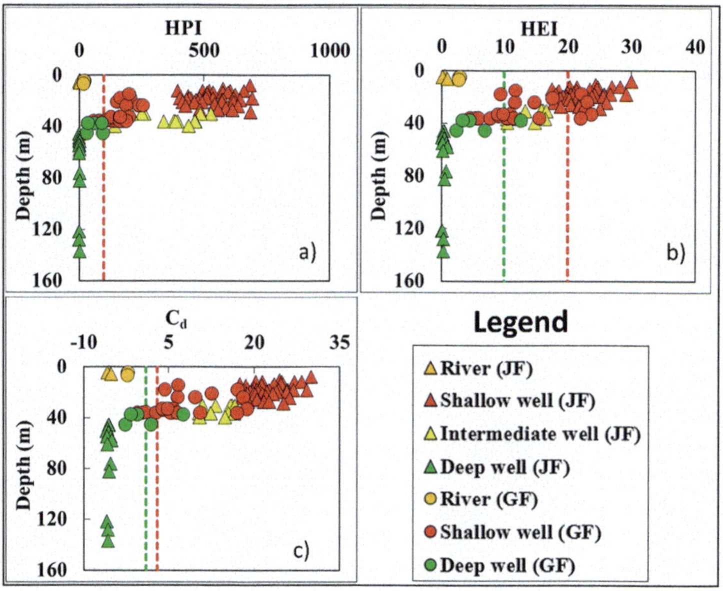

Fig. 2.9 Pollution index assessment **a** HPI, **b** HEI, and **c** degree of contamination (Cd)

2.3.7 Evaluation of Human Health Risk

Long-term ingestion of contaminated groundwater may have negative health effects. The health risk assessment is considered the most trustworthy technique for evaluating both non-carcinogenic and carcinogenic health risks [96]. This applies to both adults and children. The statistics of cancer risk and non-carcinogenic health concerns for adults and children through oral and dermal consumption are displayed in Table 2.7.

2.3.7.1 Non-carcinogenic Health Risk (NCR)

The non-carcinogenic risk (NCR) values for the JF area in shallow, intermediate, and deep samples range from 10.29–18.32, 3.10–13.13, and 0.01–0.35, with mean values for adults of 13.52 ± 2.08, 8.97 ± 3.21, and 0.09 ± 0.10; for children, the range is 13.09–23.31, 3.94–16.71, and 0.01–0.45, with mean values of 17.21 ± 2.64, 11.42 ± 4.08, and 0.11 ± 0.12 respectively. Additionally, for groundwater from shallow wells and deep wells in the GF area, the NCR values ranged from 2.25 to 13.79 and 1.25 to 4.56, with mean concentrations of 8.46 ± 2.84 and 2.49 ± 1.35 for adults. For children, the NCR values in shallow wells and deep wells ranged from 2.87 to 17.56 and 1.59 to 5.81, with mean concentrations of 10.77 ± 3.62 and 3.17 ± 1.71, respectively (Fig. 2.10a–d).

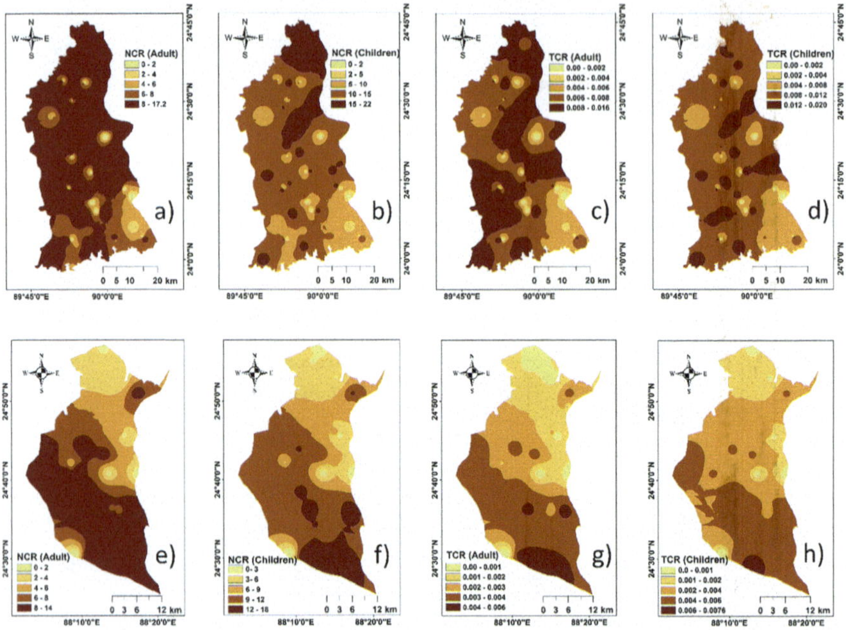

Fig. 2.10 The non-carcinogenic and the total carcinogenic health risk assessment of adults, and children

2.3.7.2 Carcinogenic Health Risk

The carcinogenic risk (CR) ranges for shallow, intermediate, and deep groundwater in the JF area were found to be 0.006–0.017, 0.004–0.010, and 2.7E−06–2.7E−06 for adults, with mean values of 0.011 ± 0.002, 0.007 ± 0.006, and 2.7E−06 ± 8.7E−22, respectively. The corresponding CR ranges for children in the JF area were 0.008–0.022, 0.005–0.012, and 3.4E−06–3.4E−06, with mean values of 0.014 ± 0.003, 0.009 ± 0.002, and 3.4E−06 ± 8.7E−22, respectively. On the other hand, the CR ranges for shallow and deep groundwater in the GF area for adults were 6.7E−04–6.0E−03 and 4.7E−04–1.9E−03, with mean values of 3.2E−03 ± 1.3E−03 and 9.3E−04 ± 5.8E−04, respectively. The corresponding CR ranges for children in the GF area were 8.5E−04–7.6E−03 and 6.0E−04–2.4E−03, with mean values of 4.0E−03 ± 1.6E−03 and 1.2E−03 ± 7.4E−04, respectively. The outcomes indicate that the concentrations of Fe, Mn, Cr, Pb, As, and Ni in both shallow and intermediate samples surpass the acceptable range of 1×10^{-6} to 1×10^{-4} for cancer risk, as recommended by USEPA [81]. These results underscore the potential adverse health effects on the local population through the consumption of contaminated drinking water. Notably, the JF areas exhibit higher cancer risk values compared to the GF area, attributed to increased irrigation and industrial activities in the former. The elevated accumulation of these contaminants in the study area poses significant cancer risks for the local residents. Spatial distribution analysis depicted in Fig. 2.10e–h reveals lower risks in the GF compared to the JF area, with risks being higher for children than adults. However, it's essential to acknowledge uncertainties in the risk assessment due to the utilization of parameters from USEPA [81], which may not be entirely suitable for the demographics of Bangladesh. Future studies should refine these assessments for more accurate evaluations, considering population-specific factors and exposure variations.

2.4 Conclusion

The groundwater plays a crucial role as the primary water source, supporting various industries, agriculture, and the daily lives of residents in the study area. The study conducted a comprehensive understanding of the groundwater quality, its suitability for drinking, groundwater sources, and the potential health risks of the Ganges and Jamuna floodplain aquifers. In this study, 105 samples were collected in the floodplain areas of Ganges floodplain and Jamuna floodplain in Bangladesh. These samples were analyzed to assess the chemistry, recharge mechanism, quality, pollution index, and potential health risks using hydrochemical methods, environmental isotope techniques, and multivariate statistical analysis. The Ca–HCO_3^- and Na–HCO_3^- are the predominant varieties of water in these aquifers. Shallow and intermediate wells in the Ganges and Jamuna floodplain aquifers have greater levels of K^+, NO_3^-, Fe, Mn, As, Cr, Ni, and Pb enrichment. The isotopic compositions indicate recent recharge from precipitation, with re-evaporation and mixing processes. IWQI reveals that groundwater from the Ganges and Jamuna floodplain (shallow and intermediate wells) aquifers is unsafe for human consumption. The Heavy Metal Evaluation Index (HEI), Heavy Metal Pollution Index (HPI), and Degree of Contamination (Cd) are higher for shallow and intermediate aquifers. Moreover, the HEI, and HPI both are higher at the Jamuna floodplain than the Ganges floodplain. The non-carcinogenic risk evaluation suggested that the mean hazard index (HI) of most groundwater samples from shallow and intermediate aquifers exceed the permissible limit and the carcinogenic risk ranges for children via oral consumption and adults and children via dermal exposure. The release of untreated heavy metal-containing effluents agricultural and industrial pollutants is dominant in the Jamuna floodplain aquifer whereas agricultural and livestock farming activities are dominant in the Ganges floodplain aquifer. The release of these contaminants to the environment should be reduced and properly managed. The study implies that proper water management will ensure safe drinking water in the Ganges and Jamuna floodplain aquifers for future generations.

References

1. Adimalla N (2021) Application of the entropy weighted water quality index (EWQI) and the pollution index of groundwater (PIG) to assess groundwater quality for drinking purposes: a case study in a rural area of Telangana State, India. Arch Environ Contam Toxicol 80(1):31–40
2. Ahmed N, Bodrud-Doza M, Islam SDU, Choudhry MA, Muhib MI, Zahid A, Hossain S, Moniruzzaman M, Deb N, Bhuiyan MAQ (2019) Hydrogeochemical evaluation and statistical analysis of groundwater of Sylhet, north-eastern Bangladesh. Acta Geochim 38:440–455
3. Ahmed N, Bodrud-Doza M, Islam ARMT, Hossain S, Moniruzzaman M, Deb N, Bhuiyan MAQ (2019) Appraising spatial variations of As, Fe, Mn and NO_3 contaminations associated health risks of drinking water from Surma basin, Bangladesh. Chemosphere 218:726–740
4. Akbor MA, Rahman MM, Bodrud-Doza M, Haque MM, Siddique MAB, Ahsan MA, Bondad SEC, Uddin MK (2020) Metal pollution in water and sediment of the Buriganga River, Bangladesh: an ecological risk perspective. Desalin Water Treat 193:284–301
5. Al-Ani MY, Al-Nakib SM, Ritha NM, Nouri AH (1987) Water quality index applied to the classification and zoning of Al-Jaysh canal, Baghdad-Iraq. J Environ Sci Health Part A 22(4):305–319
6. Al-Asad H, Moniruzzaman M, Sarker AK, Quaiyum Bhuiyan MA, Ahsan MA (2023) Hydrogeochemical evaluation, groundwater contamination and associated health risk in southern Tangail, Bangladesh. Chemosphere 332:138806
7. Anonna TA, Moniruzzaman M, Hadi AHANK, Khan AN, Sarker AK, Samanta P, Al Asad H (2022) Why pandemic coronavirus (SARS-CoV-2) hit different age groups of people in Southeast Asia? A case study in Bangladesh. J Ideas Health 5(2):655–663
8. Backman B, Bodiš D, Lahermo P, Rapant S, Tarvainen T (1998) Application of a groundwater contamination index in Finland and Slovakia. Environ Geol 36:55–64
9. Bahar MM, Reza MS (2010) Hydrochemical characteristics and quality assessment of shallow groundwater in a coastal area of Southwest Bangladesh. Environ Earth Sci 61:1065–1073
10. Balderacchi M, Benoit P, Cambier P, Eklo OM, Gargini A, Gemitzi A, Gurel M, Kløve B, Nakic Z, Trevisan M (2013) Groundwater pollution and quality monitoring approaches at the European level. Crit Rev Environ Sci Technol 43(4):323–408
11. Barzegar R, Asghari Moghaddam A, Soltani S, Fijani E, Tziritis E, Kazemian N (2019) Heavy metal (loid)s in the groundwater of Shabestar area (NW Iran): source identification and health risk assessment. Exposure Health 11:251–265
12. Bhattacharya P, Jacks G, Ahmed KM, Routh J, Khan AA (2002) Arsenic in groundwater of the Bengal delta plain aquifers in Bangladesh. Bull Environ Contam Toxicol 69(4)
13. Bhuiyan MAH, Bodrud-Doza M, Islam AT, Rakib MA, Rahman MS, Ramanathan AL (2016) Assessment of groundwater quality of Lakshimpur district of Bangladesh using water quality indices, geostatistical methods, and multivariate analysis. Environ Earth Sci 75:1–23
14. Bhuiyan MAH, Islam MA, Dampare SB, Parvez L, Suzuki S (2011) Evaluation of hazardous metal pollution in irrigation and drinking water systems in the vicinity of a coal mine area of northwestern Bangladesh. J Hazard Mater 179(1–3):1065–1077. https://doi.org/10.1016/j.jhazmat.2010.03.114
15. Bodrud-Doza MD, Islam AT, Ahmed F, Das S, Saha N, Rahman MS (2016) Characterization of groundwater quality using water evaluation indices, multivariate statistics and geostatistics in central Bangladesh. Water Sci 30(1):19–40
16. Bodrud-Doza M, Islam SDU, Hasan MT, Alam F, Haque MM, Rakib MA, Rahman MA (2019) Groundwater pollution by trace metals and human health risk assessment in central west part of Bangladesh. Groundw Sustain Dev 9:100219
17. Bortey-Sam N, Nakayama SM, Ikenaka Y, Akoto O, Baidoo E, Yohannes YB, Baidoo E, Mizukawa H, Ishizuka M (2015) Human health risks from metals and metalloid via consumption of food animals near gold mines in Tarkwa, Ghana: estimation of the daily intakes and target hazard quotients (THQs). Ecotoxicol Environ Saf 111:160–167

18. Chakraborti D, Rahman MM, Ahamed S, Dutta RN, Pati S, Mukherjee SC (2016) Arsenic contamination of groundwater and its induced health effects in Shahpur block, Bhojpur district, Bihar state, India: risk evaluation. Environ Sci Pollut Res 23:9492–9504
19. El-Rawy M, Fathi H, Abdalla F, Alshehri F, Eldeeb H (2023) An integrated principal component and hierarchical cluster analysis approach for groundwater quality assessment in Jazan, Saudi Arabia. Water 15(8):1466
20. Fallahzadeh RA, Ghaneian MT, Miri M, Dashti MM (2017) Spatial analysis and health risk assessment of heavy metals concentration in drinking water resources. Environ Sci Pollut Res 24:24790–24802
21. Freeze RA, Cherry JA (1979) Groundwater. Prentice-Hall Inc., Englewood Cliffs, p 604
22. Gotelli NJ, Ellison AM (2004) A primer of ecological statistics, 1st edn. Sinauer Associates, Sunderland
23. Habib MA, Islam ARMT, Bodrud-Doza M, Mukta FA, Khan R, Siddique MAB, Phoungthong K, Techato K (2020) Simultaneous appraisals of pathway and probable health risk associated with trace metals contamination in groundwater from Barapukuria coal basin, Bangladesh. Chemosphere 242:125183
24. Hassan HB, Moniruzzaman M, Majumder RK, Ahmed F, Bhuiyan MAQ, Ahsan MA, Al-Asad H (2023) Impacts of seasonal variations and wastewater discharge on river quality and associated human health risks: a case of northwest Dhaka, Bangladesh. Heliyon 9(7)
25. He X, Li P, Wu J, Wei M, Ren X, Wang D (2021) Poor groundwater quality and high potential health risks in the Datong Basin, northern China: research from published data. Environ Geochem Health 43(2):791–812
26. He S, Wu J (2019) Hydrogeochemical characteristics, groundwater quality, and health risks from hexavalent chromium and nitrate in groundwater of Huanhe Formation in Wuqi county, northwest China. Exposure Health 11:125–137
27. Hem JD (1985) Study and interpretation of the chemical characteristics of natural water, 3rd edn. U.S. Geological Survey Water-Supply Paper 2254. Retrieved from https://pubs.er.usgs.gov/publication/wsp2254
28. Hsu-Kim H, Kucharzyk KH, Zhang T, Deshusses MA (2013) Mechanisms regulating mercury bioavailability for methylating microorganisms in the aquatic environment: a critical review. Environ Sci Technol 47(6):2441–2456. https://doi.org/10.1021/es304370g
29. Islam ARMT, Ahmed N, Bodrud-Doza M, Chu R (2017) Characterizing groundwater quality ranks for drinking purposes in Sylhet district, Bangladesh, using entropy method, spatial autocorrelation index, and geostatistics. Environ Sci Pollut Res 24:26350–26374
30. Islam ARMT, Islam HT, Mia MU, Khan R, Habib MA, Bodrud-Doza M, Siddique MAB, Chu R (2020) Co-distribution, possible origins, status and potential health risk of trace elements in surface water sources from six major river basins, Bangladesh. Chemosphere 249:126180
31. Islam ARMT, Kabir MM, Faruk S, Al Jahin J, Bodrud-Doza M, Didar-ul-Alam M, Choudhury TR (2021) Sustainable groundwater quality in southeast coastal Bangladesh: co-dispersions, sources, and probabilistic health risk assessment. Environ Dev Sustain 1–30
32. Islam IR, Rahman M, Reza AHMS, Rahman M (2013) Groundwater geochemistry and its implication for arsenic enrichment and mobilization in shallow alluvial aquifers of Pakshi Union, Ishwardi, Pabna, Bangladesh. Int J Chem Mater Sci 1(4):069–078
33. Islam AT, Shen S, Haque MA, Bodrud-Doza M, Maw KW, Habib MA (2018) Assessing groundwater quality and its sustainability in Joypurhat district of Bangladesh using GIS and multivariate statistical approaches. Environ Dev Sustain 20:1935–1959
34. Jackson JE (2005) A user's guide to principal components. Wiley
35. Janardhana Raju N (2012) Arsenic exposure through groundwater in the middle Ganga plain in the Varanasi environs, India: a future threat. J Geol Soc India 79:302–314
36. Kabir MH, Kormoker T, Islam MS, Khan R, Shammi RS, Tusher TR, Proshad R, Islam MS, Idris AM (2021) Potentially toxic elements in street dust from an urban city of a developing country: ecological and probabilistic health risks assessment. Environ Sci Pollut Res 28(40):57126–57148

37. Karim Z (2011) Risk assessment of dissolved trace metals in drinking water of Karachi, Pakistan. Bull Environ Contam Toxicol 86:676–678
38. Kraiem Z, Zouari K, Chkir N, Agoune A (2014) Geochemical characteristics of arid shallow aquifers in Chott Djerid, south-western Tunisia. J Hydro Environ Res 8(4):460–473
39. Krishna AK, Mohan KR (2014) Risk assessment of heavy metals and their source distribution in waters of a contaminated industrial site. Environ Sci Pollut Res 21(5):3653–3669
40. Kumar M, Ramanathan AL, Mukherjee A, Sawlani R, Ranjan S (2019) Delineating sources of groundwater recharge and carbon in Holocene aquifers of the central Gangetic basin using stable isotopic signatures. Isot Environ Health Stud 55(3):254–271
41. Kumar M, Ramanathan AL, Tripathi R, Farswan S, Kumar D, Bhattacharya P (2017) A study of trace element contamination using multivariate statistical techniques and health risk assessment in groundwater of Chhaprola Industrial Area, Gautam Buddha Nagar, Uttar Pradesh, India. Chemosphere 166:135–145
42. Li P, He X, Li Y, Xiang G (2019) Occurrence and health implication of fluoride in groundwater of loess aquifer in the Chinese loess plateau: a case study of Tongchuan, Northwest China. Exposure Health 11(2):95–107
43. Li P, Wang D, Li W, Liu L (2022) Sustainable water resources development and management in large river basins: an introduction. Environ Earth Sci 81(6):179
44. Mohan SV, Nithila P, Reddy SJ (1996) Estimation of heavy metals in drinking water and development of heavy metal pollution index. J Environ Sci Health Part A 31(2):283–289
45. Moldovan A, Hoaghia MA, Kovacs E, Mirea IC, Kenesz M, Arghir RA, Moldovan OT et al (2020) Quality and health risk assessment associated with water consumption—a case study on karstic springs. Water 12(12):3510
46. Moniruzzaman M, Asad HA, Sarker AK, Bhuiyan MAQ, Ahsan MA, Majumder RK, Hassan HB (2024) Hydrogeochemical appraisal, sources, quality and potential health risk assessment in Holocene and Pleistocene aquifers in Bangladesh. Environ Sci Pollut Res 31(38):50261–50282
47. Moniruzzaman M, Lee JH, Jung KM, Kwon JS, Kim KH, Yun ST (2018) Lithologic control of the hydrochemistry of a point-bar alluvial aquifer at the low reach of the Nakdong River, South Korea: implications for the evaluation of riverbank filtration potential. Water 10(12):1763
48. Mosaffa M, Nazif S, Amirhosseini YK, Balderer W, Meiman HM (2021) An investigation of the source of salinity in groundwater using stable isotope tracers and GIS: a case study of the Urmia Lake basin, Iran. Groundw Sustain Dev 12:100513
49. Mushtaq N, Masood N, Khattak JA, Hussain I, Khan Q, Farooqi A (2021) Health risk assessment and source identification of groundwater arsenic contamination using agglomerative hierarchical cluster analysis in selected sites from upper Eastern parts of Punjab province, Pakistan. Hum Ecol Risk Assess Int J 27(4):999–1018
50. Nalbantçılar MT, Pınarkara D (2015) Impact of industry on ground water contamination: a case study in Konya city, Turkey
51. Naz A, Chowdhury A, Mishra BK, Gupta SK (2016) Metal pollution in water environment and the associated human health risk from drinking water: a case study of Sukinda chromite mine, India. Hum Ecol Risk Assess Int J 22(7):1433–1455
52. Panhwar AH, Kazi TG, Afridi HI, Shaikh HR, Arain SA, Arain SS, Brahman KD (2013) Evaluation of calcium and magnesium in scalp hair samples of population consuming different drinking water: risk of kidney stone. Biol Trace Elem Res 156:67–73
53. Piper AM (1944) A graphic procedure in the geochemical interpretation of water analyses. Trans Am Geophys Union 25(6):914–928. https://doi.org/10.1029/TR025i006p00914
54. Rafique N, Tariq SR (2016) Distribution and source apportionment studies of heavy metals in soil of cotton/wheat fields. Environ Monit Assess 188:1–10
55. Rahman MM, Islam MA, Bodrud-Doza M, Muhib MI, Zahid A, Shammi M, Kurasaki M (2018) Spatio-temporal assessment of groundwater quality and human health risk: a case study in Gopalganj, Bangladesh. Exposure Health 10:167–188
56. Rahman A, Rahaman H (2018) Contamination of arsenic, manganese and coliform bacteria in groundwater at Kushtia District, Bangladesh: human health vulnerabilities. J Water Health 16(5):782–795

57. Rahman MS, Reza AS, Siddique MAB, Akbor MA, Hasan M (2023) Accumulation of arsenic and other metals in soil and human consumable foods of Meherpur district, southwestern Bangladesh, and associated health risk assessment. Environ Sci Eur 35(1):47
58. Rani A, Parashar K, Meena R, Sharma SK, Tiwari KK, Ajaykumar V, Mondal NC (2023) Hydrochemical characteristics and potential health risks of nitrate, fluoride, and uranium in Kota district, Rajasthan, India. Environ Sci Pollut Res 30(34):82485–82505
59. Rashid A, Ayub M, Ullah Z, Ali A, Khattak SA, Ali L, Gao X, Li C, Khan S, Kaushik P (2022) Geochemical modeling source provenance, public health exposure, and evaluating potentially harmful elements in groundwater: Statistical and human health risk assessment (HHRA). Int J Environ Res Public Health 19(11):6472
60. Rezaei A, Hassani H, Hassani S, Jabbari N, Mousavi SBF, Rezaei S (2019) Evaluation of groundwater quality and heavy metal pollution indices in Bazman basin, Southeastern Iran. Groundw Sustain Dev 9:100245
61. Rezaei H, Jafari A, Kamarehie B, Fakhri Y, Ghaderpoury A, Karami MA, Salimi M (2019) Health-risk assessment related to the fluoride, nitrate, and nitrite in the drinking water in the Sanandaj, Kurdistan County, Iran. Hum Ecol Risk Assess Int J 25(5):1242–1250
62. Saha N, Rahman MS, Ahmed MB, Zhou JL, Ngo HH, Guo W (2017) Industrial metal pollution in water and probabilistic assessment of human health risk. J Environ Manage 185:70–78
63. Sappa G, Ergul S, Ferranti F (2014) Geochemical modeling and multivariate statistical evaluation of trace elements in arsenic contaminated groundwater systems of Viterbo Area, (Central Italy). Springerplus 3:1–19
64. Selvakumar S, Ramkumar K, Chandrasekar N, Magesh NS, Kaliraj S (2017) Groundwater quality and its suitability for drinking and irrigational use in the Southern Tiruchirappalli district, Tamil Nadu, India. Appl Water Sci 7:411–420
65. Shamsudduha M, Joseph G, Haque SS, Khan MR, Zahid A, Ahmed KMU (2020) Multi-hazard groundwater risks to water supply from shallow depths: challenges to achieving the sustainable development goals in Bangladesh. Exposure Health 12(4):657–670
66. Shannon CE (1948) A mathematical theory of communication. Bell Syst Tech J 27
67. Sheikhi S, Faraji Z, Aslani H (2021) Arsenic health risk assessment and the evaluation of groundwater quality using GWQI and multivariate statistical analysis in rural areas, Hashtroud, Iran. Environ Sci Pollut Res 28(3):3617–3631
68. Shi G, Chen Z, Bi C, Wang L, Teng J, Li Y, Xu S (2011) A comparative study of health risk of potentially toxic metals in urban and suburban road dust in the most populated city of China. Atmos Environ 45(3):764–771
69. Siddique MAB, Alam MK, Islam S, Diganta MTM, Akbor MA, Bithi UH, Chowdhury AI, Ullah AA (2020) Apportionment of some chemical elements in soils around the coal mining area in northern Bangladesh and associated health risk assessment. Environ Nanotechnol Monit Manag 14:100366
70. Singh S, Anil AG, Kumar V, Kapoor D, Subramanian S, Singh J, Ramamurthy PC (2022) Nitrates in the environment: a critical review of their distribution, sensing techniques, ecological effects and remediation. Chemosphere 287:131996
71. Singh DD, Thind PS, Sharma M, Sahoo S, John S (2019) Environmentally sensitive elements in groundwater of an industrial town in India: spatial distribution and human health risk. Water 11(11):2350
72. Singha S, Pasupuleti S, Singha SS, Singh R, Kumar S (2021) Prediction of groundwater quality using efficient machine learning technique. Chemosphere 276:130265
73. Smedley PL, Kinniburgh DG (2002) A review of the source, behaviour and distribution of arsenic in natural waters. Appl Geochem 17(5):517–568. https://doi.org/10.1016/S0883-2927(02)00018-5
74. Solangi GS, Siyal AA, Babar MM, Siyal P (2020) Groundwater quality evaluation using the water quality index (WQI), the synthetic pollution index (SPI), and geospatial tools: a case study of Sujawal district, Pakistan. Hum Ecol Risk Assess Int J
75. Su Z, Wu J, He X, Elumalai V (2020) Temporal changes of groundwater quality within the groundwater depression cone and prediction of confined groundwater salinity using Grey Markov model in Yinchuan area of northwest China. Exposure Health 12(3):447–468

76. Subba Rao N, Sunitha B, Adimalla N, Chaudhary M (2020) Quality criteria for groundwater use from a rural part of Wanaparthy District, Telangana State, India, through ionic spatial distribution (ISD), entropy water quality index (EWQI) and principal component analysis (PCA). Environ Geochem Health 42:579–599
77. Uddin A, Shamsudduha M, Saunders JA, Lee MK, Ahmed KM, Chowdhury MT (2011) Mineralogical profiling of alluvial sediments from arsenic-affected Ganges-Brahmaputra floodplain in central Bangladesh. Appl Geochem 26(4):470–483
78. United States Environmental Protection Agency (USEPA) (1989) Risk assessment guidance for superfund volume I human health evaluation manual (part A)
79. United States Environmental Protection Agency (USEPA) (1999) A risk assessment–multi way exposure spread sheet calculation tool. United States Environmental Protection Agency, Washington, DC
80. United States Environmental Protection Agency (USEPA) (2001) Baseline human health risk assessment Vasquez Boulevard and I-70 superfund site. Denver CO. U.S. Environmental Protection Agency. Assessed from. http://www.epa.gov/region8/superfund/sites/VB-170-Risk.pdf
81. United States Environmental Protection Agency (USEPA) (2004) Guidelines for water reuse. EPA/625/R-04/108
82. United States Environmental Protection Agency (USEPA) (2009) National Primary and Secondary Drinking Water Standards. U.S. Environmental Protection Agency. Assessed from. http://www.epa.gov/safewater/cnsumer/pdf/mcl.pdf. Accessed 20 Sept 2017
83. United States Environmental Protection Agency (USEPA) (2013) National primary drinking water regulations, drinking water contaminants. EPA 816-F-09–004, pp 141–142
84. Vetrimurugan E, Brindha K, Elango L, Ndwandwe OM (2017) Human exposure risk to heavy metals through groundwater used for drinking in an intensively irrigated river delta. Appl Water Sci 7:3267–3280
85. Vinnarasi F, Srinivasamoorthy K, Saravanan K, Gopinath S, Prakash R, Ponnumani G, Babu C (2021) Chemical weathering and atmospheric carbon dioxide (CO_2) consumption in Shanmuganadhi, South India: evidences from groundwater geochemistry. Environ Geochem Health 43:771–790
86. WHO (2011) Guidelines for drinking-water quality. World Health Organization, Geneva
87. WHO and UNICEF (2017) Progress on drinking water, sanitation and hygiene: 2017 update and SDG baselines. World Health Organization (WHO), Geneva and the United Nations Children's Fund (UNICEF), New York
88. Wagh VM, Panaskar DB, Mukate SV, Aamalawar ML, Laxman Sahu U (2020) Nitrate associated health risks from groundwater of Kadava river basin Nashik, Maharashtra, India. Hum Ecol Risk Assess Int J 26(3):654–672
89. Wagh VM, Panaskar DB, Muley AA, Mukate SV (2017) Groundwater suitability evaluation by CCME WQI model for Kadava River Basin, Nashik, Maharashtra, India. Model Earth Syst Environ 3:557–565
90. Wang X, Tang Z (2020) The first large-scale bioavailable Sr isotope map of China and its implication for provenance studies. Earth Sci Rev 210:103353
91. Wang D, Wu J, Wang Y, Ji Y (2020) Finding high-quality groundwater resources to reduce the hydatidosis incidence in the Shiqu County of Sichuan Province, China: analysis, assessment, and management. Exposure Health 12(2):307–322
92. Wu B, Zhao DY, Jia HY, Zhang Y, Zhang XX, Cheng SP (2009) Preliminary risk assessment of trace metal pollution in surface water from Yangtze River in Nanjing Section, China. Bull Environ Contam Toxicol 82:405–409
93. Xiao Y, Yin S, Hao Q, Gu X, Pei Q, Zhang Y (2020) Hydrogeochemical appraisal of groundwater quality and health risk in a near-suburb area of North China. J Water Suppl Res Technol AQUA 69(1):55–69
94. Xu Y, Xue X, Dong L, Nai C, Liu Y, Huang Q (2018) Long-term dynamics of leachate production, leakage from hazardous waste landfill sites and the impact on groundwater quality and human health. Waste Manage 82:156–166

95. Yang M, Fei Y, Ju Y, Ma Z, Li H (2012) Health risk assessment of groundwater pollution—a case study of typical City in North China plain. J Earth Sci 23(3):335–348
96. Zhang Q, Xu P, Qian H (2020) Groundwater quality assessment using improved water quality index (WQI) and human health risk (HHR) evaluation in a semi-arid region of northwest China. Exposure Health 12(3):487–500
97. Zhou Y, Ning J, Li L, Long Q, Wei A, Liu Z (2020) Health risk assessment of groundwater in Gaobeidian, north China: distribution, source, and chemical species of the main contaminants. Exposure Health 12:427–446

Chapter 3
Assessment of Pollution and Water Quality of Coastal Seawater of Saint Martin's Island

Imtiaz Ahmed Sakib, Ferdousi Begum, Adiba Mosharraf, Farhana Akter, Md. Arman Hossain, S. K. Rahat Rezwon, and Md. Abu Bin Hasan Susan

Abstract Saint Martin's island, located at the southernmost tip of Bangladesh, has significant ecological value for being most favourite tourist spot and acting as a nesting site for many internationally vulnerable marine species. This island's marine biodiversity is exceptionally wealthy due to its favorable environment, but its marine environment is facing threats day by day due to natural calamities, various types of pollution, and other anthropogenic activities. This study was conducted on Saint Martin's Island, through assessing the quality of seawater collected from 9 stations, determined using the Global Positioning System. Various physicochemical properties: pH, conductivity, total dissolved solids, turbidity, salinity, density, viscosity, refractive index and dissolved oxygen with concentration of phosphate and nitrate were measured. Concentration of arsenic, lead, mercury, zinc, sodium and magnesium were also investigated by Atomic Absorption Spectrophotometer and the trend of average concentration of mineral and heavy metals was Na > Mg and Zn > Pb > Hg > As, respectively. Furthermore, several water quality indices were calculated for physicochemical parameters and concentration of heavy metals where category of seawater quality was found "Good" according to Canadian Council of Ministers of Environment (CCME) Water Quality Index (WQI) and Weighted Arithmetic (WA) Water Quality Index (WQI) revealed "poor" quality. Thus, seawater quality parameters have been critically analysed, correlated and essential findings were noted and. Obtained results highlight variation in most of the seawater quality parameters of Saint Martin and have a good correlation with each other.

I. A. Sakib · A. Mosharraf · Md. A. Hossain · S. K. R. Rezwon
Department of Oceanography and Hydrography, Bangladesh Maritime University, Dhaka, Bangladesh

F. Begum (✉) · F. Akter
Department of Chemistry, Bangladesh Maritime University, Dhaka, Bangladesh
e-mail: lovely_ferdousi@yahoo.com

Md. A. B. H. Susan
Department of Chemistry, University of Dhaka, Dhaka, Bangladesh
e-mail: susan@du.ac.bd

G. M. Tarekul Islam et al. (eds.), *Sustainable Water Management and Environmental Resilience*, Springer Water, https://doi.org/10.1007/978-3-031-93350-9_3

Keywords Seawater · Physicochemical properties · Water quality index · Contamination factor · Heavy metal pollution index

3.1 Introduction

Physicochemical properties of coastal seawater are crucial to a healthy marine ecology and sustainable fisheries. Variability of productivity in coastal seawater is a result of fluxes of nutrients that may occur from both anthropogenic and natural sources, as well as changes in the physicochemical characteristics of the water, which can ultimately affect seawater quality (Idrus et al. 2017). Temperature, rainfall, pH, salinity, and dissolved oxygen (DO) are significant physical and chemical factors affecting the aquatic environment. Others include total alkalinity and acidity, total suspended and dissolved particles, and heavy metal pollutants. The survival of aquatic species is constrained by these conditions (flora and fauna) [19].

Saint Martin's island in Bangladesh (Scheme 3.1) is a distinctive coral-bearing island with significant ecological significance and an increasingly popular tourist destination in Bangladesh because of its geographical location and ideal environmental conditions [23]. This island's extensive area of mangrove formations and dunes provide some of the handful of surviving habitats for a number of marine species that are vulnerable or globally very rare. Due to its abundance of marine biotic resources, this island serves as a flyway and wintering place for migrating birds from East and Australasia [5]. Any work particular to physicochemical parameters of coastal seawater surrounding Saint Martin's island is not yet to be done. To sustain the seawater quality and its standard, the marine aquatic ecosystem needs to be carefully protected.

Although many attempts have already been made to conserve the numerous endangered species of turtles and corals, etc., the number of tourists has recently increased rapidly, causing the island's natural ecosystem to deteriorate [6]. Many works have been done on physicochemical properties of the ocean all over the world. An attempt has been made to establish a baseline scenario based on the studies of six physicochemical parameters: temperature, conductivity, salinity, total dissolved solids (TDS), pH, and dissolved oxygen (DO) on Saint Martin's island [13]. Environmental impact was also monitored to make a current status of Saint Martin's island coral resources and environmental assessment using pH, salinity, turbidity, and temperature (Ahammed et al. 2016).

Saint Martin's Island was designated as an ecologically critical area under the Bangladesh Environment Conservation Act, according to the proposal of the National Conservation Society (Ahammed et al. 2016). There have been several initiatives to safeguard the island's biodiversity and ecosystem, but the lack of environmental law, long-term sustainable policy implementation, and monitoring make it unlikely that an effective protected area will be established [5]. Despite several studies, a detailed physicochemical profiling along with heavy metal and nutrient analysis is required to establish a strong ground for further research for the well-being of the aquatic

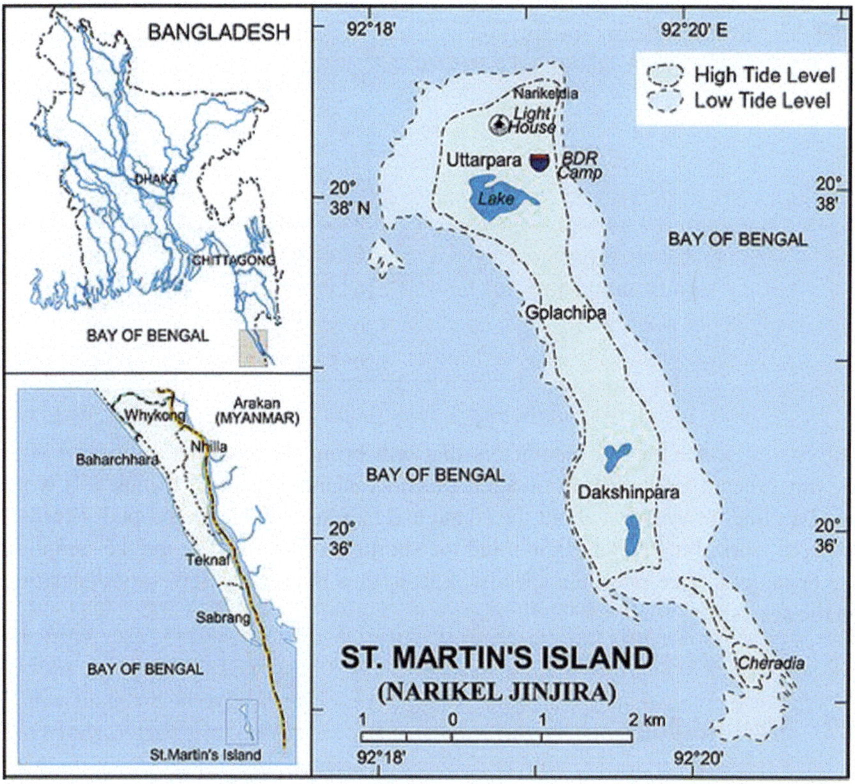

Scheme 3.1 Map of Saint Martin's island in Bangladesh [33]

environment of the island. Water Quality Index (WQI), Heavy Metal Pollution Load Index (HPI), Contamination Factor (CF), Correlation Coefficient (CC), and other indexes through the studies of physicochemical properties with determination of concentration of heavy metals or minerals will provide an idea about the current status of aquatic environment so that required initiatives can be taken. In this study, an extensive work has been detailed to establish a basic scenario of pollution status of Saint Martin's island. WQI was calculated using the measured physicochemical properties for coastal seawater of Saint Martin's island. While doing so, this index needed expert judgment and assumptions along with recognized standards, which can vary from person to person and also an absence of established standards can slightly alter the outcome. In this case, previous research works were considered. Additionally, results obtained from the studies of physicochemical properties have been correlated with each other through statistical analysis. This study primarily focuses on the physicochemical parameters along with the concentration of nutrients and heavy metals and minerals. The ultimate goal has been to give information about the pollution condition of Saint Martin's island's sorrounding coastal seawater through calculation of WQI, HPI, and CF as well as a correlation based on statistical

Table 3.1 Geographical co-ordinates of the sample sites

Sample No.	Longitude	Latitude
1	20° 36′ 56″	92° 19′ 53″
2	20° 35′ 58″	92° 20′ 1″
3	20° 35′ 19″	92° 20′ 19″
4	20° 37′ 90″	92° 19′ 58″
5	20° 34′ 20″	92° 20′ 26″
6	20° 38′ 5″	92° 19′ 11″
7	20° 37′ 57″	92° 18′ 56″
8	20° 37′ 32″	92° 18′ 60″
9	20° 37′ 13″	92° 19′ 24″

analysis on water quality parameters that will bring the research up-to-date with the most recent investigation of Saint Martin's island. Obtained results will work as a baseline information about the island and help to indicate the scope for further research work. Finally, taking in mind the shortage of time and scope the sampling has been done once only in a specific season, thus the results may vary depending on the season.

3.2 Methodology

3.2.1 Study Areas

Saint Martin's island was selected as the area for this study and it is located at the north-eastern side of the BoB, about 9 km south to Teknaf peninsula. As it is a small island, the study will cover 9 stations around the island, having latitudinal range from 20° 34′ 20″ N to 20° 38′ 5″ N latitude and longitudinal range from 92° 18′ 56″ E to 92° 20′ 19″ E longitude at the Bangladeshi territory of Bay of Bengal. The geographic coordinates for each sampling station on Saint Martin's island from which coastal saltwater was collected can be seen in Table 3.1 and Scheme 3.2.

3.2.2 Justification of the Sampling Site

As the main objective of this research was to get information about the current scenario of pollution of Saint Martin's Island's surrounding seawater, the sampling stations were chosen as possible to give priority to the main target (Scheme 3.2; Table 3.1). Coastal seawater were collected surrounding the island so that a full picture of the quality parameters of adjacent water of the island could be observed.

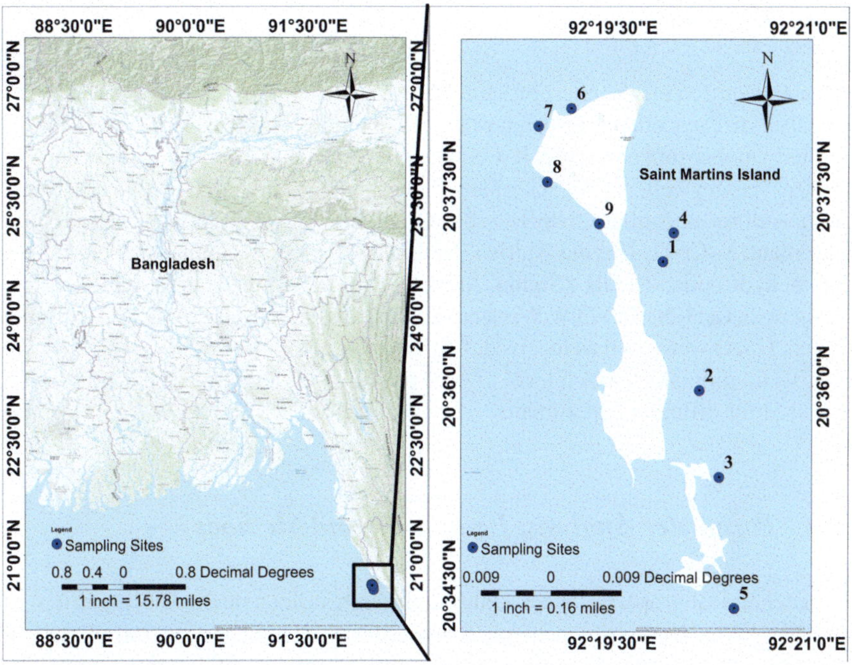

Scheme 3.2 Geographical co-ordinates of all sampling stations of Saint Martin's island (satellite view of Bangladesh and Saint Martin's island with its adjacent neighbors (left) and sampling stations around Saint Martin's island (right)

Station no.1, 2, 3, 4, and 5 cover the island's eastern side from the central point to the tip of Cheradwip. Station no. 6, 7, 8, and 9 cover the north-west side of the island.

3.2.3 Sample Collections and Preservation

Coastal seawater were collected into properly cleaned bottles with no air bubbles from the surrounding of Saint Martin's island in March 2022. Then sample bottles were labeled with station no., sealed in air tight condition, preserved in an adiabatic ice box at 4 °C temperature and fixed with nitric acid for further study. Co-ordinates of the sample locations were measured correctly and precisely through handheld Global Positioning System (GPS).

3.2.4 Materials

Seawater acquired off the coast of Saint Martin's island in March of 2022 were used for analysis to carry out this research work. Silver nitrate, $AgNO_3$, (Honeywell, Fluka, Germany), potassium chromate, K_2CrO_4 (Sigma Aldrich, USA), manganese sulfate, $MnSO_4$, (Honeywell, Fluka, Germany), potassium iodide, KI, (Sigma Aldrich, Chile), sodium thiosulfate, $Na_2S_2O_3.5H_2O$ (Sigma Aldrich, Germany), potassium dichromate, $K_2Cr_2O_7$ (Sigma Aldrich, Netherlands), starch (Sigma Aldrich, India), sodium hydroxide, NaOH (Sigma Aldrich, Sweden), nitrification inhibitor (*N*-allythiourea, $C_4H_8N_2S$), WTW Wissenschaftlich, and silver sulfate, Ag_2SO_4, (Sigma Aldrich, USA) were used as received. Ultra-pure water with resistivity 18.2 MΩ cm at 25 °C, total organic carbon level to < 10 ppb and flow rate (max.) are 1.50 L/min were used for different experiments.

3.2.5 Parameter Analysis, Equipment and Methods

Physicochemical properties were analyzed through various methods using different equipment. Analyzed physicochemical properties were temperature, turbidity, total dissolved solids (TDS), pH, salinity, electrical conductivity (EC), density, viscosity, refractive index (RI), and dissolved oxygen (DO). Temperature and pH were observed with Handheld Thermometer and WTW Lab-pH Meter inoLab® pH 7110, respectively. Salinity, EC and TDS were measured by WTW inoLab Multiparameter 9310 IDS (P) using electrometric method. DO was measured with Portable DO meter: EcoSence DO® 200A, Xylem by polarographic electrode with convenient screw-on cap membrane. Density, viscosity, turbidity, and refractive index were measured using AntanPaar vibrating tube density meter DMA 4500 ME following oscillating U-tube method, AntonPaar-Lovis 2000 M/ME microviscometer with rolling ball method, Nepthelometric method by EPA Compliant Benchtop Turbidity Meter—HI88703, HANNA and Abbemat 300 refractometer including high-resolution optical sensor with a resolution and limit error $\pm 10^{-5}$, Range nD: 1.26 to 1.72, respectively. Concentration of nutrients was measured by Single beam UV-spectrophotometer (DR 3900) using HACH Powder Pillows 8048 and 8039. Concentration of different heavy metals/minerals: As, Hg, Zn, Pb, Na and Mg were measured by Atomic Absorption Spectrophotometer (AAS) using different methods: electric hydride for As (hollow cathode lamp of wavelength 193.7 nm), cold vapor for Hg (in the light path of wavelength of 253.7 nm) and direct flame absorption for Zn, Pb, Na and Mg at 213.9, 217.0, 589.0 and 589.0 nm, respectively. Furthermore, different reagent solutions were prepared with ultrapure water using Ultrapure Water Purification System, aquaMAXTM—Ultra 370. These reagents were weighed and mixed using ATY224 Analytical Balance, Shimadzu, and Biobase Hotplate Magnetic Stirrer, BS-2H, respectively during removal of halides from these seawater for the determination of concentration of heavy metals/minerals.

3.2.6 Sample Preparation for AAS: Elimination of Halides from Seawater Samples

Halides were removed using the following method to correctly determine concentration of minerals and heavy metals in seawater samples by Atomic AbsorptionSpectrophotometer (AAS) [1]. 50.0 mL of seawater sample was poured into a 100 mL Erlenmeyer flask and then 5.0 g of $AgSO_4$ was added. A small magnetic rod was placed inside the flask before it was placed on the magnetic stirrer plate. Different samples from different stations were mixed for a period of time ranging from 60 to 90 min at room temperature, and the precipitation that was white at first changed to a faded lilac-colored mixture. At this point, mixing was stopped and the flask was taken from the magnetic stirrer and placed in a 40° inclined position on a suitable inclined supporting frame for a rest period of 5–10 min. Sedimentation of colored precipitate was very quick and after filtration ca. 45.0 mL of the clear seawater was collected in 60 mL sample bottles.

3.2.7 Tools/Methods for Data Analysis

Spatial distribution of different seawater quality parameters was observed by Inverse Distance Weighted (IDW) interpolation method [2]. Several software packages were used to obtain, manage, process, analyze, visualize and interpret data. ArcGIS 10.8 was used for mapping of study area and Microsoft Excel 2013 was used for data profiling, linear correlation plots [17, 23] and Pearson Correlation [32]. In this study, Canadian Council of Ministers of the Environment (CCME) Water Quality Index (CCME WQI) [4] and Weighted Arithmetic Water Quality Index (WA-WQI) [10] methods were used to calculate WQI.

3.3 Results and Discussion

3.3.1 Assessment of Seawater Quality Parameters and Their Spatial Distribution Using GIS Based Inverse Distance Weighted Interpolation

3.3.1.1 Temperature

Figure 3.1 shows temperature and spatial distribution temperature of seawater acquired off the coast of Saint Martin's island. Temperature varied between 26.09 and 30.01 °C with average temperature of 27.45 °C. Highest and lowest temperatures were recorded at station no. 9 and 2, respectively.

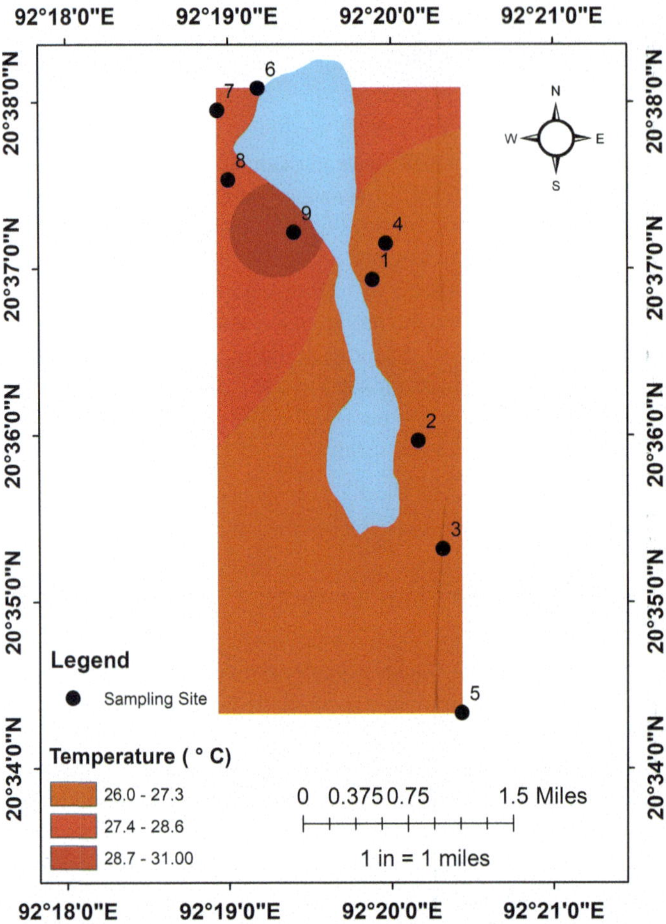

Fig. 3.1 Temperature (left) and spatial distribution of temperature (right) in coastal seawater around Saint Martin's island [31]

Interpolated distribution (Fig. 3.1, right) examined that lower range (26.0–27.3 °C) in the central to southern side and northern side of the island had highest range (28.7–31.0 °C) of temperature, where most of tourism activities takes place (Mian 2005). With the exception of one site, all temperatures were within the normal range; nonetheless, the difference was too little to be significant. Variation in temperature was observed as the seawater was collected in day time in dry season, the adjacent air temperature was high. Furthermore, Ocean currents and wind patterns also affect the temperature of seawater.

3.3.1.2 pH

Figure 3.2 shows pH and spatial distribution of pH in seawater acquired off the coast of Saint Martin's island. Seawater was slightly alkaline at all the stations (Fig. 3.2) and pH of various stations ranged between 8.21 and 8.31 where highest and lowest pH were found at station no. 8 and 1, 7, respectively with average pH of 8.26. The analysis of the pH interpolation (Fig. 3.2, right) revealed that the pH of the seawater at almost every station followed a similar pattern, with very little variation. Maybe because it was winter, the pH that was obtained was within the standard range for seawater. The pH levels at every site were within the acceptable range, indicating that the island's coastline saltwater is not yet being impacted by ocean acidification [24].

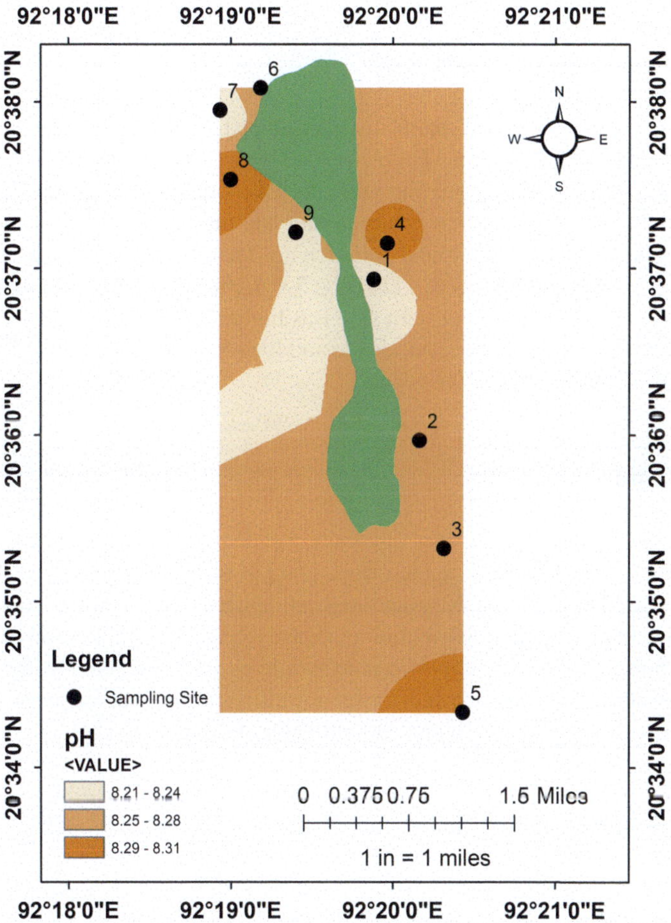

Fig. 3.2 PH (left) and spatial distribution of pH (right) in coastal seawater around Saint Martin's island [31]

3.3.1.3 Turbidity

Figure 3.3 gives turbidity and spatial distribution of turbidity in coastal seawater around Saint Martin's Island. Observed turbidity of all seawater was quite low (range from 0.2 to 1.67 NTU) which is desired for a sustainable coral reef environment (Fig. 3.3). Highest and lowest turbidity were found at stations 5 and 3, respectively with average turbidity of 0.65 NTU. Interpolation of the turbidity (Fig. 3.3, right) showed that The northern portion of the island, where there is a lot of tourist, has slightly higher turbidity (1.17–1.67 NTU) might be due to the massive amount of silt that has been deposited there from the deltaic system [15]. A healthy, sustainable coral reef habitat during the dry season depends on turbidity levels between 0.15 and 0.70 NTU, which have been observed throughout the island, with the exception of the northern half [29].

3.3.1.4 Total Dissolved Solids (TDS)

Figure 3.4 represents TDS and spatial distribution of TDS in coastal seawater around Saint Martin's Island. Range of TDS in seawater was quite lower (24.95–25.63 g/L) than average TDS (35 g/L) of seawater reflects good quality of seawater. This could be because there is a decrease in water mixing and the sea is calmer in the winter. Lowest and highest TDS were found at station no. 7 and 8, respectively with average TDS of 25.25 g/L. Distribution of TDS (Fig. 3.4, right) clearly showed that TDS of coastal seawater on the north western side was higher (25.41–25.62 g/L) as well as gradually became lower on the central side and lowest value (24.95–25.17 g/L) was found on the southern side of the island.

3.3.1.5 Density

Figure 3.5 shows density and spatial distribution of density in coastal seawater around Saint Martin's Island. Density ranged from 1.02334 to 1.02396 g/cm^3 with average density of 1.02369 g/cm^3. Lowest and highest density were found at station no. 6 and 9, respectively. Spatial distribution of density (Fig. 3.5, right) showed that the island's central part had the highest density, while its southern and northern regions had the lowest. Surface fluxes, wind-driven upwelling, advection, temperature, TDS, salinity, and other variables all interact to affect surface density [27].

3.3.1.6 Viscosity

Figure 3.6 gives viscosity and spatial distribution of viscosity in seawater acquired off the coast of Saint Martin's island. Viscosity was ranged from 0.942 to 0.949 mPa's with average viscosity of 0.944 mPa's. Lowest and highest viscosity were found at station no. 1 and 8, respectively andit is apparent from interpolation plot (Fig. 3.6,

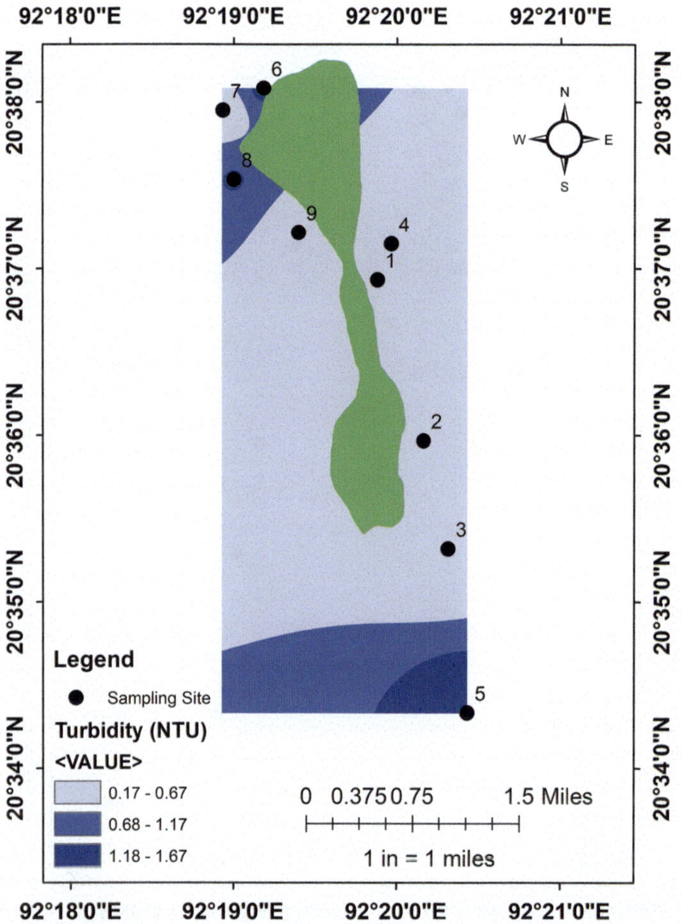

Fig. 3.3 Turbidity (left) and spatial distribution of turbidity (right) in coastal seawater around Saint Martin's island [31]

right) that viscosity was lower (0.942–0.944 mPa's) on the southern side and comparatively higher (0.947–0.948 mPa's) in the island's north western part. In general, viscosity rises as temperature decreases and the ease with which molecules can move with one another is related to a liquid's viscosity [28].

3.3.1.7 Refractive Index (RI)

RI and spatial distribution of RI in coastal seawater around Saint Martin's island are shown in Fig. 3.7. RI ranged from 1.34998 to 1.35074 nD with average RI of 1.351 nD and Lowest and highest RI were monitored at station no. 1 and 8, respectively. Distribution of RI (Fig. 3.7, right) clearly showed that range of RI

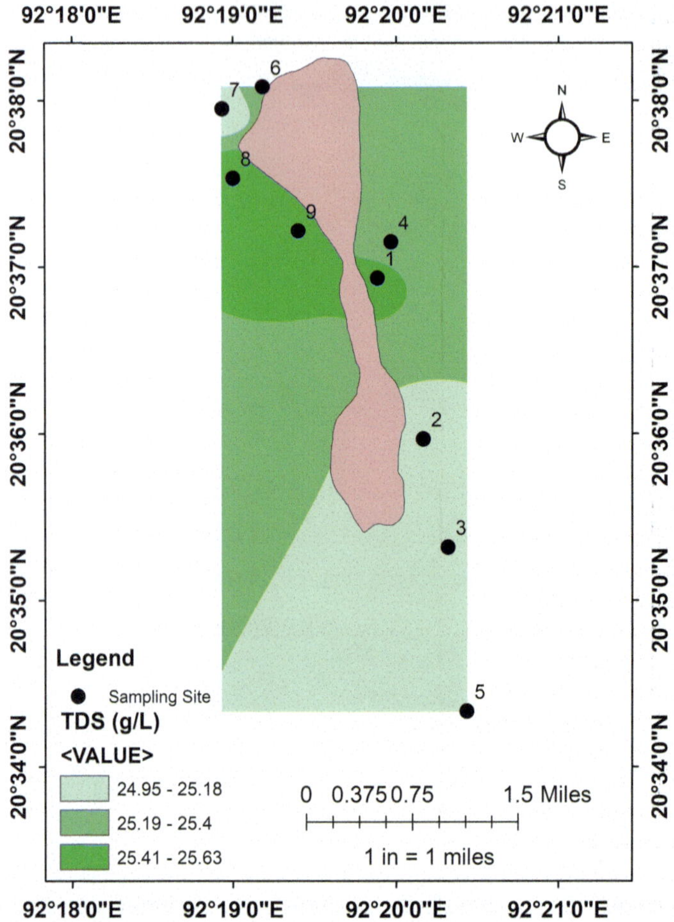

Fig. 3.4 TDS (left) and spatial distribution of TDS (right) in coastal seawater around Saint Martin's island [31]

was low (1.3499–1.3500 nD) starting from the center to the southern part, mid-range (1.3502–1.3505 nD) on north eastern and western part and high (1.305–1.307 nD) on the island's north eastern tip. RI increases as salinity, TDS, and temperature decrease [16]. Therefore, variations in salinity, TDS, and temperature may be the cause of the RI changes observed around the island.

3.3.1.8 Salinity

Salinity and spatial distribution of salinity in coastal seawater around Saint Martin's island are given in Fig. 3.8. Salinity ranged from 33.58 to 35.54 ppt where highest and lowest salinity were observed at station no. 8 and 5.7 respectively, with average

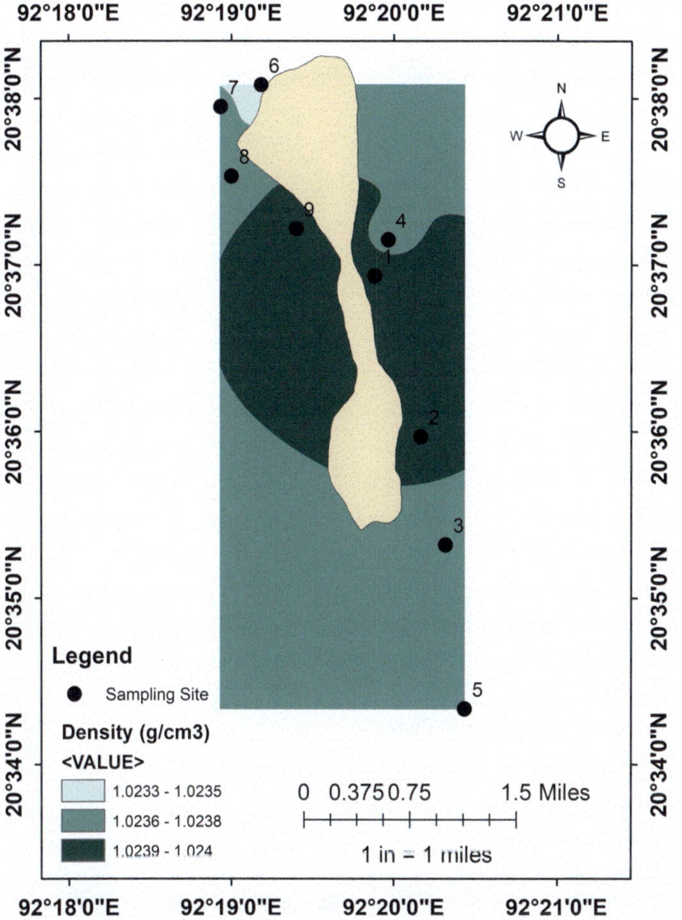

Fig. 3.5 Density (left) and spatial distribution of density (right) in coastal seawater around Saint Martin's island [31]

salinity of 33.04 ppt. From spatial distribution (Fig. 3.8, right), it was observed that salinity was in lower (32.54–32.88 ppt) on the southern side, mid-range (32.23–33.89 ppt) on island's north east and central part and highest (33.24–33.58 ppt) island's north western part. It is commonly known that temperature, evaporation, and precipitation all affect salinity [7]. Salinity may therefore fluctuate throughout the island as a result of temperature variations and seasonal changes brought on by precipitation, evaporation, freshwater discharge, and other factors.

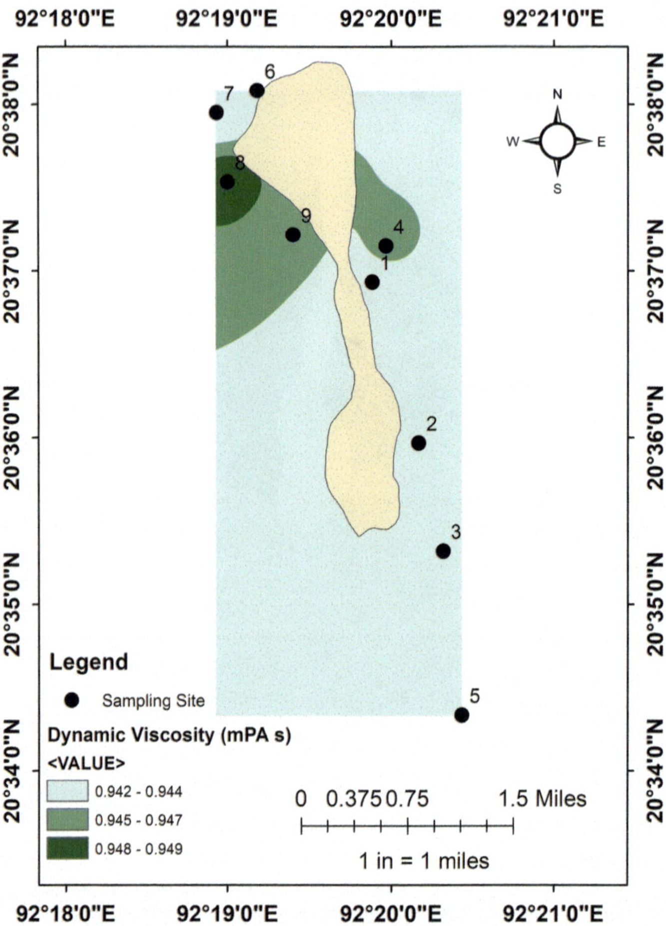

Fig. 3.6 Viscosity (left) and spatial distribution of viscosity (right) in coastal seawater around Saint Martin's island [31]

3.3.1.9 Electrical Conductivity (EC)

EC and spatial distribution of EC in coastal seawater around Saint Martin's island are shown in Fig. 3.9. EC was ranged between 49.91 to 51.22 mS/cm and lowest as well as highest EC were found at station no. 7 and 8, respectively with average EC of 50.54 mS/cm.

From spatial distribution of EC (Fig. 3.9, right), it was noticed that EC was lowest on the island's south-eastern side and highest on the north-eastern side. EC is correlated with Salinity [35] and increases as salinity increases and salinity might have affected the EC observed of seawater around the island.

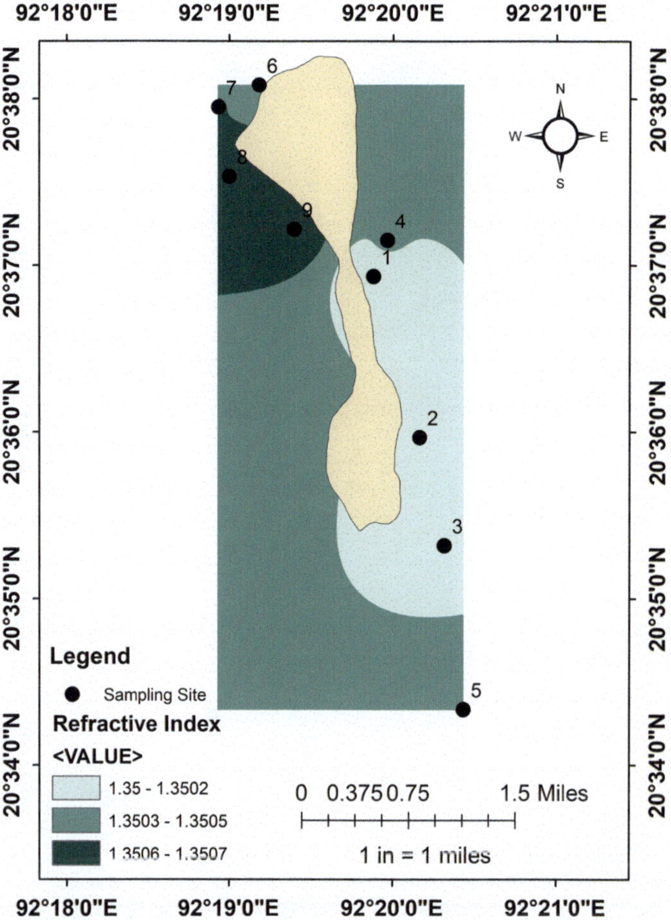

Fig. 3.7 RI (left) and spatial distribution of RI (right) in coastal seawater around Saint Martin's island [31]

3.3.1.10 Dissolved Oxygen (DO)

DO of seawater acquired off the coast of Saint Martin's island was shown in Fig. 3.10. DO ranged between 6.19 and 8.08 mg/L where lowest and highest DO were found at station no. 6 and 4, respectively with average DO of 6.97 mg/L over the island.

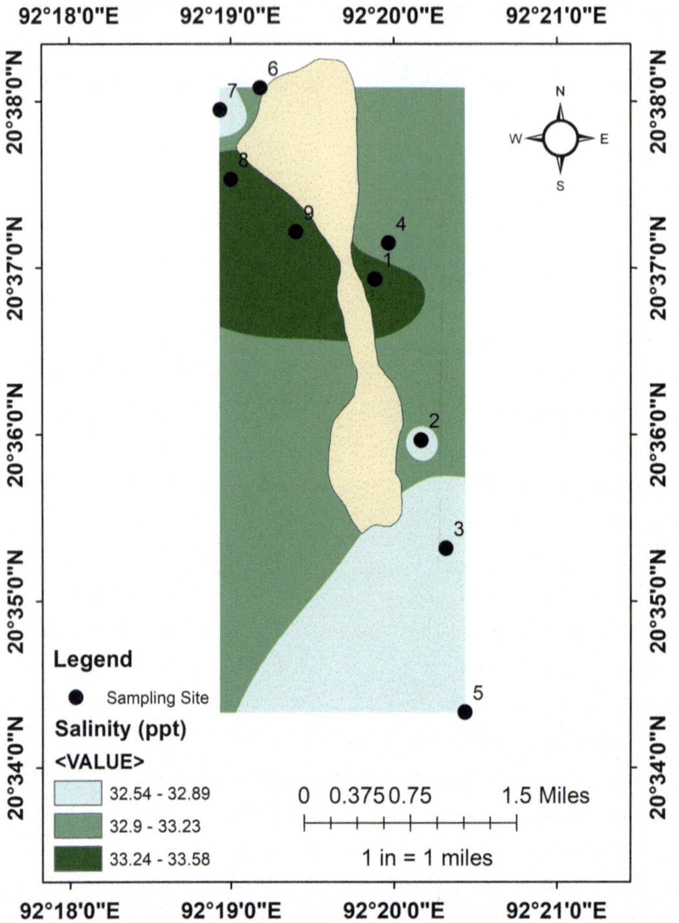

Fig. 3.8 Salinity (left) and spatial distribution of salinity (right) in coastal seawater around Saint Martin's island [31]

3.3.1.11 Concentration of Nutrients

Phosphate

Concentration of phosphate, [PO_4^{3-}] and spatial distribution of [PO_4^{3-}] in coastal seawater around Saint Martin's island are shown in Fig. 3.11. [PO_4^{3-}] ranged from 0.04 to 0.53 mg/L with average of 0.1711 mg/L. It was observed from spatial distribution (Fig. 3.11, right) that [PO_4^{3-}] was comparatively lower in the southern side and in the northwestern side, [PO_4^{3-}] was maximum (Fig. 3.11), since the tourism activities affect this part most frequently. Inappropriate practices in agriculture, runoff from landscapes and cities, septic system leaks, sewage treatment, and plant emissions all contribute to the high [PO_4^{3-}] levels that are seen [18].

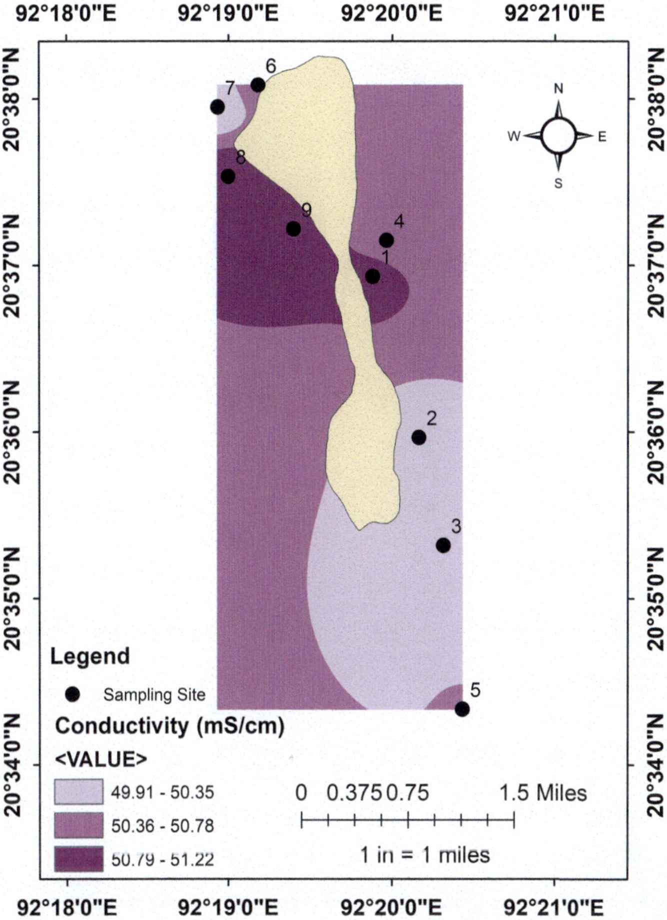

Fig. 3.9 EC (left) and spatial distribution of EC (right) in coastal seawater around Saint Martin's island [31]

Nitrate

Figure 3.12 shows concentration of nitrate, $[NO_3^-]$ ranged from 1.2 to 2.5 mg/L with average of 1.6556 mg/L. On the contrary to $[PO_4^{3-}]$, $[NO_3^-]$ was highest in the islands central area and a slightly less in the island's southern part (Fig. 3.12, right). Most of $[NO_3^-]$ in the surface seawater is formed by river runoff, which includes all of the river activities: agricultural, farming, industrial, and waste [30].

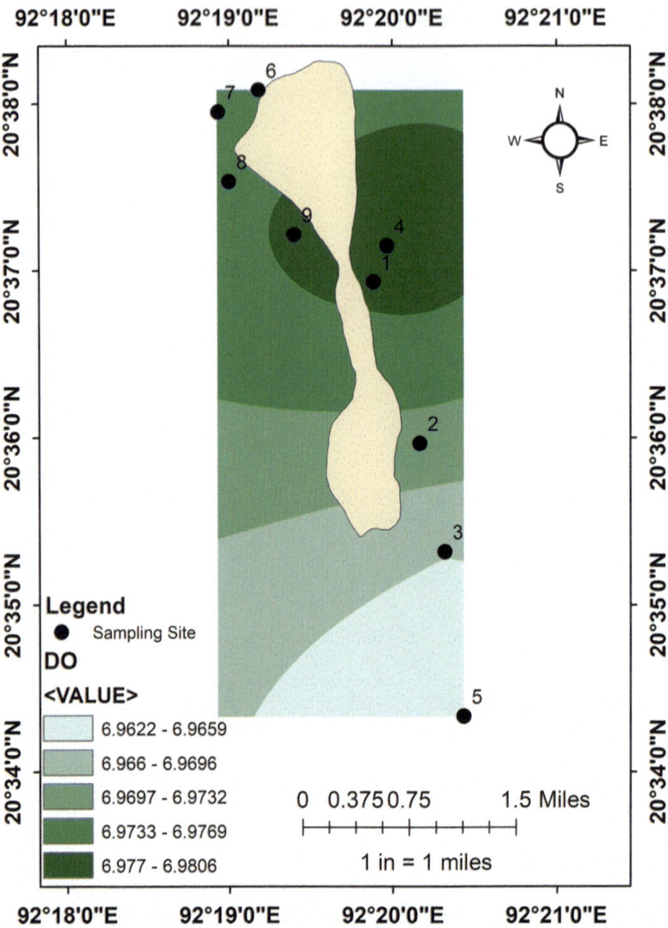

Fig. 3.10 DO in coastal seawater around Saint Martin's island

3.3.2 Determination of Concentration of Heavy Metals and Minerals

3.3.2.1 Removal of Halides for Determination of Concentration of Heavy Metals

Salinity of seawater were lower than their initial values after using the procedure outlined in Sect. 2.6, displayed in Fig. 3.13 [1]. Concentrations of various heavy metals/minerals were determined after removal of halides from coastal seawater using AAS.

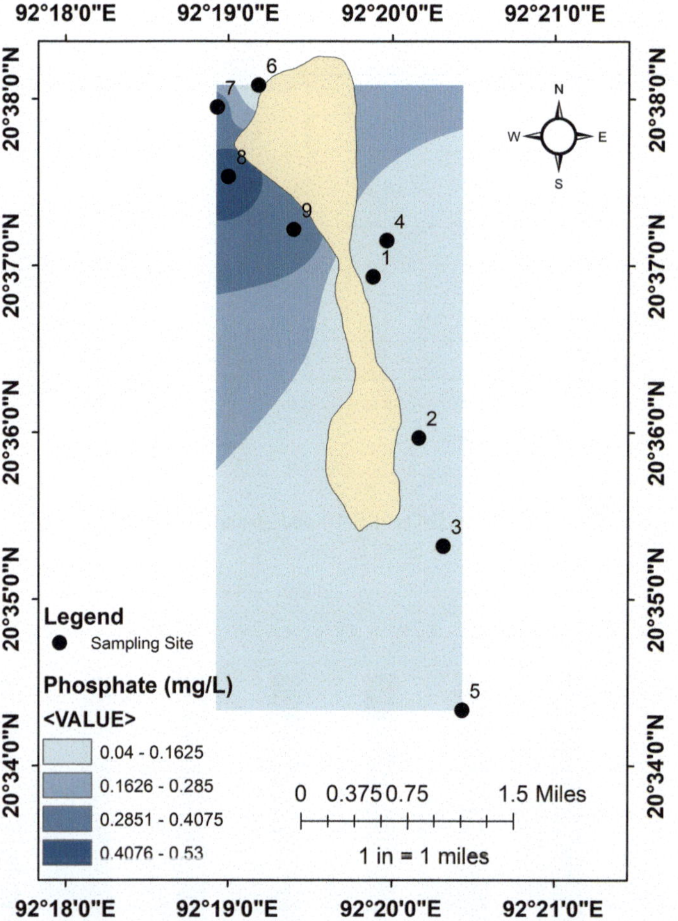

Fig. 3.11 Concentration of phosphate, [PO_4^{3-}] (left) and spatial distribution of [PO_4^{3-}] (right) in coastal seawater around Saint Martin's island [31]

3.3.2.2 Concentration of Heavy Metals

Arsenic (As)

Concentration of arsenic, [As] in coastal seawater of Saint Martin's island is shown in Fig. 3.14. [As] was found in a measurable range for only 3 stations out of 7 preselected stations in which highest and lowest [As] was 0.09 and 0.03 μg/L for station no. 2, 7 and 3, respectively. [As] was < 0.5 μg/L which is lower than the standard permissible limit of [As] in seawater, 50 μg/L [34].

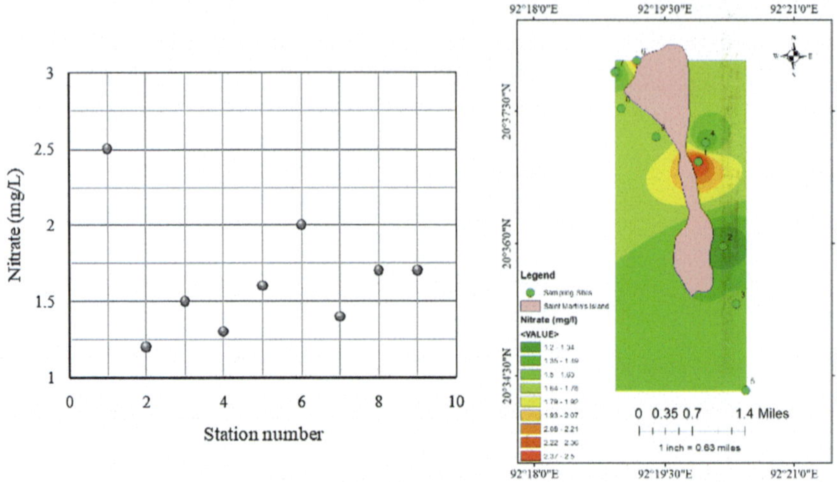

Fig. 3.12 Concentration of nitrate, [NO_3^-] (left) and spatial distribution of [NO_3^-] (right) in coastal seawater around Saint Martin's island [31]

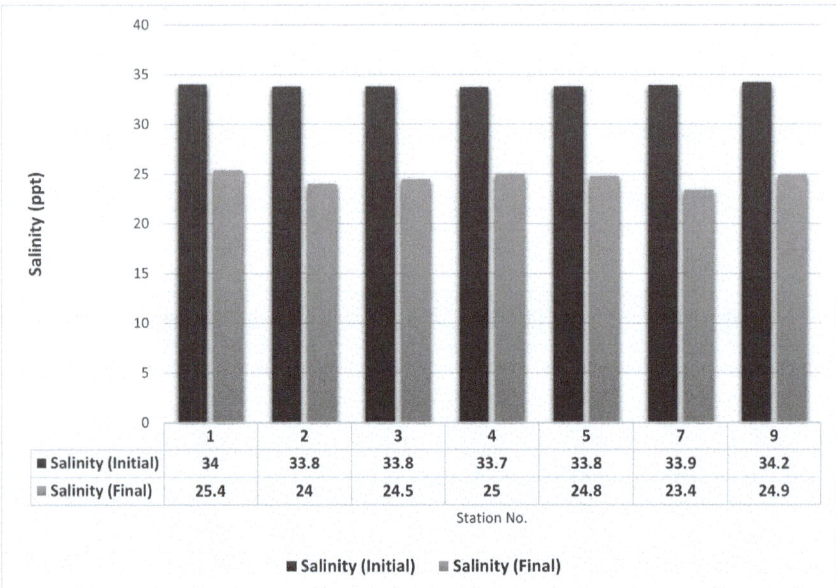

Fig. 3.13 Initial and final salinity of coastal seawater after removal of halides collected from Saint Martin's island

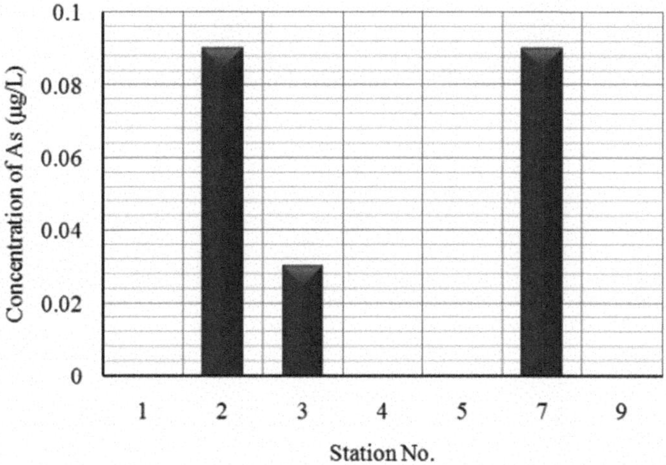

Fig. 3.14 Concentration of As, [As] in coastal seawater around Saint Martin's island

Fig. 3.15 Concentration of Hg, [Hg] in coastal seawater around Saint Martin's island

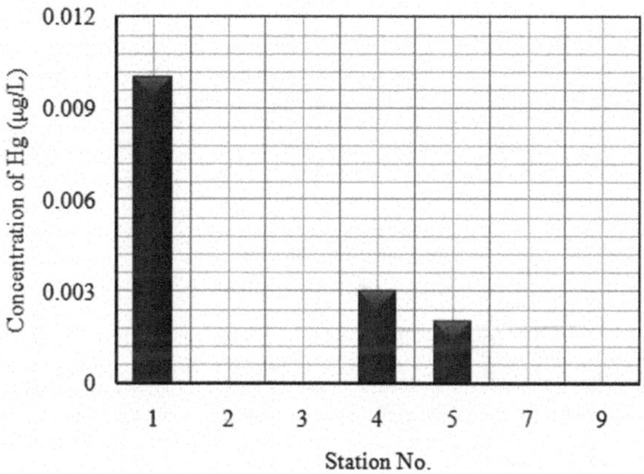

Mercury (Hg)

Concentrations of mercury, [Hg] of some preselected seawater were determined by AAS with cold vapor method [25], is shown in Fig. 3.15. Among 7 stations, only 3 stations had [Hg] in a measurable range in which highest and lowest [Hg] were measured in station no. 1 (0.01 µg/L) and 5 (0.002 µg/L) indicating [Hg] is within the standard permissible limit of [Hg] in seawater, 1 µg/L [34].

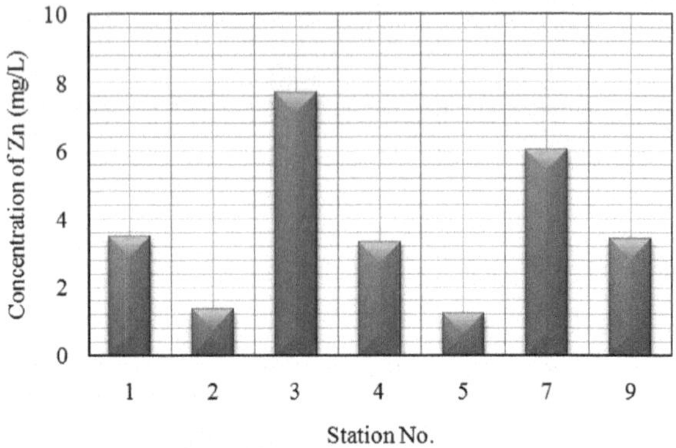

Fig. 3.16 Concentration of zinc, [Zn] in coastal seawater around Saint Martin's island

Zinc (Zn)

Concentration of zinc, [Zn] (Fig. 3.16) in 7 preselected stations around Saint Martin's island ranged from 1.23 to 7.72 mg/L. Highest and lowest [Zn] were found at station no. 3 (7.72 mg/L) and 5 (1.23 mg/L). Average [Zn] was 3.787 mg/L which was high as a large portion of total Zn in seawater is absorbed by particles or complexes with dissolved organic matter through aerial deposition and average residence time of Zn is 1400 years [8]. But the observed Zinc Concentration was within the standard permissible limit 5000 μg/L of [Zn] in Seawater [34].

Lead (Pb)

Figure 3.17 represents concentration of lead, [Pb] in coastal around Saint Martin's island and measured [Pb] ranged from 0.05 to 0.09 mg/L in which highest and lowest [Pb] were found at station no. 4 and 9, respectively with average [Pb] of 0.06 mg/L in the coastal seawater in Saint Martin's island which is lower than the ideal limit (50 μg/L) of lead in seawater [26].

3.3.2.3 Concentration of Minerals

Sodium (Na)

Figure 3.18 represents concentration of sodium, [Na] in coastal seawater around Saint Martin's island and [Na] ranged from 11,334 to 12,948 mg/L where lowest and highest [Na] were found at station no. 1 and 9, respectively with average [Na] of 11,933 mg/L in coastal seawater in Saint Martin's island. As the salinity of seawater was higher, [Na] were also high.

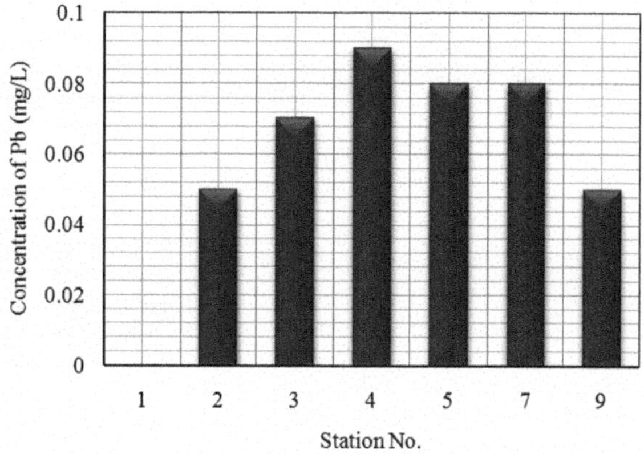

Fig. 3.17 Concentration of lead, [Pb] in coastal seawater around Saint Martin's island

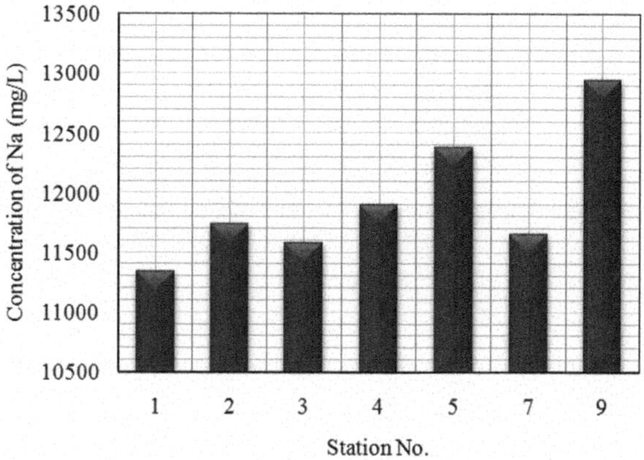

Fig. 3.18 Concentration of sodium, [Na] in coastal seawater around Saint Martin's island

Magnesium (Mg)

Figure 3.18 represents concentration of magnesium, [Mg] of preselected seawater in Saint Martin's island and [Mg] ranged from 784 to 1268 mg/L. Highest and lowest [Mg] were found at stations no. 6 and 4, respectively, with average [Mg] of 1099 mg/L due to sediment discharge occurring near the shore and terrestrial sediments are main source of Mg in seawater [12].

3.3.3 Correlation Between Physicochemical Properties

3.3.3.1 Pearson Correlation

Relationship of two variables was measured by Pearson's correlation coefficient (r) and the correlation coefficient ranges in value from − 1 to + 1. Positive correlation

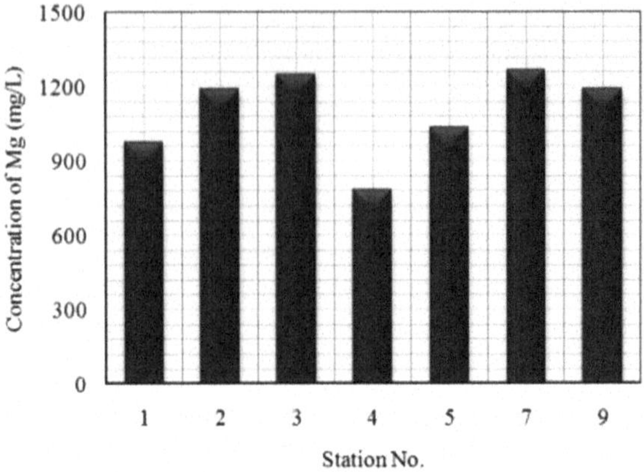

Fig. 3.19 Concentration of magnesium, [Mg] in coastal seawater around Saint Martin's island

	pH	*Conductivity*	*TDS*	*DO*	*Salinity*	*Temperature*	*Density*	*SG*	*RI*	*Turbidity*	*Phosphate*	*Alkalinity*	*Viscosity*	*Nitrate*
pH	1.000													
Conductivity	0.192	1.000												
TDS	0.145	0.976	1.000											
DO	0.099	-0.198	-0.155	1.000										
Salinity	0.130	0.924	0.979	-0.142	1.000									
Temperature	-0.049	0.437	0.456	-0.127	0.343	1.000								
Density	-0.261	0.368	0.358	0.310	0.393	-0.048	1.000							
Specific Gravity	-0.261	0.377	0.370	0.293	0.407	-0.042	1.000	1.000						
Refractive Index	0.258	0.392	0.380	0.173	0.264	0.868	0.025	0.024	1.000					
Turbidity	0.490	0.286	0.133	-0.505	0.016	0.191	-0.432	-0.432	0.284	1.000				
Phosphate	0.071	0.480	0.519	0.211	0.449	0.766	0.206	0.208	0.881	0.013	1.000			
Alkalinity	-0.127	0.224	0.057	-0.298	0.001	-0.348	0.231	0.225	-0.281	0.508	-0.301	1.000		
Viscosity	0.635	0.713	0.701	0.203	0.639	0.478	0.115	0.117	0.724	0.348	0.730	-0.108	1.000	
Nitrate	-0.368	0.576	0.551	-0.646	0.524	0.089	-0.002	0.009	-0.230	0.261	-0.088	0.492	-0.045	1.000

Fig. 3.20 Pearson correlation coefficient between physicochemical properties in coastal seawater around Saint Martin's island

coefficient values show a tendency for one variable's increase to be correlated with the other variable's drop, and vice versa. Correlation coefficient values near zero show little association between variables, whereas those near − 1 or + 1 show a significant linear relationship between two variables [14]. Figure 3.20 shows Pearson correlation coefficient between all the observed physicochemical properties used in this study. All physicochemical properties have good correlation between them, a few have negative correlation between them and others have a somewhat neutral relation.

3.3.4 Pollution Status of Saint Martin's Island Surrounding Seawater

3.3.4.1 Canadian Council of Ministers of Environment (CCME) WQI of Physicochemical Parameters

CCME WQI [20] was evaluated for physicochemical parameters and Table 3.2 shows the result obtained from during the calculation WQI.

Table 3.2 Calculated CCME WQI of Seawater acquired off the coast of Saint Martin's island

	Temp (°C)	Salinity (ppt)	pH	DO (mg/L)	TDS (g/L)	EC (mS/cm)	Nutrients (mg/L)		Turbidity (NTU)
							PO_4^{3-}	NO_3^-	
Std.	25–30	15–36	7.42–8.65	5.24–7.92	0.5–30	> 15	0.005–2	0.3–3.0	30
St. No.									
1	26.09	33.42	8.21	6.54	25.48	50.96	0.04	2.5	0.41
2	26.01	32.88	8.26	7.1	25.05	50.09	0.06	1.2	0.21
3	26.8	32.8	8.27	6.67	25.07	50.13	0.07	1.5	0.2
4	26.11	33.06	8.3	8.08	25.26	50.54	0.06	1.3	0.34
5	26.9	32.54	8.28	6.64	25.01	50.35	0.04	1.6	1.67
6	28.29	33.09	8.27	6.19	25.29	50.56	0.04	2	1.29
7	28.15	32.54	8.21	7.56	24.95	49.91	0.31	1.4	0.17
8	28.68	33.58	8.31	6.94	25.63	51.22	0.53	1.7	1.21
9	30.01	33.43	8.24	7.04	25.56	51.1	0.39	1.7	0.35
CCME WQI calculation		No. of failed variables		2	F_1 (scope)		25	F_1^2	625
		Total variables = 8							
		Total no. of tests		72	F_2 (frequency)		2.7777778	F_2^2	7.716049383
		Total failed tests = 2							
		$\sum$*excursion*		0.020535354	F_3 (amplitude)		0.02851	F_3^2	0.000813002
		$\sqrt{(F_1^2 + F_2^2 + F_3^2)} = 25.15386377$							
		CCME WQI = 85.5							

Table 3.3 Calculated WA-WQI of seawater acquired off the coast of Saint Martin's island

Parameters	Standards	Stations (WiQi)	Quality
pH	7.42–8.65	Station 1 = 79.0434	Very Poor
DO (mg/L)	5.24–7.92	Station 2 = 57.0271	Poor
EC (mS/cm)	> 15	Station 3 = 63.1874	Poor
TDS (g/L)	0.5–30	Station 4 = 58.2731	Poor
Salinity (ppt)	15–36	Station 5 = 63.6936	Poor
Temperature (°C)	25–30	Station 6 = 71.1729	Poor
Phosphate	0.005–2	Station 7 = 71.5721	Poor
Nitrate (mg/L)	0.3–3.0	Station 8 = 88.7831	Very Poor
Silicate (mg/L)	< 16	Station 9 = 81.71732	Very Poor
Nitrite (mg/L)	3		
Turbidity (NTU)	30		

According to CCME WQI, it was observed that physicochemical properties of seawater acquired off the coast of Saint Martin's island yields a result of 85.5 which fell under the category "Good" means water quality has minor degree of threat or impairment; conditions rarely depart from natural or desirable levels. Occasional departure from ideal parameters might take a short time due to natural and anthropogenic activities. Continuous monitoring and source analysis of these deviations might be required to save the island from further contamination.

3.3.4.2 Weighted Arithmetic Water Quality Index (WA-WQI)

Weighted Arithmetic Water Quality Index (WA-WQI) [10] was also evaluated for physicochemical parameters and Table 3.3 shows the result obtained from during the calculation WA-WQI for each stations.

WA-WQI shows the water quality of coaster seawater to be "poor" in average, where 6 out of the 9 stations were marked "poor" which means that seawater of these stations can only be used for irrigation and industrial usage. And other three stations were marked "very poor" which means that water of these stations can be used for irrigation only. Further detailed investigation will provide a clear and concise picture.

3.3.4.3 Heavy Metal Pollution Index (HPI)

Table 3.4 represents calculated HPI values of measured heavy metals and HPI score of heavy metals is 208.8945 in coastal seawater of Saint Martin's island.

From the calculation of HPI, it was observed that the coastal seawater around Saint Martin's island has a HPI score of 208.8945 and any score above 100 is indicative of 'good' water quality and that of less exposure to heavy metal and pollution [21].

Table 3.4 HPI values in coastal seawater of Saint Martin's island

Metal	Average value in water sample (M) ppb	Standard permissible limit in water sample (S) ppb [34]	Ideal limit (I) ppb [26]	Unit weight (W)	Sub-index Q	W * Q	HPI
Zn	3787.14	5000	3000	0.0002	39.357145	0.007871	208.8945
Pb	60	50	10	0.02	125	2.5	
As	0.03	50	10	0.02	24.925	0.4985	
Hg	2.1428	1		1	0	214.2857	
			Total	1.0402	Total	217.2921	

Table 3.5 CF for heavy metals in seawater of Saint Martin's island

Metal	St. 1	St. 2	St. 3	St. 4	St. 5	St. 7	St. 9	Avg CF
Zn (mg/L)	– 0.76	– 0.91	– 0.48	– 0.77	– 0.91	– 0.59	– 0.77	– 0.74
Pb (mg/L)	– 1	0	0.4	0.8	0.6	0.6	0	0.2
As (μg/L)	– 1	– 0.99	– 0.99	– 1	– 1	– 0.99	– 1	– 0.99
Hg (μg/L)	9	– 1	– 1	2	1	– 1	– 1	1.14
CD	1.55	– 0.72	– 0.52	0.25	– 0.08	– 0.49	– 0.69	– 0.10

Thus, score of 208.8945 indicates that coastal seawater of Saint Martin's island is not yet polluted by heavy metals and was in "Good" quality during March 2022.

3.3.4.4 Contamination Factor (CF)

Table 3.5 and Fig. 3.21 represent CF, degree of contamination (CD) and category of water quality depending on CF for heavy metals and CD has been found as – 0.10. Figure 3.21 (right) shows that average contamination factor for seawater acquired off the coast of Saint Martin's island was – 0.10, which is less than 1 and described as low contamination factor, which is an indication of little or no contamination.

3.4 Conclusions

A detailed analysis was conducted on the physicochemical properties and concentration of minerals and heavy metals found in coastal seawater from Saint Martin's Island. The sea surface temperature was affected by the low air temperature of winter season. All station had a slightly alkaline pH while no significant variations in pH were observed. Due to decreased evaporation during the winter season, salinity and electrical conductivity were high. Low levels of turbidity and total dissolved

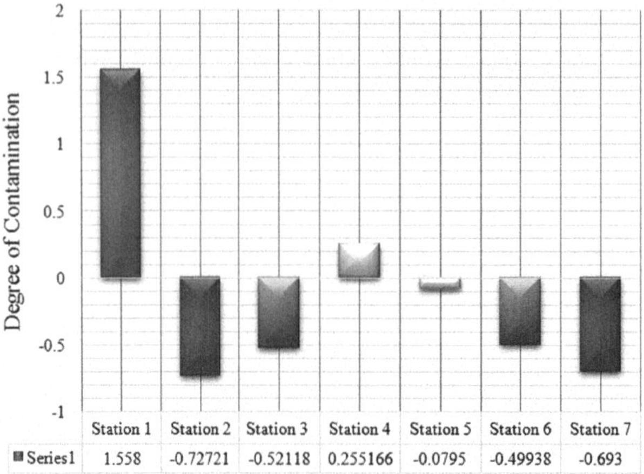

Contamination Factor	Category
CF < 1	Low
1 < CF <3	Moderate
3 < CF <6	Considerable
6 < CF	Very High

Fig. 3.21 CD at each sampling station around Saint Martin's island (left) and contamination factor category (right) [9]

solids (TDS) were observed which is beneficial for the coral reefs of Saint Martin's Island. DO were more than 6 mg/L, which is necessary for fish populations, phytoplankton growth, primary production, and sustainable marine life. Density, viscosity, and refractive index (RI) were higher due to the high salt content of seawater. The amounts of arsenic, mercury, lead, and zinc among the heavy metals examined differed. Mercury and Lead levels in some stations were over the permissible limit. However, the concentrations of magnesium, zinc, sodium, and arsenic were within the permitted limits at all station. Phosphate and nitrate concentrations were well within the accepted range, suggesting nutrient-enriched seawater. Overall, the collected seawater's physicochemical characteristics were satisfactory, and showed a strong correlation between the parameters. The island's water quality was rated as "Good" by CCME WQI, used in this study which indicates that more research and monitoring are needed to ensure the island's sustainability. However, the water quality was rated as "Poor" by the WA-WQI, suggesting that it should only be used for industrial and irrigation purposes. According to the Heavy Metal Pollution Index, heavy metal pollution has not yet had a major effect on the island. Different result on two indexes expresses the need for more detailed study. The impact of heavy metals on specific sample stations was brought into focus by the Contamination Factor. The analysis's findings will provide valuable insights for future management and provide the groundwork for future physicochemical study. This study recommends future studies to be conducted with seasonal varieties and added properties for a more enhanced and wider study. The island which is already declared a ecologically critical area needs strong legislation and monitoring by the authority to prevent any further degradation.

References

1. Abbassi M, Banaoui A, Charioui I, Abderrazak K, Elkhou A, Nadir M, Agnaou M, Lefrere L, Hamidi F (2017) Physico-chemical characterization of the coastal waters of the city of Sidi Ifni (Morocco). J Mater Environ Sci 8:3112–3120
2. Achilleos GA (2011) The inverse distance weighted interpolation method and error propagation mechanism—creating a DEM from an analogue topographical map. J Spat Sci 56(2):283–304
3. Ahammed SS, Hossain MA, Abedin MZ, Khaleque A (2016) A study of environmental impacts on the coral resources in the vicinity of the Saint Martin Island, Bangladesh. Int J Sci Technol Res 5(1):37–39. www.ijstr.org
4. Ahmed AAM (2013) A Canadian water quality guideline-water quality index (CCME-WQI) based assessment study of water quality in Surma River. J Civ Eng Constr Technol 4:81–89
5. Akther S (2019) Marine biodiversity and ecosystem of the St. Martin's Island in Bangladesh: perceived threats and vulnerabilities to environmental management. Orient Geogr 60:61–74
6. Alam O, Deng TL, Uddin MN, Alamgir M (2015) Application of environmental ethics for sustainable development and conservation of Saint Martin's Island in Bangladesh. J Environ Sci Nat Resourc 8:19–27
7. Blunden J, Arndt DS, Richter-Menge JA, Sánchez-Lugo A, Sharon S, Thorne PW, Willett KM et al (2014) American meteorological society state of the climate in 2014
8. Chester R, Bradshaw GF, Corcoran PA (1994) Trace metal chemistry of the North Sea particulate aerosol; concentrations, sources, and seawater fates. Atmos Environ 28(17):2873–2883
9. Chinedu I, Roseline N-T, Tochukwu E, Uzochukwu E, Lucy I, Darlinton A, Ngozi V (2016) Anthropogenic impact and geo-accumulation of heavy metal levels of soils in Owerri, Nigeria. Br J Appl Sci Technol 12:1–9
10. Hyarat T, Mustafa AK (2021) Comparison between weighted arithmetic and Canadian council of ministers of the environment water quality indices performance in Amman-Zarqa area, Jordan. Jordan J Earth Environ Sci 12:295–305
11. Idrus FA, Chong MD, Rahim NSA, Basri MM, Musel J (2017) Physicochemical parameters of surface seawater in Malaysia exclusive economic zones off the coast of Sarawak. Borneo J Resourc Sci Technol 7:1–10
12. Jeandel C, Peucker-Ehrenbrink B, Jones MT, Pearce CR, Oelkers EH, Godderis Y, Lacan F, Aumont O, Arsouze T (2011) Ocean margins. the missing term in oceanic element budgets?. Eos, Trans Am Geophys.Union. 92(26):217–218. https://doi.org/10.1029/2011EO260001
13. Kashem M, Sultan N, Mahbub K (2019) Assessment of physico-chemical status of coastal seawater of the Saint Martin's Island, Bangladesh. Int J Sci Eng Res 10:84–91
14. Kirch W (2008) Pearson's correlation coefficient. In: Encyclopedia of public health. Springer Netherlands, pp 1090–1091
15. Krishna KS, Mohammad I, Karlapati S, Rao D, Mishra J, Saha D (2016) Sediment pathways and emergence of Himalayan source material in the Bay of Bengal. Curr Sci 110:363–372
16. Kroiß A, Sebastian E, Kuczaty J, Cornelius T, Wolf S, Alexander P, Markus S, Thomas S (2015) Optical methods for simultaneous measurement of temperature and concentration polarization. In: The international desalination association world congress on desalination and water reuse 2015/San Diego, CA, USA, DAWC15-Kroiß_51500, pp 1–24
17. Kumari K, Suniti Y (2018) Linear regression analysis study. J Pract Cardiovasc Sci 4:33
18. Lang M, Ping L, Xiaoyuan Y (2013) Runoff concentration and load of nitrogen and phosphorus from a residential area in an intensive agricultural watershed. Sci Total Environ 458–460:238–245
19. Lawson O, Emmanuel L (2011) Physico-chemical parameters and heavy metal contents of water from the mangrove swamps of Lagos Lagoon, Lagos, Nigeria. Adv Biol Res 5:8–21
20. Lumb A, Doug H, Tribeni S (2006) Application of CCME water quality index to monitor water quality: a case study of the Mackenzie River Basin, Canada. Environ Monit Assess 113(1):411–429

21. Majhi A, Subhra KB (2016) Application of HPI (heavy metal pollution index) and correlation coefficient for the assessment of groundwater quality near ash ponds of thermal power plants. Int J Sci Eng Adv Technol 4(8):395–405
22. Mian SM (2005) Program development office for integrated coastal zone management plan (PDO-ICZMP): proceedings of the round table discussion on holistic approach for sustainable management of St. Martin's Island. Working paper WP038
23. Muhibbullah, Mukaka MM (2012) Statistics corner: a guide to appropriate use of correlation coefficient in medical research. Malawi Med J 24(3):69–71
24. National Research Council (2010) Ocean acidification: a national strategy to meet the challenges of a changing ocean. National Academies Press
25. Oda CE, Ingle JD Jr (1981) Speciation of mercury with cold vapor atomic absorption spectrometry by selective reduction. Anal Chem 53(14):2305–2309
26. Okoro HK, Fatoki OS, Adekola FA, Ximba BJ, & Snyman R, (2013). Physico-chemical characteristics and 1-year monitoring of heavy metal pollution and its seasonal variation in seawater of Cape Town Harbour, South Africa. Fresenius Environ Bulletin, 22(10):2855–2866
27. Petit T, Susan ML, Simon AJ, Stuart AC (2021) Role of air-sea fluxes and ocean surface density in the production of deep waters in the eastern subpolar gyre of the North Atlantic. Ocean Sci 17(5):1353–1365
28. Petravic J (2004) Influence of strain on transport in dense Lennard-Jones systems. J Chem Phys 120(15):7041–7049
29. Raj N, Kumaraswamy K (1999) Status of the coral reef environment of the Gulf of Mannar, South India. J Natcon Nat Conserv India 11:161–168
30. Rustiah W, Alfian N, Muhammad L, Andi TF (2019) Distribution analysis of nitrate and phosphate in coastal area: evidence from Pangkep River, South Sulawesi. Int J Agric Syst 7(1):9–17
31. Sakib IA, Begum F, Akter F, Arman MH, Susan MABH (2024) Spatial distribution of physicochemical parameters for quality of coastal seawater at Saint Martin's Island, Bangladesh. Bangladesh Marit J 8(1):153–168
32. Sedgwick P (2012) Pearson's correlation coefficient. BMJ 345:e4483
33. Siddiqui MSI, Ahasan A, Islam N, Kundu P, Munshi MN, Chowdhury EH (2016) Peste des petits ruminants (PPR) virus antibodies in goats and cattle of the Saint Martin Island in Bangladesh. Bangladesh Vet 31(2):55–59
34. Sowrav SF, Rahman SM, Hossain MA, Hafiz KB, Hossain NI, Alam MN, Rahman MS, Choudhury TR (2023) Multi-dimensional approach for an environmental health assessment of a deltaic mangrove ecosystem, Sundarbans. Environ Adv 12:100377. https://doi.org/10.1016/j.envadv.2023.100377
35. Tyler RH, Tim PB, Takuto M, Melissa MZ, James RR (2017) Electrical conductivity of the global ocean. Earth Planet Space 69(1):156

Chapter 4
Assessment of Potential to Improve Water Quality in Natural Wetlands: A Case Study in Kotagala Natural Wetland, Nuwara Eliya, Sri Lanka

J. M. A. U. Jayasekara, N. D. K. Dayawansa, and M. I. M. Mowjood

Abstract Wetlands are essential for purifying water, but human activity poses serious risks to many of them. This study investigates the removal efficiency dynamics of the Kotagala natural wetland in Nuwara-Eliya, Sri Lanka, which is increasingly endangered by encroachment and untreated effluent discharges. There is a need to better understand the factors influencing the efficacy of wetlands in contaminant removal under varying environmental conditions. To address this, we utilized a general removal efficiency equation to calculate removal efficiency values. Then descriptive statistics and P values were used to analysis the data statistically. Temporal analysis of removal efficiencies shows significant deviations from the general trend, particularly in the removal of phosphate and nitrate. Analysing the impact of monthly precipitation on removal efficiencies reveals a positive relationship between wet months on EC, TSS, Nitrate and Phosphate. On the other hand, salinity and TDS exhibit negative correlations with precipitation. All the p values associated with the different parameters shows the less values than significant level which is 0.05. It concludes, there is a significant relationship in between the removal efficiencies and the rainfall fluctuations. Regardless challenges, the Kotagala wetland exhibits potential for improving water quality, highlighting its importance as a natural remedy further study reveals that rainfall is not only factor influence on the nutrient removal efficiencies.

Keywords Removal efficiency · Wetland · Water quality · Rainfall

J. M. A. U. Jayasekara (✉)
Postgraduate Institute of Agriculture, University of Peradeniya, Peradeniya, Sri Lanka
e-mail: awanthi.jayasekara@gmail.com

N. D. K. Dayawansa · M. I. M. Mowjood
Department of Agricultural Engineering, Faculty of Agriculture, University of Peradeniya, Peradeniya, Sri Lanka

G. M. Tarekul Islam et al. (eds.), *Sustainable Water Management and Environmental Resilience*, Springer Water, https://doi.org/10.1007/978-3-031-93350-9_4

4.1 Introduction

One of the most beneficial and biologically diverse ecosystems on Earth is the wetlands [22]. Widely called "the kidneys of Earth," wetlands act as natural filters for water, removing pollutants through a complex interplay of chemical, biological, and physical processes. Wetlands stand out as essential nature-based solutions for the purification of water [11, 20]. As water passes through the complex root systems of wetland plants, it naturally filters out sediments and other pollutants [14].

As described by the Ramsar Convention in [16], a wetland is either permanently or temporarily saturated with static or flowing fresh, salty, or salt water, often found in marshes, fens, and peatlands. Wetlands are mainly classified into natural and artificial categories. Both types perform essential functions involving physical, chemical, and biological processes for natural water purification by filtering and absorbing a variety of pollutants from surface water [9], through phytoremediation [13] and bioremediation [9]. Additionally, wetlands provide habitats for flora and fauna [15], grazing areas for animals, and tourism opportunities [10]. They also play roles in flood control [18] and carbon sequestration [6].

However, over nutrient and other contaminant inputs could overwhelm wetlands, causing the ecosystem to change from being a sink to a source of pollutants. This could lead to a change in diversity and a loss of ecological quality. The effectiveness of removing nutrients and pollutants from wetlands is influenced by several factors, including the watershed characteristics, precipitation, the state of the wetland, and other internal mechanisms operating within the system [3].

Furthermore, by decomposing organic matter and converting pollutants into innocuous substances, the diverse microbial communities found in wetlands play a critical role in the purification of water. Plants and microbes have a symbiotic relationship that serves as a strong defence against a variety of water pollution sources, including industrial discharges and agricultural runoff [7]. Not only do these ecosystems maintain water quality within wetland boundaries but also contribute to downstream water bodies, demonstrating the interplay of their impact [25].

In addition to that, wetlands provide main ecosystem services, including climate regulation [19], supporting healthy ecosystems and maintaining biodiversity [2], and water purification and waste treatment. Considering the human benefits, wetlands cater ecosystem services such as providing food [8], medical resources [23], recreational opportunities and economic benefits [1].

Due to the growing worldwide problems of water scarcity and declining water quality, it is essential to understand the various roles of wetlands in water purification. The Secretariat of the Ramsar Convention [16] states that in order to ensure a sustainable future for these essential ecosystems, conservation efforts must acknowledge and tackle the complex relationships that exist between wetlands, water purification, and human well-being.

The Kotagala wetland is located in Nuwara-Eliya, Central province of Sri Lanka. It is crucial for maintaining the ecosystem's hydrology and serving the community in the vicinity and downstream. Areas of the marsh that belongs to the wetland are being encroached by the adjacent agricultural land. The wetland is threatened by untreated domestic and small-scale industrial effluents discharges through the inflowing waterways; however, its water quality is affected by these causes. To implement effective conservation and management strategies and protect the integrity of the ecosystem as well as the welfare of communities dependent on its services, it is essential to evaluate the potential of Kotagala natural wetlands to improve water quality. Therefore, the objective of this study is to assess the potential of Kotagala natural wetlands to improve the water quality.

4.2 Methodology

4.2.1 Study Area

The study was conducted in Sri Lanka's central hill country, in the Kotagala natural wetland (53° 30′ N, 80° 34′ E, and 57° N, 80° 38′ E). In the Hatton Valley, it is one of the largest and infrequently accessible wetlands. Locally referred to as "Kotagala Oya," the stream originates from the Kotagala wetlands and supplies water for the glorious waterfalls located downstream. Over the wetland's catchment area, the average yearly rainfall is 2929 mm. The region has recorded mean maximum and lowest temperatures of 270° C and 90° C, respectively. The wetland has a catchment area of 13 km^2 and is located at 1724 m above sea level. As shown in Fig. 4.1, tea plantations comprise the majority of the land use types in the catchment region.

4.2.2 Data Collection and Sample Measurements

In case of water sample collection, monthly basis water samples were taken from one (S1) outlet and six (S2–S7) inlets (Fig. 4.2) between November 2021 and August 2022. The analysis of six distinct physicochemical characteristics was conducted to examine the variations in water quality in specific 7 sampling points. The samples were subjected for the measurements of nitrate, phosphate, total suspended solid (TSS), total dissolved solid (TDS), electrical conductivity (EC), and salinity. Parameters: nitrate (Spectrophotometric method), phosphate (Spectrophotometric method), TSS (conventional gravimetric laboratory procedure), TDS (Multi-parameter), EC (Multi-parameter), and salinity (Multi-parameter) were determined after bringing the water samples into Joint Research and Development Centre (JRDC) and Department of Agricultural Engineering laboratories at the University of Peradeniya, Sri Lanka.

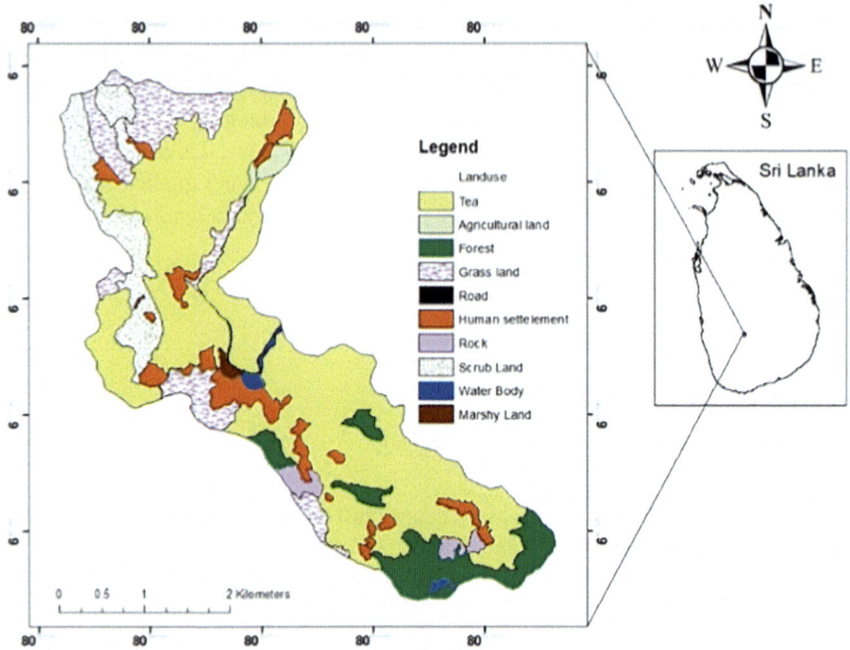

Fig. 4.1 Catchment area of the wetland

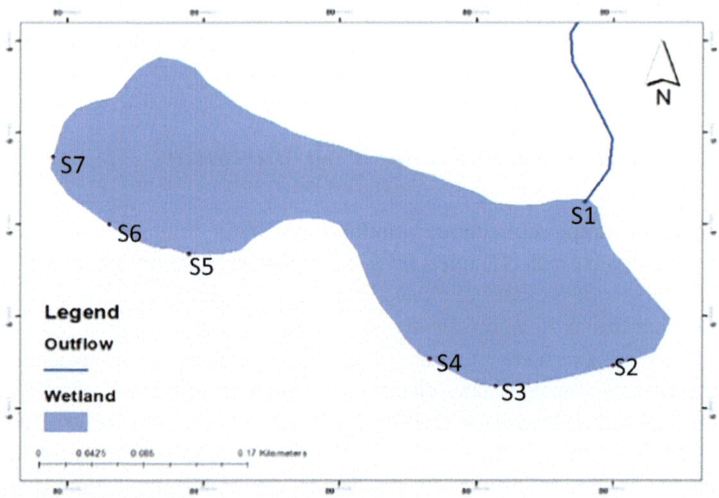

Fig. 4.2 Sampling points distribution within the wetland

4.2.3 Data Analysis

Data was analysed initially by descriptive statistics by calculating mean, maximum, minimum and standard deviation. After that, removal efficiencies of EC, TDS, TSS, salinity, nitrate and phosphate were calculated using the following removal efficiency Eq. (4.1).

$$(Ci - Ce)/Ci = E \tag{4.1}$$

where Ci = influent concentration of a pollutant Ce = effluent concentration of a pollutant, and E = wetland's removal efficiency (%). For the statistical analysis Excel 2013 statistical package was used.

4.3 Results

4.3.1 Descriptive Variation of Water Quality Parameters

The overall water quality status of Kotagala Wetland throughout the study period is summarized in Table 4.1. EC ranged from 75.90 to 138.56 μS/cm with a standard deviation of 19.45. Salinity, measured in Practical Salinity Units (PSU), varied from 0.09 to 0.13, averaging 0.11. TDS concentration fluctuated between 40.41 and 75.28 mg/l, with a mean of 53.34 mg/l. TSS ranged narrowly from 773.60 to 844.12 mg/l, with a mean of 810.9 mg/l. Phosphate concentrations averaged 0.37 mg/l, ranging from 0.07 to 0.9 mg/l, while nitrate values ranged from 0.79 to 2.94 mg/l, averaging 1.44 mg/l. These findings underscore the dynamic nature of Kotagala Wetland's water quality, providing crucial insights for environmental management and conservation efforts.

Table 4.1 Descriptive statics of the water quality parameters

Parameter	Mean	Max	Min	SD
EC (μS/cm)	102.12	138.56	75.90	19.45
Salinity (PSU)	0.11	0.13	0.09	0.01
TDS (mg/l)	53.34	75.28	40.41	11.62
TSS (mg/l)	810.9	844.12	773.60	18.44
Nitrate (mg/l)	1.44	2.94	0.79	0.68
Phosphate (mg/l)	0.37	0.9	0.07	0.26

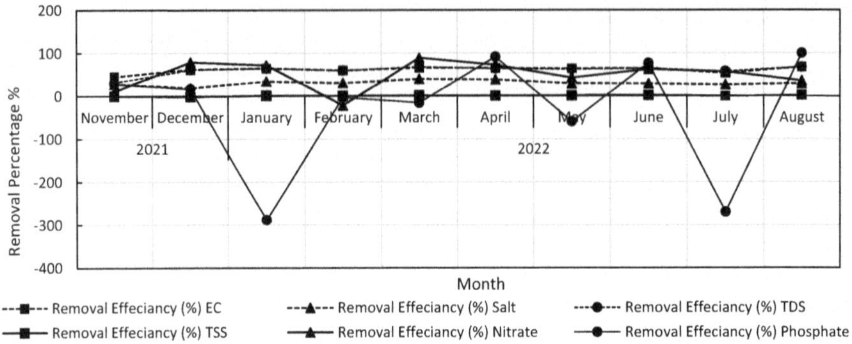

Fig. 4.3 Temporal variation of removal efficiencies

4.3.2 Temporal Variation in Removal Efficiency

The removal efficiency percentages for various water quality parameters from November 2021 to August 2022 are showed in Fig. 4.3, illustrating the system's capacity to mitigate environmental stressors over time. While most parameters demonstrated consistent removal trends, notable fluctuations were observed in EC, salinity, TDS, and TSS removal efficiencies. EC removal ranged from 45.01 to 67.82%, with peak removal in August. Salinity removal varied from 18.43 to 40.00%, with the lowest in December. TDS removal fluctuated between 30.64 and 67.95%. TSS removal exhibited a range of reactions, from − 1.71 to 1.88%, including negative efficiencies. Nitrate removal ranged widely from − 21.67 to 89.27%, and phosphate removal showed substantial variation from − 288.09 to 100.00%.

4.3.3 Effect of Rainfall to Variation of Removal Efficiencies in Water Quality Parameters

Monthly rainfalls from November 2021 to August 2022 are shown in Fig. 4.4. Highest rainfall was found in June and August. This shift in rainfall patterns over time could have influence the water quality ensuing differences in removal efficiency.

Table 4.2 presents the distinct patterns of each parameter's removal effectiveness in relation to varying rainfall amounts throughout the November 2021–August 2022 study period.

Figure 4.5 shows an increasing trend in EC with rising rainfall, suggesting that the wetter months may have higher removal efficiency. The effectiveness of EC removal peaked in August 2022 at 67.82%, which was correlated with 408.5 mm of monthly rainfall. A substantial correlation between rainfall and EC clearance effectiveness is demonstrated by a p-value of 0.004 ($p < 0.05$).

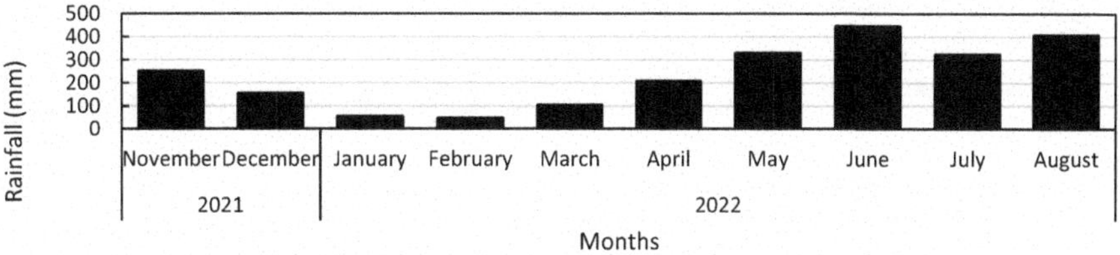

Fig. 4.4 Monthly rainfall for study duration

Table 4.2 Removal efficiencies of parameters with the rainfall

Year	Month	Monthly RF (mm)	Removal efficiency (%)					
			EC	Salinity	TDS	TSS	Nitrate	Phosphate
2021	November	253.1	45.01	27.55	30.64	− 0.71	9.88	27.47
	December	156.9	62.02	18.43	61.60	− 1.71	79.09	16.22
2022	January	56.6	64.85	34.16	64.52	0.84	71.96	− 288.09
	February	47.8	60.01	30.64	59.65	0.43	− 21.67	− 3.21
	March	105.4	66.20	40.00	67.95	0.83	89.27	− 15.75
	April	209.4	65.38	38.66	64.26	0.56	72.83	91.50
	May	330.8	64.21	29.59	62.36	1.09	42.71	− 59.48
	June	446.2	65.07	29.07	64.57	1.65	62.58	76.60
	July	324.0	53.44	26.65	57.57	− 0.49	56.88	− 269.86
	August	408.5	67.82	29.41	67.28	1.88	35.63	100.00

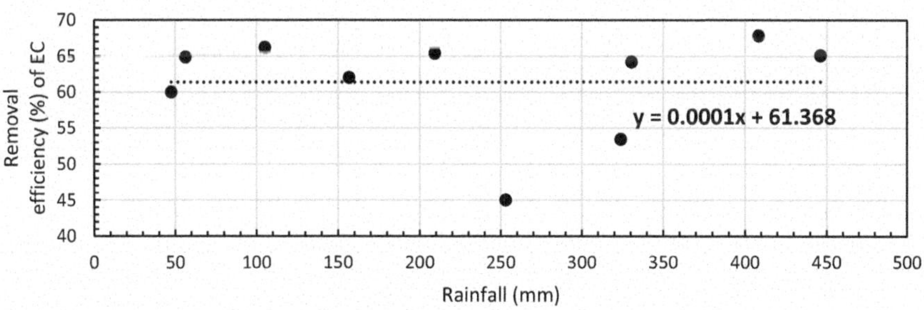

Fig. 4.5 Relationship between removal efficiency of EC with rainfall

The effectiveness of removing salinity exhibited a more variable response, peaking at 40.00% in March 2022. Overall, removal efficiency of salinity shows the negative trend with the rainfall as shown in the Fig. 4.6. *P* value was calculated as 0.004 for the relationship between removal efficiency of salinity and rainfall which value has fallen under the significant level ($p = 0.05$).

Rainfall and TDS removal efficiency also showed a negative association, with rainfall amount as shown in Fig. 4.7. This pattern implies that precipitation could

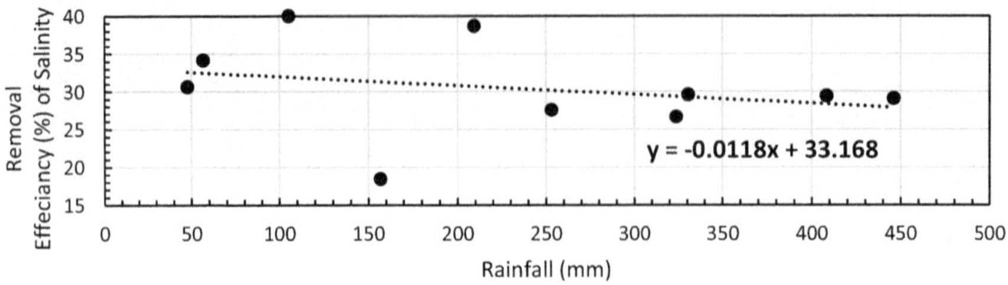

Fig. 4.6 Relationship between removal efficiency of salinity with rainfall

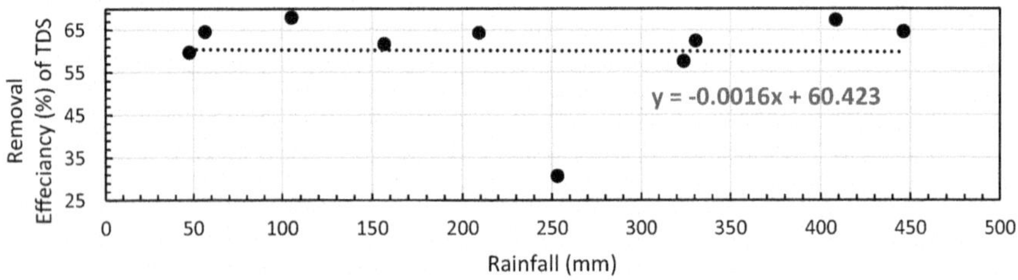

Fig. 4.7 Relationship between removal efficiency of TDS with rainfall

help with reduce TDS removal. However, calculated p value ($p = 0.004$) is less than the significant level ($p = 0.05$).

According to Fig. 4.8 the removal efficiency of TSS showed a positive pattern in response to variations in rainfall, showing both rises and declines.

A significant peak of 79.09% efficiency was recorded in December 2021 with a rainfall of 156.9 mm, as shown in Fig. 4.9, which shows a relationship between monthly rainfall and nitrate removal efficiency. This suggests that there is a positive correlation between increased rainfall and improved nitrate removal efficiency. The substantial p-value of 0.004 ($p < 0.05$) supports the idea that wetter months may help wetlands remove nitrate more successfully.

Phosphate removal efficiency fluctuated noticeably with changes in monthly rainfall, peaking at 100.00% in August 2022 alongside substantial rainfall (408.5 mm). Conversely, efficiency plummeted to − 288.09% in January 2022 during a period

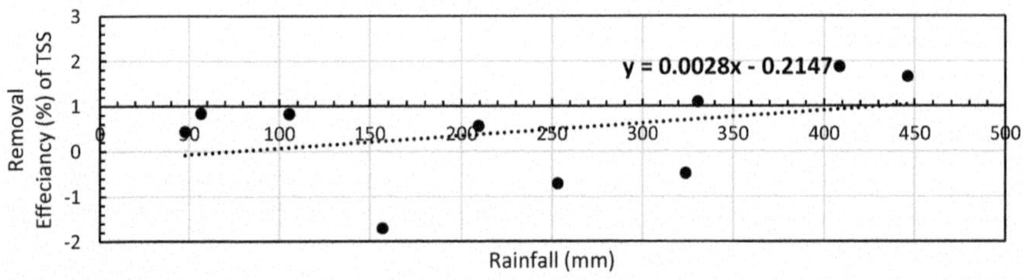

Fig. 4.8 Relationship between removal efficiency of TSS with rainfall

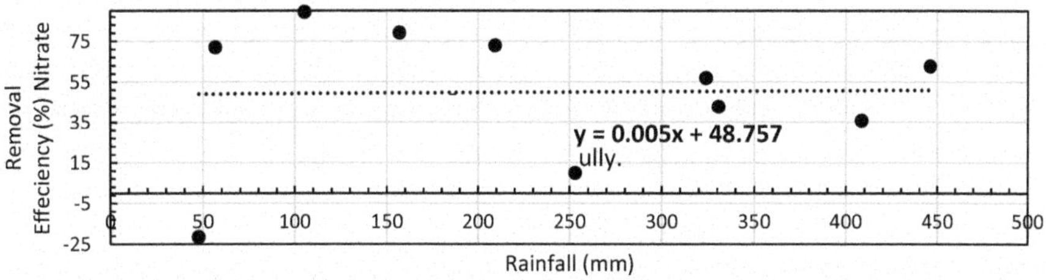

Fig. 4.9 Relationship between removal efficiency of Nitrate with rainfall

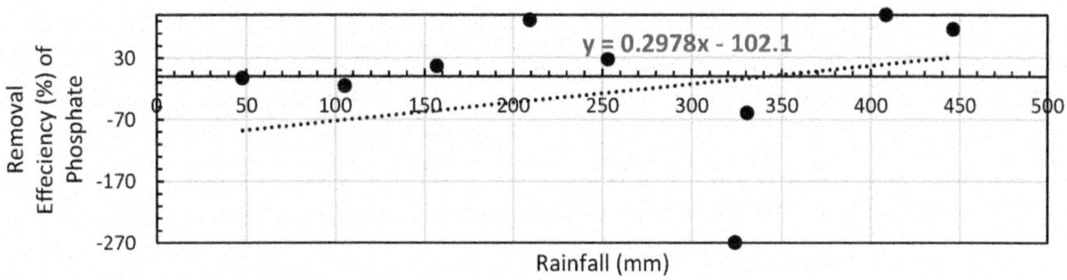

Fig. 4.10 Relationship between removal efficiency of phosphate with rainfall

of minimal rainfall (56.6 mm), suggesting clearer phosphate removal during drier months. This trend underscores a positive correlation between rainfall and phosphate removal, validated by a significant *p*-value of 0.001 ($p < 0.05$), as depicted in Fig. 4.10.

4.4 Discussion

In the descriptive statics, the mean, maximum, minimum and standard deviation was calculated. A variety of studies have underscored the significance of taking temporal variations and descriptive statistics into account when evaluating the water quality of wetlands. The wetland's capacity to gradually lessen environmental stressors is demonstrated by the temporal dynamics of removal efficiencies. The salinity level variation shows the least standard deviation and narrow range, indicating some stability in the wetland. But EC and TSS shows the higher standard deviation and wider range when compare to the other water quality parameters, suggest possible environmental stressors and the requirement for focused management approaches.

However, for the TSS mean value was less than the maximum permissible level as TSS 2100 mg/l. Phosphate concentrations have been measured, ranging from 0.09 to 7 mg/l, with an average of 0.37 mg/l. It is crucial to understand that, according to the Central Environmental Authority (CEA) in 2019, this mean value is below the allowable threshold standards (0.13 mg/l) for inland waters. The aforementioned data highlights the necessity of additional focus and actions to lower phosphate levels to within allowable bounds in order to safeguard inland water resources. Phosphate and nitrate, on the other hand, displayed distinct patterns that deviated from the overall pattern observed in the other measurements. There was a regular and ordered pattern in the overall distribution of the parameters that were being observed. But there was a noticeable deviation in the phosphate and nitrate removal efficiency. This indicates that there are large differences in the system's ability to reduce phosphate levels; negative values suggest that there may have been periods in which the system act as source.

The collected data and accompanying analysis show complex correlations between monthly rainfall and the removal efficiencies of different water quality parameters. Figure 4.5 clearly illustrates the positive correlation between rainfall and the efficiency of EC removal. August 2022 in particular shows an upward trend in EC removal efficiency in tandem with higher monthly rainfall. A low p-value (0.004), indicating a strong correlation between increased precipitation and enhanced EC removal, supports the statistical significance. This is consistent with the hypothesis that the presence of higher inflows during the wetter months leads to improved removal efficiency because of increased dilution. Wetter months may result in lower salinity removal efficiencies, as indicated by the negative trend for salinity removal efficiency in Fig. 4.6 in response to rainfall. Figure 4.6's depiction of the negative relationship between rainfall and TDS removal efficiency suggests that more precipitation may help with TDS removal. The correlation is strengthened by the statistically significant *p*-value (0.004), indicating that the removal of dissolved solids from the water might be facilitated by wetter conditions.

Significant positive correlation between rainfall and Total Suspended Solids (TSS) removal was observed, particularly evident in August 2022, which exhibited the highest value. This suggests that increased rainfall levels during specific months can influence variations in TSS removal efficiency. However, Birch et al. [4] reported contrasting results in their investigation of constructed wetland systems. They noted that during the most intense rainfall event, TSS removal efficiency was notably lower compared to other events. This discrepancy highlights the nuanced impact of rainfall characteristics, such as intensity, on TSS removal processes within wetland systems.

Figure 4.9 and the peak in December 2021 confirm the positive correlation between increased rainfall and nitrate removal efficiency. This observation is consistent with the exceptionally high nitrate removal efficiency in December 2021. The statistical significance of this relationship is highlighted by the low *p*-value of 0.004. The complex relationship between phosphate removal efficiency and monthly rainfall is demonstrated by the distinct fluctuations in the latter, as illustrated in Fig. 4.10. The greatest removal efficiency in August 2022, which also happened to coincide with an abundance of rain, confirms the positive correlation. On the other hand, the January 2022 removal efficiency that was negative points to a possible negative impact during drier conditions. The statistical significance of this relationship is supported by the low *p*-value (0.001).

It is reported by Yano et al. [24] and Gurung et al. [12] that increased rainfall intensity reduces the effectiveness of nutrient removal in wetlands. Our latest research, however, provides a more complex picture and suggests that this trend is not consistent across all parameters. Remarkably, the only materials in our study that followed the expected trend of decreasing removal efficiency with increasing rainfall were salinity and TDS. This disparity suggests that the effects of rainfall may differ based on the particular characteristics of the constituents involved and requires for a closer investigation of the complex dynamics governing nutrient removal processes in natural wetlands. The underlying mechanisms causing these variations should be further investigated in order to provide important information for better wetland management and conservation strategies, especially in areas vulnerable to heavy precipitation events. Certain points of individual removal efficiencies have negative values in the graphs, despite the fact that the overall patterns of the removal efficiencies exhibit rising and downward tendencies. The intricate biogeochemical dynamics of natural wetland systems are reflected in the study's negative removal efficiencies for some parameters. These kinds of incidents happen when these compounds are produced or released by internal processes in the marsh, which raises the concentrations at the outlet relative to the intake. For example, sediment resuspension from wind-induced mixing, changing hydrodynamic conditions, or bioturbation by aquatic species can all be responsible for increases in TSS. Phosphate concentrations can increase as a result of mineralization and desorption processes or the anaerobic release of phosphorus from sediments [17]. Similarly, nitrification in aerobic wetland microenvironments can lead to the generation of nitrate [5]. These results demonstrate the dual function of natural wetlands as suppliers and sinks of nutrients and sediments, highlighting the necessity of taking these dynamics into account when assessing their potential to improve water quality and developing sustainable management strategies.

4.5 Conclusion

The water quality and purification potential of the Kotagala natural wetland in Nuwara-Eliya, Sri Lanka, was studied. Water quality parameters varied, with salinity exhibiting the least variation and TSS and EC suggesting potential environmental stressors. Even though phosphate mean values are below allowable limits, targeted management strategies are still required. The wetland's dynamic reaction to external stressors was shown by the temporal variation in removal efficiencies. The overall pattern was altered by the phosphate and nitrate removal efficiencies, highlighting the need for further investigation into the wetland's nutrient removal mechanisms.

The effect of monthly rainfall on removal efficiencies was also investigated in this study. The idea that wetter months improve removal efficiency is supported by the positive correlation found between rainfall and EC removal efficiency. In contrast, the removal efficiencies of TDS and salinity demonstrated a negative correlation with rainfall, indicating that these parameters may be removed less effectively during wetter periods. Rainfall and TSS removal showed a positive correlation, suggesting possible gains during the rainy season. The results highlight the necessity for sophisticated management approaches by adding to our understanding of the complex dynamics of wetland water purification processes. For effective wetland conservation, the study emphasizes the significance of considering particular water quality parameters, their variations, and the intricate interactions with environmental factors.

The study emphasizes the critical role of wetlands as naturally occurring solutions for water purification in light of the increasing worldwide scarcity of water and the deteriorating quality of water. The Kotagala wetland has the potential to improve water quality, despite threats from encroachment and untreated effluent discharges. The insights gained from this research can inform conservation efforts and sustainable management practices to ensure the continued ecological and societal benefits of wetlands.

Acknowledgements This work was carried out with the aid of a grant from the International Development Research Center (IDRC), Ottawa, Canada. Their financial support is greatly appreciated.

References

1. Aazami M, Shanazi K (2020) Tourism wetlands and rural sustainable livelihood: the case from Iran. J Outdoor Recreat Tour 30:100284
2. Baker C, Thompson JR, Simpson M (2009) Hydrological dynamics I: surface waters, flood and sediment dynamics. The wetlands handbook 2:120–168
3. Bastviken S (2006) Nitrogen removal in treatment wetlands—factors influencing spatial and temporal variations. Linkopings Universitet (Sweden)
4. Birch GF, Matthai C, Fazeli MS, Suh JY (2004) Efficiency of a constructed wetland in removing contaminants from stormwater. Wetlands 24(2):459–466
5. Burgin AJ, Hamilton SK (2007) Have we overemphasized the role of denitrification in aquatic ecosystems? A review of nitrate removal pathways. Front Ecol Environ 5(2):89–96
6. Byun C, Lee SH, Kang H (2019) Estimation of carbon storage in coastal wetlands and comparison of different management schemes in South Korea. J Ecol Environ 43:1–12
7. Churchill E (2005) Environmental degradation and human well-being: report of the millennium ecosystem assessment. Popul Dev Rev 31:389–398
8. De Groot D, Brander L, Finlayson CM (2018) Wetland ecosystem services. In: The wetland book: I: structure and function, management, and methods. Springer Netherlands, pp 323–333
9. Dias S, Correia B, Fraga-Santiago P, Silva C, Baptista PC, Gomes CR, Almeida CMR (2021) Potential of an estuarine salt marsh plant (*Phragmites australis* (Cav.) Trin. Ex Steud 10751) for phytoremediation of bezafibrate and paroxetine. Hydrobiologia 848:3291–3304
10. Fernando SLJ, Shariff NM (2017) Site suitability analysis for ecotourism development at the Kirala Kele partial-nature-based wetland of southern Sri Lanka. Int J Sci 32(3):89–104
11. Ferreira CS, Kašanin-Grubin M, Solomun MK, Sushkova S, Minkina T, Zhao W, Kalantari Z (2023) Wetlands as nature-based solutions for water management in different environments. Curr Opin Environ Sci Health 33:100476
12. Gurung SB, Geronimo FKF, Choi H, Hong J, Kim LH (2018) Analysis of the factors affecting nutrients removal in hybrid constructed wetland treating stormwater runoff. J Wetlands Res 20(1):54–62
13. Jabłońska E, Winkowska M, Wiśniewska M, Geurts J, Zak D, Kotowski W (2021) Impact of vegetation harvesting on nutrient removal and plant biomass quality in wetland buffer zones. Hydrobiologia 848:3273–3289
14. Mitsch WJ, Gosselink JG (2015) Wetlands. Wiley

15. Rahimi L, Malekmohammadi B, Yavari AR (2020) Assessing and modeling the impacts of wetland land cover changes on water provision and habitat quality ecosystem services. Nat Resour Res 29(6):3701–3718
16. Ramsar Convention (2014) The convention on Wetlands, https://www.ramsar.org/publications. Accessed on 23 Aug 2023
17. Reddy KR, D'angelo EM (1997) Biogeochemical indicators to evaluate pollutant removal efficiency in constructed wetlands. Water Sci Technol 35(5):1–10
18. Rojas O, Soto E, Rojas C, López JJ (2022) Assessment of the flood mitigation ecosystem service in a coastal wetland and potential impact of future urban development in Chile. Habitat Int 123:102554
19. Salimi S, Almuktar SA, Scholz M (2021) Impact of climate change on wetland ecosystems: a critical review of experimental wetlands. J Environ Manage 286:112160
20. Sileshi A, Awoke A, Beyene A, Stiers I, Triest L (2020) Water purifying capacity of natural riverine wetlands in relation to their ecological quality. Front Environ Sci 8:39
21. Skov H (2018) UN convention on Wetlands (Ramsar): implications for human health. In: Encyclopedia of the anthropocene, pp 479–485. https://doi.org/10.1016/b978-0-12-809665-9.09347-2
22. Smardon R (2014) Wetland ecology principles and conservation, 2nd edn. Water 6(4):813–817. https://doi.org/10.3390/w6040813
23. Yang TX, Sheng LX, Zhuang J, Lv XG, Cai YP (2019) Function, restoration, and ecosystem services of riverine wetlands in the temperate zone. Ecol Eng 137:46–52
24. Yano KAV, Geronimo FKF, Reyes NJDG, Jeon M, Kim L (2019) Comparison of nutrient removal efficiency of an infiltration planter and an infiltration trench. J Wetlands Ress 21(4):384–391
25. Zedler JB, Kercher S (2005) Wetland resources: status, trends, ecosystem services, and restorability. Annu Rev Environ Resour 30(1):39–74

Chapter 5
The Bangladesh Erosion Monitor: Automatic Detection of Riverbank Erosion and Its Potential Use for Project Planning

Imran Khan, Arjen V. Haag, James Lilly, Kymo Slager, Md Shahadat Hossain, Raqubul Hasib, Morsheda Begum, William Oliemans, and Zahirul Haque Khan

Abstract The Bangladesh Erosion Monitor (BEM), developed through the Google Earth Engine platform using JavaScript, is a novel tool designed to assess erosion and accretion in Bangladesh's major river systems. Utilizing the extensive historical records of the Joint Research Centre (JRC), this approach conducts a comprehensive planform change analysis. It intricately blends data covering administrative boundaries, essential infrastructures, the dynamics of water flow, and patterns of global land use. This integration offers a multi-faceted perspective on how river systems

I. Khan (✉) · M. S. Hossain · R. Hasib · M. Begum · Z. H. Khan
Institute of Water Modelling, Dhaka, Bangladesh
e-mail: imk@iwmbd.org

M. S. Hossain
e-mail: sht@iwmbd.org

R. Hasib
e-mail: rqb@iwmbd.org

M. Begum
e-mail: mbm@iwmbd.org

Z. H. Khan
e-mail: zhk@iwmbd.org

A. V. Haag · J. Lilly · K. Slager · W. Oliemans
Deltares, MH Delft, Netherlands
e-mail: arjen.haag@deltares.nl

J. Lilly
e-mail: jim.lilly@deltares.nl

K. Slager
e-mail: kymo.slager@deltares.nl

W. Oliemans
e-mail: william.oliemans@deltares.nl

G. M. Tarekul Islam et al. (eds.), *Sustainable Water Management and Environmental Resilience*, Springer Water, https://doi.org/10.1007/978-3-031-93350-9_5

evolve and interact with both the natural environment and human-developed landscapes. This approach aids in understanding erosion impacts and supports informed decision-making. Including actionable information for policymakers, the tool identifies vulnerable infrastructure like embankments and ferry terminals, estimates potential damages to agriculture and households from river erosion, and can perform a scenario analysis to calculate inundation depths in breached coastal polders. Its multifaceted data layers enable users to predict and strategize against the adverse effects of riverbank erosion and flooding. Validated against real-time riverbank survey data, this tool represents an advancement in the field of environmental risk management by effectively implementing geospatial technology in novel new ways. It provides valuable insights for policy development, especially in addressing river erosion and flood hazards in Bangladesh, thus contributing meaningfully to the field.

Keywords Google Earth Engine · Remote sensing · Global Surface Water Dataset · Bankline erosion

5.1 Introduction

Located at the confluence of the Ganges, Brahmaputra, and Meghna rivers, the delta region in Bangladesh is a highly dynamic and morphologically active area, facing continuous changes in its landscape [6]. As a result, the susceptibility of these regions to riverbank and coastal erosion presents a formidable challenge with extensive impacts on communities, critical infrastructure, and diverse vulnerable ecosystems [2, 11, 15]. Riverbank and coastal erosion are dynamic phenomena influenced by various natural and anthropogenic factors, including hydrological patterns, land use change, and climatic variability [3, 5].

While these are, for the most part, natural processes with potential positive outcomes: e.g., accretion of new land [9], the most prevalent consequences are negative: e.g., erosion of agricultural land and embankments, and potentially consequent flooding of polders and communities [7]. To cope with these impacts, responsible authorities, such as the Bangladesh Water Development Board (BWDB), require innovative and efficient tools that can provide a rapid and accurate overview of riverbank and coastal erosion risks at a nation-wide scale for project planning and management.

This study is embedded in the "Joint-Cooperation between the Netherlands and Bangladesh for integrating a Geospatial Analysis Platform for Erosion Monitoring in Bangladesh in Risk Management Policy Planning," addressing the pressing need for sophisticated risk assessment mechanisms in the Ganges-Brahmaputra-Meghna delta. This collaboration, involving the Institute of Water Modelling (IWM) from Bangladesh and Deltares from the Netherlands, is directed towards enhancing the application of advanced geospatial analysis in risk management. It aims to equip the BWDB with a science-driven tool for comprehensive erosion assessment, blending technological innovation with localized expertise. This initiative, while focusing on creating the BEM, also emphasizes capacity building and fostering collaborative ties, marking a significant stride in addressing the multifaceted environmental challenges in the delta region.

In addressing the challenges of erosion in the Ganges-Brahmaputra-Meghna delta, this study investigates the capabilities of Google Earth Engine (GEE) in developing a comprehensive tool for automatic detection of riverbank erosion and monitoring. GEE's integration of satellite imagery, cloud computing, and analytical capabilities enables a deeper understanding of dynamic land use processes and their impacts [1]. By harnessing the vast array of remotely sensed data offered within GEE, the research focuses on developing an innovative and scalable approach for the automatic identification and monitoring of both coastal and riverbank erosion, effectively utilizing the extensive data capabilities of the platform.

The methodology aims to reinforce science-based decision-making, particularly in the areas of project planning and implementation. Current morphological models, while detailed, often lack comprehensive coverage and can be time-intensive to employ [10, 13]. In a similar vein, existing studies relying solely on satellite data, although ongoing, tend to have a limited scope. This research introduces a novel approach that combines readily accessible earth observation data with specific local data to conduct real-time, comprehensive analysis of erosion patterns. This new method is anticipated to guide policy decisions by providing crucial information to the BWDB for the effective planning and implementation of erosion mitigation strategies. While it offers a nationwide overview, it is not meant to entirely replace more detailed modeling or survey studies. Rather, it serves as a valuable initial assessment tool to assist in the prioritization and selection of sites for further, more detailed investigation.

5.2 Methodology

5.2.1 Workflow

The Landsat program, a joint venture between the National Aeronautics and Space Administration (NASA) and the United States Geological Survey (USGS), has been providing over 35 years of continuous satellite imagery. This extensive dataset makes it particularly suitable for long-term change analysis, a method previously employed in related research topics, as evidenced by Donchyts et al. [4]. For the preliminary dataset, the Landsat-derived Global Surface Water Dataset (GSWD) was utilized [12], which delineates water and land at a monthly time step from 1984 to 2021 at approximately 30 m spatial resolution. From this, a yearly classification was derived, which in turn is used to map erosion (land one year, water the next) and sedimentation (vice versa), providing a multi-decadal time series of these classes. In addition to this, an algorithm is under development that would enable the BEM tool to analyze Landsat images operationally and provide near-real-time surface water layer for subsequent analysis. In the next step, locally-derived information on the location of polders, embankments, and riverbank lines was used to assess the trend of erosion near them, yielding an estimate of near-future erosion probability for each 1 km stretch. The polders, embankments, and riverbank data used in the tool are a combination of field visit information and digitized data of those existing infrastructures and river bankline. Finally, for stretches with erosion potential, the adverse cascading consequences were simulated using a GIS-based approach. This approach investigates the worst-case scenario by assuming a complete failure of these embankment stretches. IWM has a nationwide validated 1D model which can provide flow and water level information at important locations for different hydrological events. Using the 1–100-year flood water levels from these nation-wide model simulations, combined with the elevation values from the Bangladesh Coastal DEM as well as other global DEM datasets (NASA DEM [8], MERIT DEM [16]), an indicative flood extent and depth are provided in real-time. These extreme flood hazard maps are combined with landcover (ESA WorldCover) [17] and world population [14] datasets to provide rough estimates on the cascading vulnerability or risk. ESA World Cover provides a global land cover classification at 10m resolution and is available for 2020 and 2021, while WorldPoP provides the spatial distribution of population for various years, both of which are vital for analysis of the impacts of erosion and flooding. The entire processing pipeline is configured in Google Earth Engine (GEE). A web-based application is created on top of this to enable viewing and user interaction, with the current prototype available at: https://arjenhaag.users.earthengine.app/view/bangladesh-erosion-monitor. The interface of the Bangladesh Erosion Monitor Tool has been shown in Fig. 5.1, while the processing pipeline of the tool is shown in Fig. 5.2.

Figure 5.2 also includes intermediate steps and products. Datasets are shown generically, specific names of those used at the time of writing are listed in the text. Elements shown in grey are being researched for near real-time purposes.

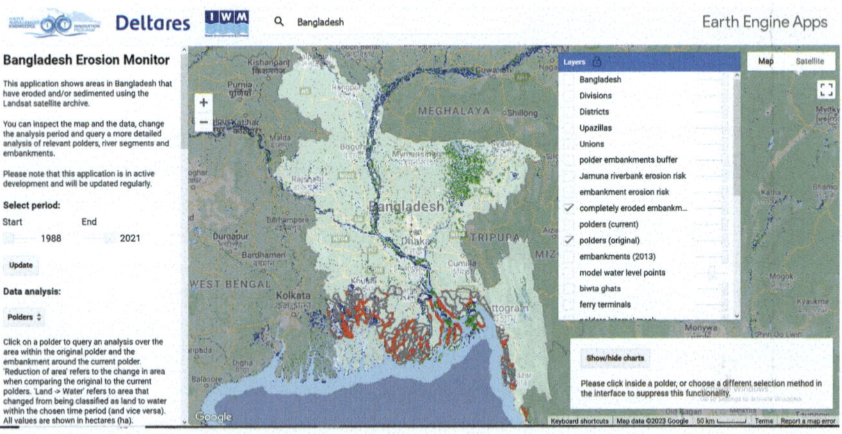

Fig. 5.1 Bangladesh Erosion Monitor Tool interface

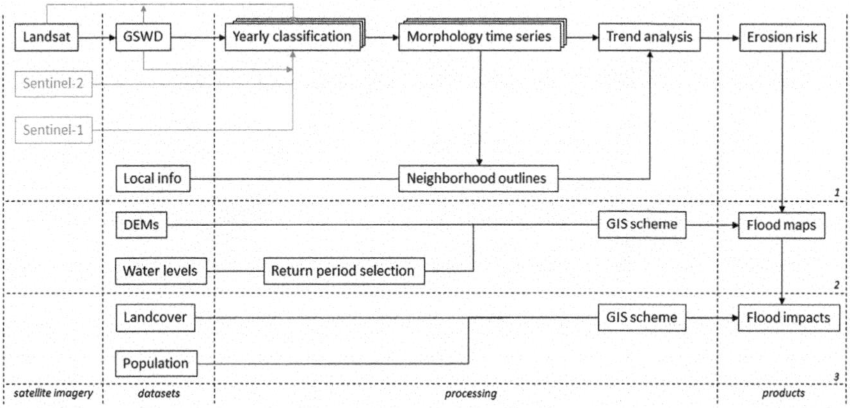

Fig. 5.2 Processing pipeline of the Bangladesh Erosion Monitor, from input (left) to output (right)

5.2.2 *Dataset*

The datasets which have been utilized in the BEM Tool is shown in Table 5.1.

Table 5.1 Dataset of the Bangladesh Erosion Monitor tool

Database	Data source
National boundary of Bangladesh	IWM
Administrative boundary (a) Division Boundary (b) District boundary (c) Upazilla boundary (d) Union boundary	IWM
Coastal embankment (polder)	IWM and Deltares
BIWTA Ghat/Ferry terminal	IWM
Land elevation	
NASA DEM	Global 1 arc second (~ 30 m) elevations [8]
MERIT DEM	Global 3 arc second (~ 90 m) elevations [16]
Coastal DEM	IWM
Global data	
Surface water	Global Surface Water Dataset [12]
Land cover	ESA WorldCover [17]
Population	WorldPop [14]
Mathematical model data	
Water level data	IWM
Flow velocity	IWM

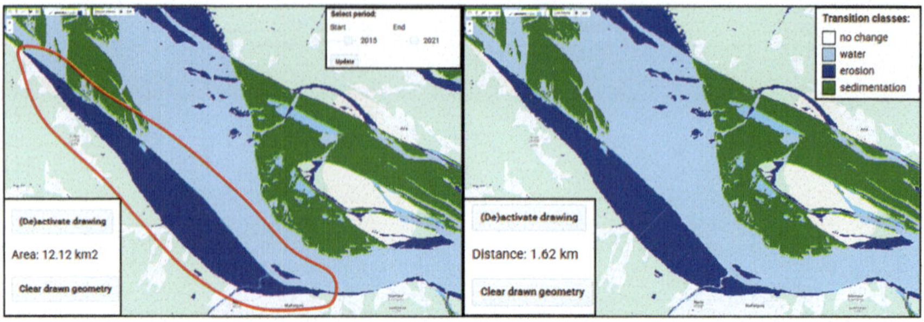

Fig. 5.3 Historical erosion near Naria, Shariatpur, Bangladesh between 2015 and 2021. **a** The picture shows the eroded area in square kilometers and **b** the picture shows the spatial shifting of the bankline using the measurement tool of the Bangladesh Erosion Monitor Tool. All these measurements can be done by the user instantaneously for any specified period

5.3 Results

The BEM serves as an analytical tool for assessing historical riverbank erosion across the nation and also incorporates features designed to predict near-future erosion of embankments and the potential cascading effects of flooding.

5.3.1 Historical Erosion

Historical erosion monitoring is one of the main objectives for the development of the BEM Tool. Currently, the tool can analyze the available imagery from 1988 to 2021 enabling users to access information on historical erosion or deposition trends. Users have the flexibility to select any start and end year between 1988 and 2021 to generate custom historical erosion and deposition trend maps, which are calculated on-the-fly in GEE. The application features a measuring tool that allows the user to measure the spatial scales, both in length and area, of erosion deposition zones. Figure 5.3 showcases this part of the interface of the BEM, illustrating its functionality in tracking and analyzing historical erosion.

5.3.2 Potential Economic Impact of Historical Erosion

With the "Admin. Bounds" option, the BEM Tool provides the user with an interactive window for generating the potential economic impact of erosion based on a user selected administrative boundary. The user has the option to click any administrative boundary (e.g., Division/District/Upaziila/Union) on the tool's interactive window and the tool will provide the potential economic impact of the historical erosion

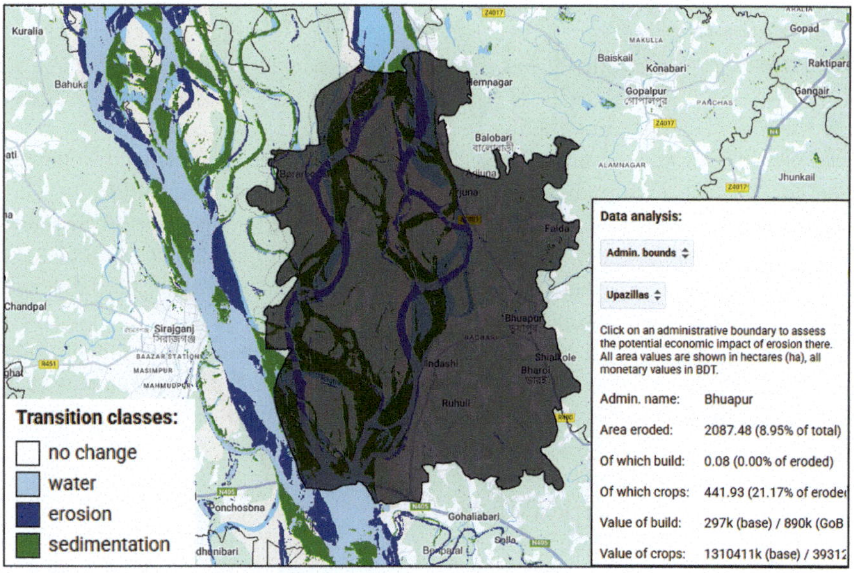

Fig. 5.4 The potential economic impact of historical erosion analyzed using the Bangladesh Erosion Monitor (BEM) tool for Bhuapur Upazila, Tangail, Bangladesh

of the selected administrative boundary. In real-time, the tool generates five key data points for the selected administrative boundary: the total area of eroded land, the total urban area lost, the total agricultural land lost, as well as the probable values of the lost urban and agricultural lands in the selected administrative boundary between 1988 and 2021 due to erosion. ESA WorldCover data [17] is used in the background to make the calculations of the impact of historical erosion and the probable value is assumed to be 2,965,200 and 3,706,500 Bangladeshi Taka for agricultural and build-up land, respectively. Figure 5.4 shows the potential economic impact of historical erosion analyzed using the Bangladesh Erosion Monitor (BEM) tool for Bhuapur Upazila, Tangail, Bangladesh. The interactive window of the BEM gives the user the freedom to click on the map to select any Division/District/Upazilla/Union (Admin Boundary) of Bangladesh and the statistics for the selected boundary will be generated spontaneously.

5.3.3 *Erosion Prediction Based on Historical Trends*

By employing historical bankline data from Bangladesh's major river systems, the tool can predict erosion susceptibility for all available banklines in the near future, such as the upcoming monsoon season(s), by conducting a trend analysis of the yearly erosion rate near the banklines. In this analysis, erosion susceptibility is calculated for every 1 km stretch of the riverbank. The BEM examines the extent of erosion within

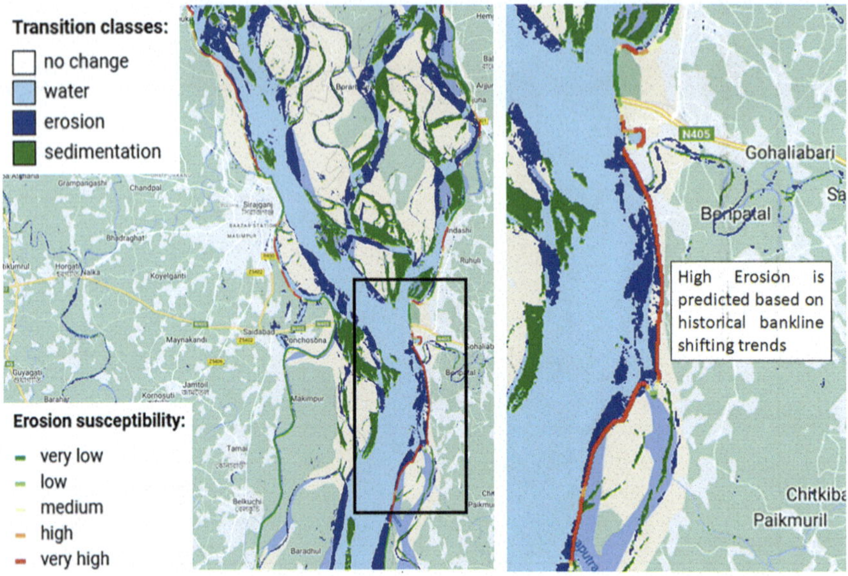

Fig. 5.5 Riverbank erosion prediction of the Bangladesh Erosion Monitor tool

a 200 m buffer region around each stretch and based on the rate of erosion within that region, assigns an erosion susceptibility rating to that stretch of the riverbank. The riverbank erosion prediction near the Bangabandhu Bridge from the Bangladesh Erosion Monitor tool has been shown in Fig. 5.5.

5.3.4 Polder Erosion Susceptibility Based on Historical Trends

Employing a similar methodology to that used for erosion prediction, the tool can predict the erosion susceptibility of different stretches of all the polders of the coastal regions of Bangladesh. Erosion susceptibility of the polders is calculated at every 2001 km stretch of the polder outline, by looking at the yearly erosion/sedimentation rates within a 200 m outwards buffer around that stretch. Figure 5.6 shows the area classified as land, seasonal and permanent water (lines, left y-axis), as well as yearly erosion and sedimentation rates (columns, right y-axis) for the total original polder area (top) and for the area near the current (2021) embankments around that polder (59/2 PROJECT). The erosion rates near the embankments show a clear increase starting around 2013, which is reflected in high erosion susceptibilities for some of the embankment stretches shown in Fig. 5.7.

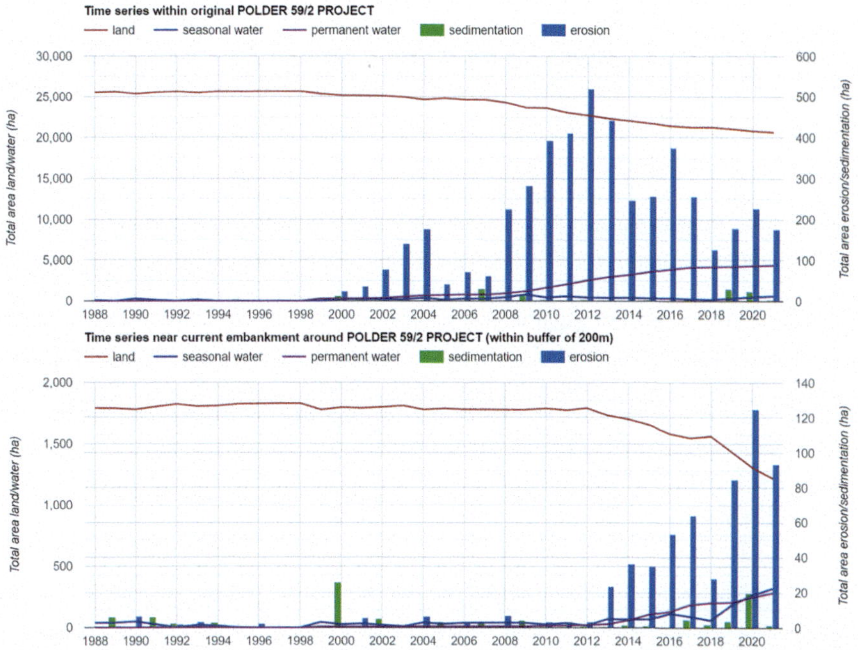

Fig. 5.6 Area classified as land, seasonal and permanent water (lines, left y-axis), as well as yearly erosion and sedimentation rates (columns, right y-axis) for the total original polder area (top) and for the area near the current (2021) embankments around that polder (59/2 PROJECT, same as in figure below)

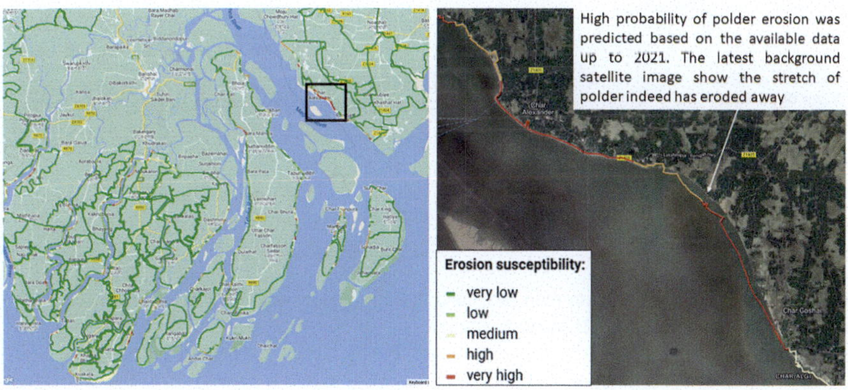

Fig. 5.7 Polder erosion susceptibility prediction by the Bangladesh Erosion Monitor tool based on data available up to 2021

5.3.5 Flood Inundation Map and Flood Risk Assessment Based on Possible Embankment Breach Due to Erosion

The tool can produce rough estimates of flood inundation assuming complete failure of a stretch of polder embankment. With the "Embankment" option activated, the user has the option to click on any stretch of the polder embankments. Once clicked, the tool will assume that the stretch of embankment has failed completely and generate a flood inundation map for a 1 in 100-year return water level from the model output point closest to that stretch of embankment. It is assumed that water can only flow to regions that are hydrodynamically connected (i.e. adjacent and with lower elevation than the water level) to the failed embankment stretch, but it should be noted that the DEMs often do not include infrastructure or other detailed features which might block the flow of water in reality. The tool will also generate graphs on the total number of people and cropland affected by the 1 in 100-year flooding due to the embankment failure. The interface of the polder flood inundation and flood risk assessment of the BEM tool is shown in Fig. 5.8.

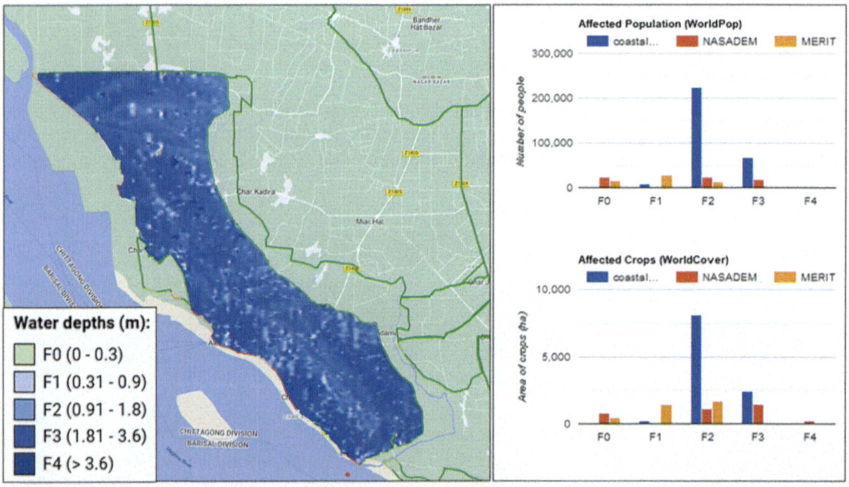

Fig. 5.8 Polder flood inundation map (left) and flood risk assessment statistics generated by the Bangladesh Erosion Monitor (BEM) tool (right)

5.4 Discussion

5.4.1 Validation of Water Surface Detection Algorithm

The water surface detection algorithm has been checked against field surveyed bankline for multiple years for proper validation of the algorithm. An example of the validation process involves comparing the algorithm-generated water surface against the field surveyed bankline for the Noakhali region has been shown in Fig. 5.9.

Figures 5.9 and 5.10 illustrate the specific locations where deviations occur between the surveyed bankline and the yearly averaged water surface data generated by the BEM. Figure 5.11 presents a graph depicting these deviations. In this graph, a negative value signifies erosion of the surveyed bankline in comparison to the BEM-generated water surface data, while a positive value indicates deposition. Notably, as shown in Fig. 5.11, deviations of up to 200 m have been observed, highlighting the variances captured by this analysis. Relatively small deviations, especially those smaller than the resolution of the data (~ 30 m), could be caused by differences in resolution between the source data and surveyed data. Larger deviations should be less affected by this and might originate from the fact that the surveyed bankline is from a precise moment in time, within a specific month, while the yearly map of

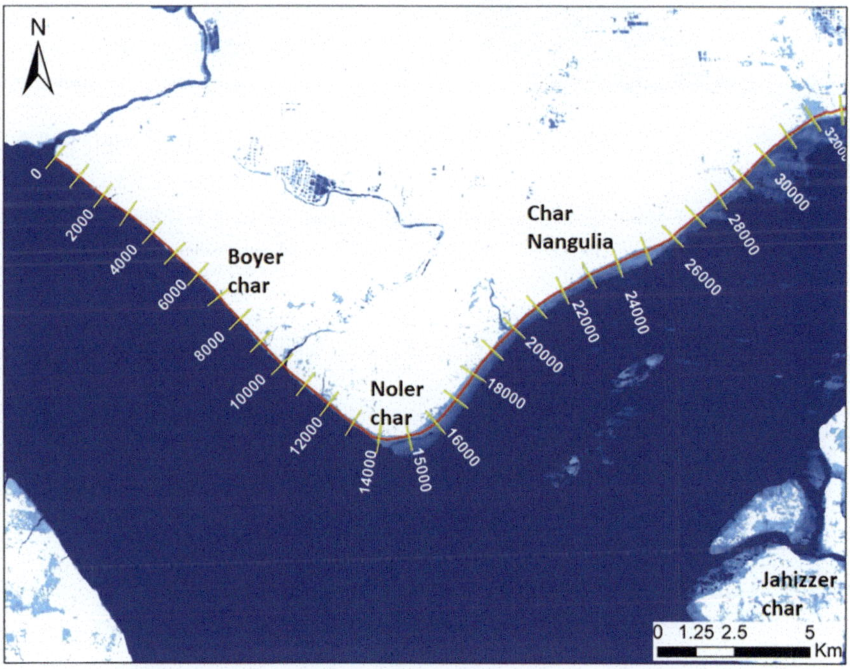

Fig. 5.9 River bankline (August, 2021) is superimposed on the Year 2021 BEM generated water surface near Boyer Char

the BEM shows aggregated values and includes months before, as well as after the survey, so the situation might already have changed.

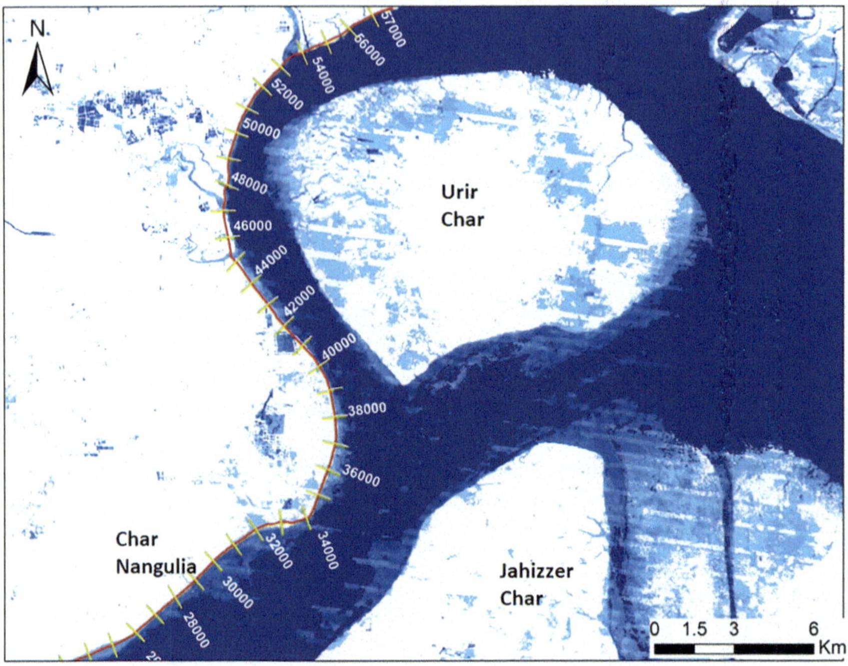

Fig. 5.10 River bankline (August, 2021) is superimposed on the Year 2021 BEM generated water surface near Urir Char

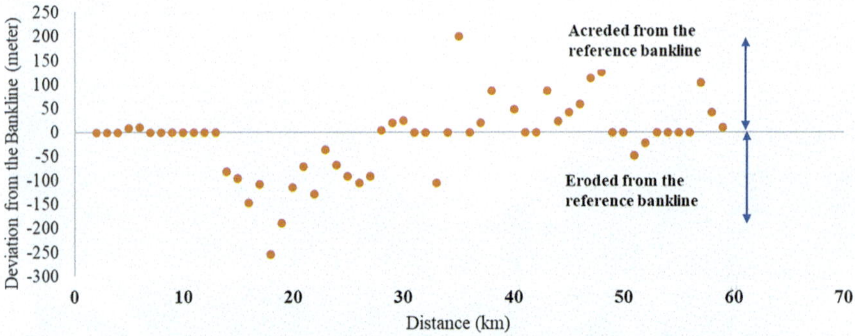

Fig. 5.11 Deviation of surveyed bankline from the reference bankline (August 2021) obtained from the BEM generated water surface slope

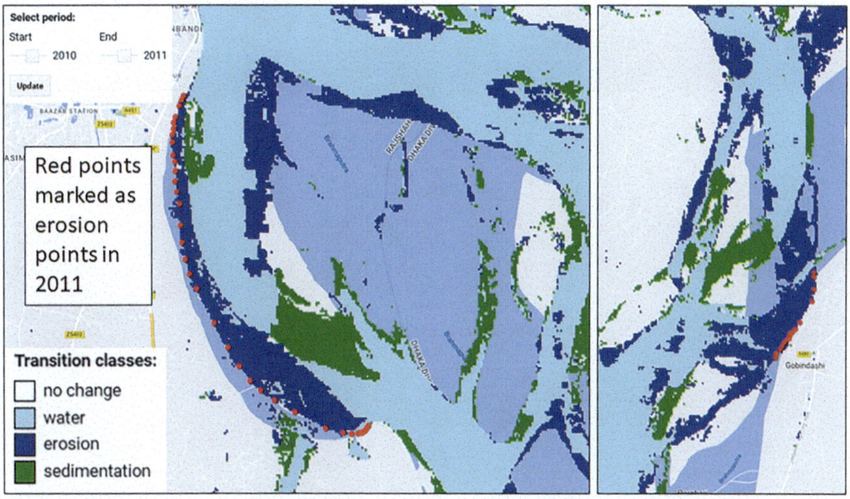

Fig. 5.12 Validation of historical erosion of downstream of Sirajganj Hard Point (left) and Bhuapur (right) for 2010–2011

5.4.2 *Validation of Historical Erosion Events*

In addition to validating the water surface algorithm, checks were conducted on the erosion detection algorithms to ascertain the BEM's effectiveness in tracking historical erosion events within Bangladesh. Archives of historical erosion events from various locations across Bangladesh, sourced from the IWM database, were utilized to construct a custom dataset. This dataset was then compared with the BEM's erosion dataset for the corresponding period and location, to validate the BEM's ability to accurately detect historical erosion. A sample historical erosion validation plot has been shown in Fig. 5.12, where the historical erosion information collected from the field in 2011, downstream of Sirajganj Hard Point and Bhuapur has been validated against BEM erosion detection. The red points indicate GPS locations recorded in Post Monsoon 2011, and the BEM's erosion-sedimentation analysis covers the 2010–2011 period. This comparison demonstrates the BEM's proficiency in accurately identifying historical erosion events, as corroborated by the field-surveyed erosion data.

5.4.3 *Erosion Prediction Validation*

The bank erosion prediction module of the BEM tool considers the historical river bankline shifting up to 2021 and provides susceptibility to erosion for the near future (e.g. 2022). The erosion prediction scheme has been validated against observed riverbank erosion that occurred in 2022 in multiple reaches of Jamuna River as well

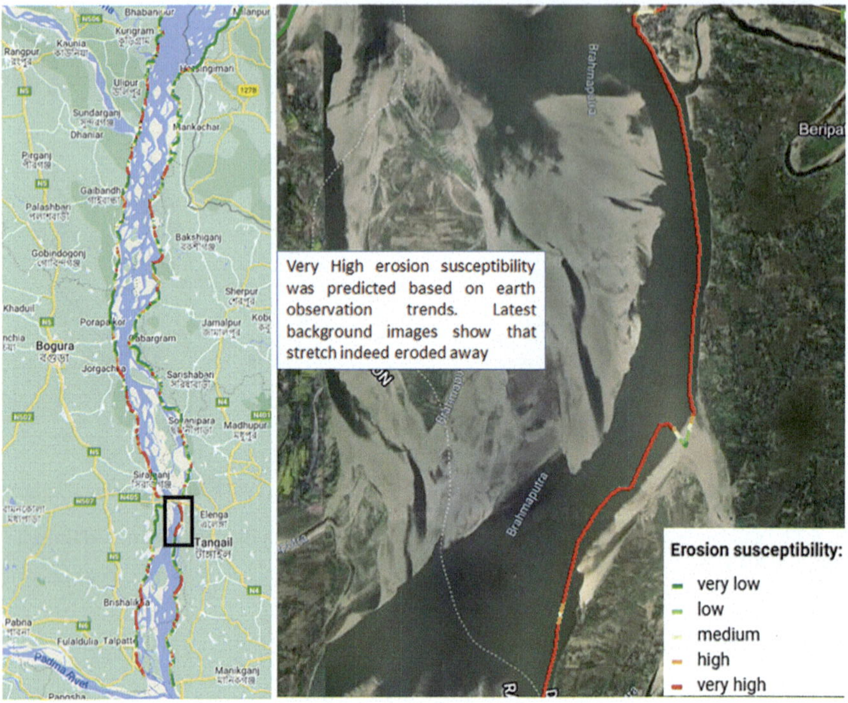

Fig. 5.13 Bangladesh Erosion Monitor predicted riverbank erosion of the Jamuna River for 2022 (left) and validation of the predicted erosion against the latest satellite image downstream of Bangabandhu Bridge (right). *Source of background image* Google Maps (within GEE)

as at multiple polders of Southern Bangladesh. The erosion predicted by BEM for 2022 has been shown in Fig. 5.13 (left). Figure 5.13 (right) shows the validation of the predicted erosion, just downstream of existing Bangabangdhu Bridge, against the latest available high-resolution satellite image. The image shows that the stretch of the Jamuna River has indeed eroded away as predicted by the BEM. Similar validation has also been carried out for Boyer Char polder in the Lower Meghna River, which has been shown in Fig. 5.14. Similar to Fig. 5.13, Fig. 5.14 also shows that the erosion predicted for the reach based on earth observation trend analysis has indeed occurred.

5.4.4 Limitations and Future Improvements

The current version of the tool uses a global dataset for images analysis up to 2021. However, work is ongoing to improve the algorithm based on recent Landsat images and provide a near-real-time surface water layer for subsequent analysis. Once fully

Fig. 5.14 Bangladesh Erosion Monitor predicted polder erosion near Boyer Char for 2022 (left) and validation of the predicted erosion against the latest satellite image near Chairman Ghat, Boyer Char (right). *Source of background image* Google Maps (within GEE)

developed, the algorithm will also be able to provide near-real-time prediction capability on a nation-wide scale. The 30 m resolution of the Landsat images cannot identify the water surface extent of the smaller rivers of Bangladesh. Hence although the BEM tool performs very well for the major rivers of Bangladesh, it has significant limitation in monitoring and predicting erosion for the smaller rivers, as well as small-scale erosion that might occur along the larger rivers and coastal region. To improve this, the inclusion of Copernicus Sentinel-1 and -2 is also being considered, which would improve spatial resolution up to 10 m. However, since these satellites do not have the long-term continuous archive Landsat has, it might introduce a shift in the yearly trend that is used for the erosion susceptibility analysis. The erosion prediction module in the current version is purely based on the earth observation trend analysis of the river bankline movement and does not consider the overall morphological process of riverbank erosion, including soil composition, flow rates and sediment loads. The bank erosion prediction module is still being developed, and in this module, more factors (such as discharge, near bank velocity, soil properties, etc.) are being considered which will improve the prediction module further. However, current results and their validation indicate that this relatively simple method can already provide an estimate of erosion susceptibility that is in line with actual erosion occurrences.

The generated flood inundation maps of the BEM are a simplified version of probable inundation as a result of complete failure of an embankment stretch and a 1-in-100-year flood event. The hydrodynamics within the region are not considered, and the land elevation of polders is not adjusted for the presence of infrastructures or other features that might block the flow of water in reality. As a result, the flood map only provides a rough estimate to enable quick insights that could drive a more detailed study. Moreover, the duration of the inundation depths associated with a 1-in-100-year flood is not currently indicated but is being considered for an upcoming update. The latter is also the case for flood events with different return periods, especially those that could occur more often, to provide information on possible impacts that might take place often and/or soon after erosion of an embankment (instead of with a major 1-in-100-year event).

5.4.5 *Field Validation and Assessing Riverbank Erosion*

The initial stage of our field validation research included visits to two distinct river sites in Bangladesh—the Jamuna River and the Arial Khan River—to gauge the extent of on-site erosion. The Jamuna River site, located near the Bangabandhu Bridge (refer to Fig. 5.15) in north-west Bangladesh, and the Arial Khan River site, near the Padma Bridge in south-west Bangladesh, were chosen due to their contrasting characteristics: the Jamuna is a vast river, averaging 12–16 km in width, while the Arial Khan is comparatively narrower, with a width of 300–500 m.

The field visits to the Jamuna and Arial Khan Rivers had multiple objectives: (i) assessing riverbank erosion in two distinct rivers; (ii) refining ground truthing methodology for the erosion monitor; (iii) testing the social survey questionnaire, a precursor to broader social surveys; (iv) exploring the social impacts of erosion. These visits were important for collecting validation data and setting the stage for comprehensive social surveys. Such surveys are imperative for achieving a holistic perspective on the broader implications of riverbank erosion, refining the BEM tool's application, and generating actionable information for policymakers.

5.5 Conclusion

This research, set against the backdrop of the Ganges-Brahmaputra-Meghna delta's complex and dynamic environment, represents a step forward in the ability to monitor erosion. Utilizing the innovative capabilities of GEE, the BEM has been developed to enable large scale (in both space and time) erosion monitoring. The tool not only traces the historical patterns of erosion but also forecasts potential future risks associated with it by providing a near-future erosion susceptibility estimate and by considering impacts that might follow from erosion. It embodies a synergy of extensive satellite data analysis, advanced algorithmic modeling, and an implementation of a deep understanding of the local geographical and hydrological nuances.

This amalgamation of data allows for a multi-decadal analysis of erosion and sedimentation, delivering insights with resolution and accuracy. This ability to predict impending erosion and its subsequent effects, such as potential flood risks, makes the information actionable for decision makers.

Aside from the previously mentioned technical developments (e.g. inclusion of near real-time data), the next phase of this project will encompass a crucial element of socio-economic ground truthing. This phase is designed to align our technical findings with the tangible socio-economic impacts observed on the ground, thereby offering a more comprehensive perspective of the erosion phenomenon. The insights

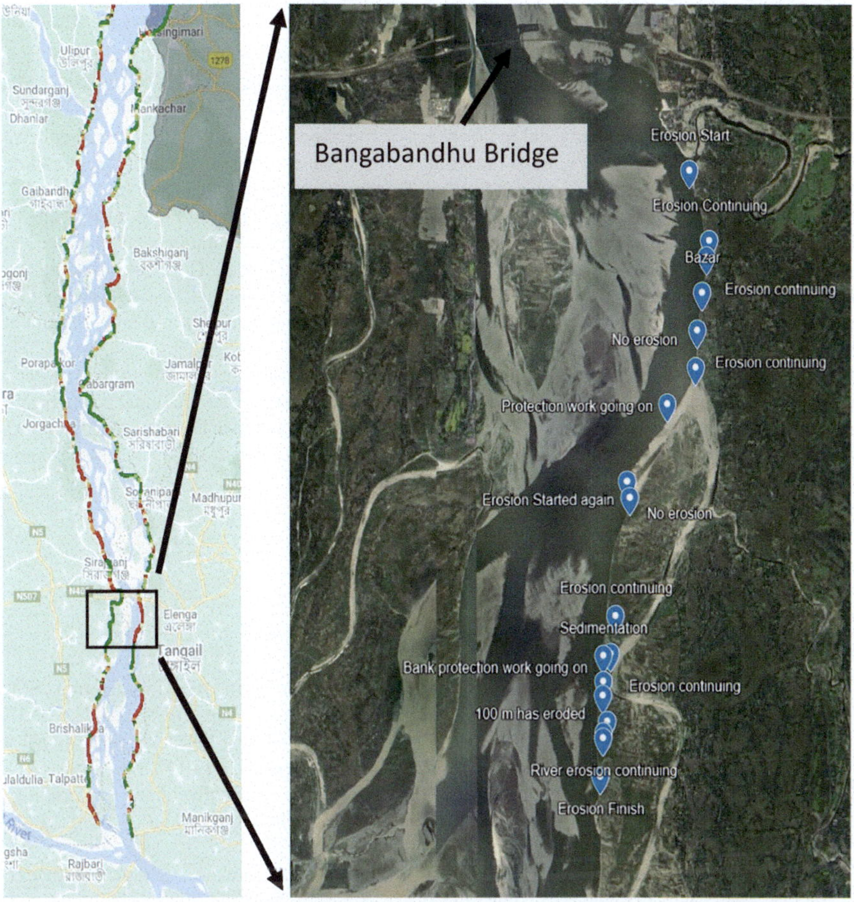

Fig. 5.15 Location visited during field trip to Jamuna River

gleaned from this phase are anticipated to augment its practical utility for local communities and policymakers.

This work could have significant implications for decision and policymakers. It has the potential to inform a broad spectrum of decisions, ranging from infrastructure planning to formulating emergency response strategies. By providing a near-real-time assessment of erosion risks, the tool can rapidly equip policymakers with the knowledge to make informed, proactive decisions, potentially reducing the adverse effects of erosion on vulnerable communities, infrastructure and ecosystems.

In essence, this research demonstrates the efficacy of integrating advanced satellite imagery analysis with the bankline survey data at the Padma River and Lower Meghna River, measured by IWM. It highlights the potential of such integrative approaches in tackling complex environmental challenges. As the project progresses, its findings are expected to make contributions to the fields of environmental science, risk management, and policy formulation, ultimately aiding in the development of resilient and adaptive strategies in response to evolving landscapes.

Acknowledgements This research is part of the "Action on the Ground" project, under the auspices of the Water and Climate Coalition and the Ministry of Infrastructure and Water Management. We extend our thanks to Robert Moree from the Ministry, for his support in the project. Our gratitude also goes to Kymo Slager and Arjen Haag for their leadership role and guidance in the project.

We acknowledge the Bangladesh Water Development Board, especially the contributions of Dr. Robin Kumar Biswas, for their significant involvement and collaboration in this initiative, which has been instrumental in enhancing climate resilience in Bangladesh. The close collaboration between Deltares and the Institute of Water Modelling (IWM) throughout this project has been vital in aligning our efforts with the specific needs and context of Bangladesh's water management challenges.

We also would like to credit William Oliemans for his essential role in shaping the project's direction. Additionally, we value and appreciate the collective efforts of international partners and financing institutions that have supported critical climate and water action in Bangladesh, leveraging data and knowledge to strengthen policy planning and implementation in the face of climate change challenges.

References

1. Amani M, Ghorbanian A, Ahmadi SA, Kakooei M, Moghimi A, Mirmazloumi SM, Moghaddam SHA, Mahdavi S, Ghahremanloo M, Parsian S, Wu Q, Brisco B (2020) Google Earth Engine cloud computing platform for remote sensing big data applications: a comprehensive review. IEEE J Sel Top Appl Earth Observ Remote Sens 13:5326–5350
2. Bhuiyan MAH, Islam SMD-U, Azam G (2017) Exploring impacts and livelihood vulnerability of riverbank erosion hazard among rural household along the river Padma of Bangladesh. Environ Syst Res 6(1):25
3. Bui LHN, Bui LT (2020) Modelling bank erosion dependence on natural and anthropogenic factors—case study of Ganh Hao estuary, Bac Lieu—Ca Mau, Vietnam. Environ Technol Innov 19:100975
4. Donchyts G, Baart F, Winsemius H, Gorelick N, Kwadijk J, Van De Giesen N (2016) Earth's surface water change over the past 30 years. Nat Clim Change 6(9):810–813
5. Huang J, Zhang Z, Feng Y, Hong H (2013) Hydrologic response to climate change and human activities in a subtropical coastal watershed of southeast China. Reg Environ Change 13(6):1195–1210
6. Islam SN (2016) Deltaic floodplains development and wetland ecosystems management in the Ganges–Brahmaputra–Meghna Rivers Delta in Bangladesh. Sustain Water Resour Manag 2(3):237–256
7. Johnson DL, Lewis LA (2007) Land degradation: creation and destruction. Rowman & Littlefield
8. Jpl N (2020) NASADEM Merged DEM Global 1 arc second V001. NASA EOSDIS Land Processes DAAC
9. Langhorst T, Pavelsky T (2023) Global observations of riverbank erosion and accretion from Landsat imagery. J Geophys Res: Earth Surf 128(2):e2022JF006774
10. Latteux B (1995) Techniques for long-term morphological simulation under tidal action. Mar Geol 126(1):129–141
11. Paszkowski A, Goodbred S, Borgomeo E, Khan MSA, Hall JW (2021) Geomorphic change in the Ganges–Brahmaputra–Meghna delta. Nat Rev Earth Environ 2(11):763–780

12. Pekel J-F, Cottam A, Gorelick N, Belward AS (2016) High-resolution mapping of global surface water and its long-term changes. Nature 540(7633):418–422
13. Schumm SA (2007) River variability and complexity. Cambridge University Press
14. Tatem AJ (2017) WorldPop, open data for spatial demography. Sci Data 4(1):1–4
15. Tri VPD, Trung PK, Trong TM, Parsons DR, Darby SE (2023) Assessing social vulnerability to riverbank erosion across the Vietnamese Mekong Delta. Int J River Basin Manag 21(3):501–512
16. Yamazaki D, Ikeshima D, Tawatari R, Yamaguchi T, O'Loughlin F, Neal JC, Sampson CC, Kanae S, Bates PD (2017) A high-accuracy map of global terrain elevations. Geophys Res Lett 44(11):5844–5853
17. Zanaga D, Van De Kerchove R, Daems D, De Keersmaecker W, Brockmann C, Kirches G, Wevers J, Cartus O, Santoro M, Fritz S (2022) ESA WorldCover 10 m 2021 v200

Chapter 6 Prediction of Stream Bank Failure at Paturia Ferry Ghat of Padma River of Bangladesh in Terms of Bank Materials

Uma Saha, Fatima Rukshana, Nayan Chandra Ghosh, Md. Moniruzzaman, Sumiya Ferdhous, Bikash Roy, and Kazi Rezaul Karim

Abstract This study has been undertaken to predict the bank failure mechanism of Padma River at Paturia ferry ghat of Bangladesh regarding bank materials. In this context, laboratory test results of soils and water levels were analyzed. From the analysis, it has been observed that the low flood level at Paturia ferry ghat varies from 1.8 to 2.0 mPWD at which the bank materials are non-cohesive. However, the high flood level varies from 7 to 9 mPWD which exceeds the bank elevation of bank line 6.667 to 6.935 mPWD, where the materials from the bank line to the ground surface are cohesive layers containing 30% clay. The presence of such clay particles creates a higher level of bonding among the other particles which are more resistant to surface erosion because of their lower permeability. These soils are more susceptible to bank failure during the rapid drawdown of water levels due to the increases in pore pressures which lead to tension cracks. The study results will inform the design

U. Saha (✉) · F. Rukshana · Md. Moniruzzaman · S. Ferdhous · K. R. Karim
Geotechnical Research Laboratory, River Research Institute, Faridpur, Bangladesh
e-mail: umasaharri@gmail.com

F. Rukshana
e-mail: frukshana@rri.gov.bd

Md. Moniruzzaman
e-mail: mmoniruzzaman@rri.gov.bd

S. Ferdhous
e-mail: sferdhous@rri.gov.bd

K. R. Karim
e-mail: krkarim@rri.gov.bd

N. C. Ghosh
Dhaka Laboratory, River Research Institute, Dhaka, Bangladesh
e-mail: ncghosh@rri.gov.bd

B. Roy
Hydraulic Research Laboratory, River Research Institute, Faridpur, Bangladesh
e-mail: broy@rri.gov.bd

G. M. Tarekul Islam et al. (eds.), *Sustainable Water Management and Environmental Resilience*, Springer Water, https://doi.org/10.1007/978-3-031-93350-9_6

engineer about the bank failure mechanism; however, a comprehensive geotechnical investigation will be needed prior to the design for the protection of the Paturia ferry ghat.

Keywords Bank materials · Failure mechanism · Paturia ferry ghat · Padma River · Bank failure

6.1 Introduction

6.1.1 Soil Formation in Bangladesh

A major part of Bangladesh is situated on the Bengal delta formed by three major rivers Ganges, Brahmaputra, and Meghna [1]. The sediments carried by the huge discharges of these rivers have built a broad delta, forming most of the area of Bangladesh and the submerged delta plain in the Bay of Bengal. These huge sediments are the major sources of formation of 80% soils of the country. The remaining 20% of soils have been formed in Tertiary and Quaternary sediments of hills (12%) and uplifted Pleistocene terraces (8%) [4].

In many areas, the soil surveys recognized active, young, and old floodplain landscapes. Active floodplains occupy land within and adjacent to the main river systems where shifting channels deposit and erode new sediments during the annual floods. Newly deposited alluvium within this floodplain is stratified into different layers. Usually, silty and clay deposits are finely stratified, whereas sandy deposits, as well as mixed sandy and silty deposits, are coarsely stratified.

The young and old floodplains are virtually stable land where the main river channel has moved away, but they are crossed by tributaries or distributaries. On these floodplains, the process of soil formation is dominated by sediment deposition, as evidenced by soil characteristics [6].

Different texture of the soil is found in different parts of Bangladesh due to the varying types of sediments deposited at the floodplains when they over-flooded the land. Sandy material is deposited near the riverbanks and silty material is further away from the banks whereas clay is the furthest away from the river. When rivers change their course, the pattern of sandy and silty materials is seen on the slightly higher ridges of the rivers, and clay materials are seen left behind in the adjoining basins [3].

6.1.2 Causes of Failure of Riverbank

Sediment gets into the river not only from the catchment but also from the contribution from the erosion of its banks. The cause of the riverbank failure includes the underwater erosion along the toe of the bank during the falling stage of the river; direct erosion of the riverbank; sloughing of the saturated bank caused by rapid drawdown; liquefaction of saturated silty and sandy bank material; erosion due to seepage from banks at low river discharge; and scour along waterline due to wind or wave wash of passing vessels. Bank failures depend more on the properties of the bank material than the direct erosive action of stream flow. Eventually, the probable causes of erosion are river instability, soil erosion, sediment transport, and deposition. Large quantities of sediment on the slopes of the Himalayan Mountains are shed due to intense monsoon rainfall, high tectonic activities, soil conditions, and human intervention [10].

The riverbank can collapse depending on several factors predominating the geotechnical characteristics. When the gravitational forces act on a bank and exceed the forces that hold the sediment together, depending on sediment type, layering, and moisture content, the banks can collapse [7].

A stream bank failure is closely related to the composition of the stream bank material. Geotechnical failures are usually the result of moisture conditions in the stream bank which create forces that exceed bank resistance. Human actions are often responsible, for example, any unnatural destruction of bank vegetation promotes erosion by hydraulic forces [9].

The guidelines regarding the failure of riverbanks suggest that when banks are destabilized by the piping of cohesion-less soil from lenses then capillary action temporarily decreases the angle of repose of the bank material to less than the existing bank slope. Liquefaction of fine-grained material causes fluid-like failures of the bank from pore pressure increase during rapid drawdown. Shrinking and swelling of clay soils during the wetting and drying cycle causes tension cracks, freezing and thawing of soil weakens the shear strength, subsurface moisture changes weaken the internal shear strength of the soil mass at the interface of different soil types and wave action is the impact of waves hitting directly on exposed soil causes bank failure. Waves vary with wind speeds and duration, water depth, and the continuous length of water over which winds blow in one direction [12]. Flow properties, bank material composition, bank geometry, bank moisture conditions, channel geometry, vegetation, and man-induced factors influence riverbank failure [13].

Terzaghi and Peck [11] explained the signs of riverbank failure and named a slide as the failure of a mass of soil located beneath a slope. It involves a downward and outward movement of the entire mass of soil that participates in the failure. The failure of a slope in a cohesive material is preceded by the formation of tension cracks behind the upper edge of the slope. They expressed that slope failure occurs along a surface of sliding that intersects the slope at or above its toe. They also explained that base failure occurs if the soil beneath the level of the toe of the slope is unable to sustain the weight of the overlying material, the failure occurs along a surface that passes at some distance below the toe of the slope.

The failure of riverbanks is a common problem in Bangladesh. Devastating floods and excessive rainfall are accelerating the failure process which results in immense damage to agriculture and infrastructures every year. In Bangladesh, riverbank failure occurs enormously in monsoon and during draw-down situations. The

Fig. 6.1 Riverbank scenario at Paturia ferry ghat

people suffered unbearable miseries wherever the riverbank failure occurred. Paturia ferry ghat is such a bank failure-prone riverbank, which has been shown in Fig. 6.1.

In this context, it is necessary to provide attention to how to control riverbank protection through proper planning and design. This bank failure phenomenon is indeed to discontinue through engineering practice. Under such circumstances, it is necessary to identify the mechanism of riverbank failure prior the riverbank protection. Although many factors influence the bank failure activities even it is necessary to prioritize the specified bank failure mechanism. In this regard, an attempt has been made to predict the bank failure mechanism of the Paturia ferry ghat of Padma River of Bangladesh through influential factors such as bank material composition, and bank moisture conditions with the help of a flow property such as water level. The main objective of this study is to predict the bank failure mechanism of the Paturia ferry ghat of the Padma River of Bangladesh with the help of geotechnical properties of the soil of the riverbank and the water level of the corresponding points.

6.2 Methodology

6.2.1 Sampling and In-Situ Testing Location

The location of the Paturia Ferry ghat of the Padma River of Bangladesh has been considered a bank failure-prone area. Baruria was the location of the Paturia side that has been selected for this study as shown in Figs. 6.2, 6.3 and 6.4 respectively. The location and the coordinates are shown in Table 6.1.

The location of the study area is adjacent to one of the ferry ghats of the Padma River at the Paturia of the Manikganj district of Bangladesh. The boring is conducted up to 22 m by the wash boring method which is shown in Fig. 6.2. In this study, both the in-situ and laboratory testing data were used as the primary data. As the phenomenon of riverbank failure is exclusively associated with soil grain sizes and their properties as well as flow properties such as water level, velocity, discharge, and stress, etc. so, some of these data have been collected from geotechnical investigation

Fig. 6.2 Riverbank scenario and boring point at Paturia ferry ghat

reports of River Research Institute of Paturia side of Padma River of Bangladesh and were used as the secondary data [8].

6.2.2 Laboratory Investigation

Saturation, pore pressure, and shear strength are the most important parameters for evaluating the riverbank failure incident, although many other soil testing parameters are interrelated for riverbank failure. The grain size of soils is mostly related to all the important parameters such as permeability as well as saturation which give direction to the natural moisture content. In the laboratory, the selected soil parameters that govern the riverbank failure are tested. Natural moisture content, specific gravity, permeability, shear strength, and grain sizes are of utmost importance and tested directly. Saturation and pore pressure are determined from consolidation tests as well as specific gravity consideration and field depth of their respective pressure. Saturation has also been considered here 100% as the soils are submerged. Grain size has been analyzed according to ASTM-D421-58 and D422-63 (Bowles 1978) and the grain sizes have been accounted for from the graph followed in the Unified Soil Classification System. Water levels and soil testing parameters are considered to predict the failure mechanism.

6.2.3 Definitions and Ranges of Different Parameters

The grain size analysis is determined through three general procedures such as (i) sieve analysis, (ii) hydrometer analysis, (iii) combined analysis. Particle sizes are determined from the Grain size distribution curve which is shown in Fig. 6.5 (Table 6.2)

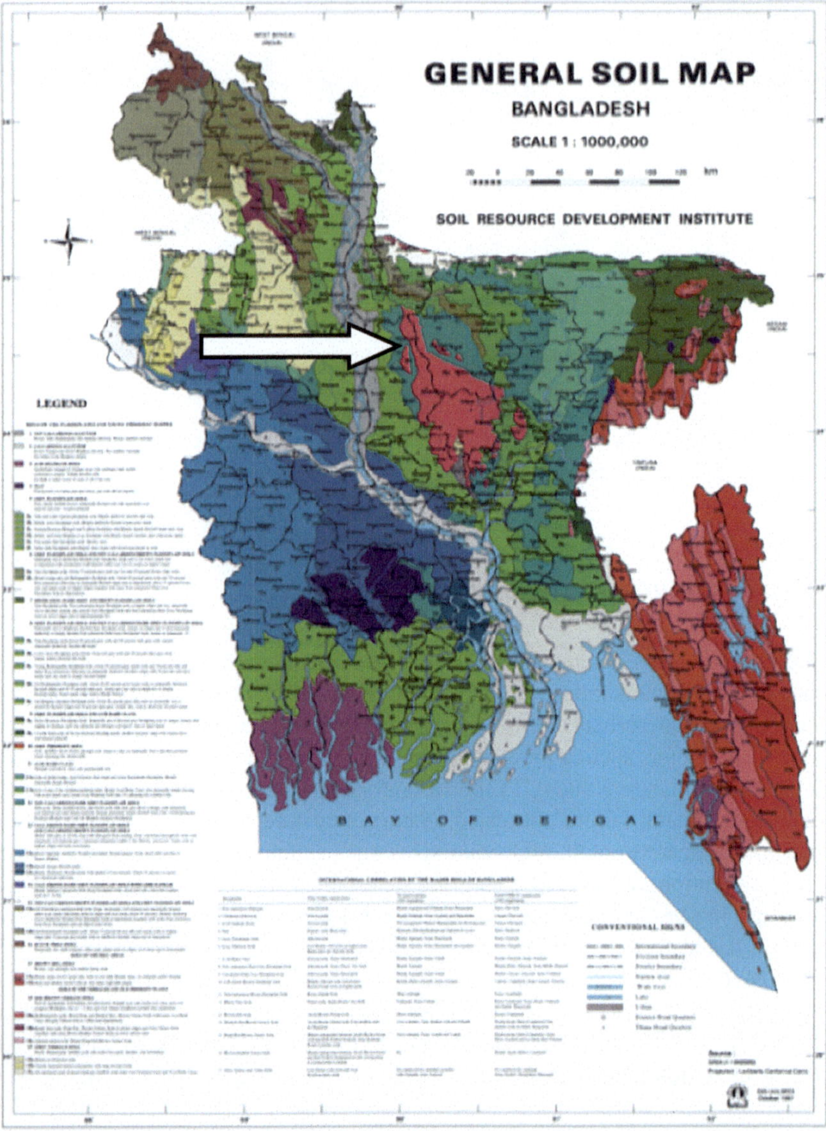

Fig. 6.3 General soil map of Bangladesh

Arora [2] stated that permeability is a very important engineering property of soils. Permeability is essential in many soil engineering problems, such as the settlement of buildings, yield of wells, and seepage through and below the earth structures. It controls the hydraulic stability of the soil masses. The permeability of soils is also required in the design of filters used to prevent piping in hydraulic structures.

Fig. 6.4 Study area map of investigation sites

Table 6.1 Coordinates of the boreholes

Location	Hole No.	Coordinate	
		Northing	Easting
Baruria	1	N-2633143	E-786834
	2	N-2633178	E-786856

The coefficient of permeability depends on the particle size and many other factors (Table 6.3).

Terzaghi and Peck [11] established a density index table regarding the number of blows SPT-N-values as shown in Table 6.4.

Arora [2] established a tabular form of typical names in terms of group symbols and their major division based on the Unified Soil Classification System (Table 6.5).

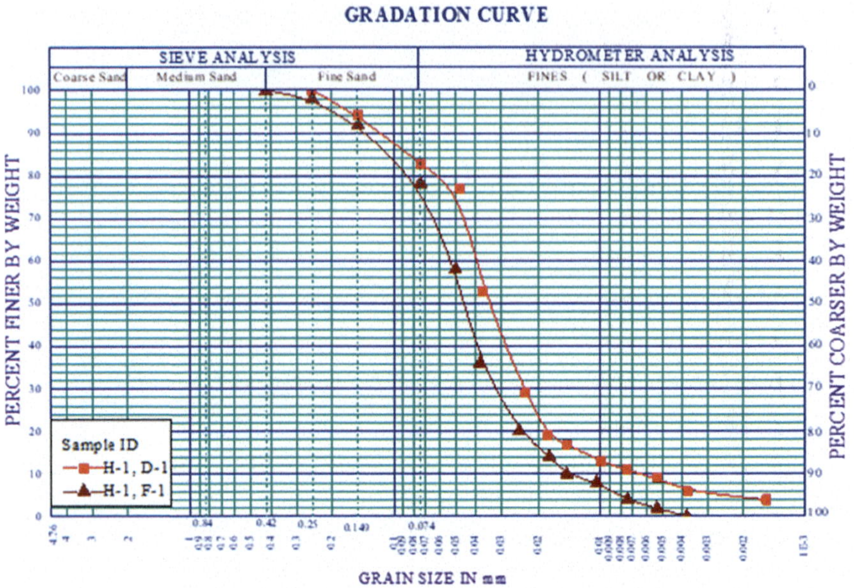

Fig. 6.5 Grain size distribution curve

Table 6.2 Sand, silt, and clay according to their range

Name of the grain	Range of values in mm
Gravel	< 76.2–4.76>
Sand	< 4.76–0.074>
Silt	< 0.074–0.002>
Clay	< 0.002

Source Garg [5]

Table 6.3 Typical values of the coefficient of permeability

Sl. No.	Soil type	Coefficient of permeability (mm/s)	Drainage properties
1	Clean gravel	10^{+1} to 10^{+2}	Very good
2	Coarse and medium sands	10^{-2} to 10^{+1}	Good
3	Fine sand, loose silt	10^{-4} to 10^{-2}	Fair
4	Dense silt, clayey silts	10^{-5} to 10^{-4}	Poor
5	Silty clay, clay	10^{-8} to 10^{-5}	Very poor

According to USBR, the soils having a coefficient of permeability greater than 10^{-3} mm/s are classified as pervious and those with a value less than 10^{-5} mm/s as impervious. The soils with a coefficient of permeability between 10^{-5} and 10^{-3} mm/s are designated as semi-pervious (source [2]).

Table 6.4 Density index (I_D) of sand

Number of blows (SPT-N values)	Density index (I_D)
0–4	Very loose
4–10	Loose
10–30	Medium dense
30–50	Dense
Over 50	Very dense

Source [11]

Table 6.5 Unified soil classification system

Major division			Group symbols	Typical names
Coarse-grained soils [more than 50% retained on No. 200 sieve (0.075 mm)]	Sand [more than 50% of coarse fraction passing No. 4 sieve (4.75 mm)]	Clean sand	SP	Poorly graded sands
			SW	Well graded sands
		Sands with fine	SM	Silty sands
Fine-grained soils [50% or more passing No. 200 sieve (0.075 mm)]	Silts and clays liquid limit 50% or less		ML	Inorganic silts of low plasticity

Source [2]

Terzaghi and Peck [11] provided a relation of consistency and compressive strength in the tabular form concerning the number of blows SPT-N-values as shown in Table 6.6.

Table 6.6 Relation of consistency of clay, number of blows N on sampling spoon and unconfined compressive strength, q_u in tons per sqft

Consistency	Very soft	Soft	Medium stiff	Stiff	Very stiff	Hard
Unconfined compressive strength, q_u (TSF)	0–0.25	0.25–0.50	0.50–1.00	1.00–2.00	2.00–4.00	> 4.00
Compressive strength (kN/m^2)	0–23.94	23.94–47.88	47.88–95.76	95.76–191.52	191.52–383.04	> 383.04
Standard penetration resistance-‘N’	0–2	2–4	4–8	8–16	16–32	> 32

Source [11]

6.3 Result and Discussion

The surface water level and groundwater level of the Paturia ferry ghat have been plotted in Fig. 6.6. From the water level graph, it has been observed that the maximum water level was about 8 mPWD in September whereas the minimum water level was about 2.5 mPWD in March. In contrast, the maximum groundwater level was observed at one point above 8 mPWD and the other point was 7 mPWD in September whereas the minimum groundwater level was observed at 3 mPWD in March.

The geotechnical parameters are specially considered as depth, SPT-N values, particle sizes, natural moisture content, permeability, and compressive strength in order of depth. The field investigation and laboratory test results are shown in tabular form in Table 6.7.

From Fig. 6.6 it is observed that very few layers contain clay particles wherever silt particles are dominant up to 7.5 m at H-1 and up to 9.0 m at H-2. However, after this depth, sand particles are dominant up to the depth of exploration. Here it is mentioned that clay particles at a 3 m depth layer contain about 30% at H-1 and a 1.5 m depth layer contains about 20% which is susceptible to resist bank erosion from seepage and leads to resistance over these layers accordingly (Fig. 6.7).

Figure 6.8a shows that the SPT values gradually increase with depth on average. However, SPT values suddenly decrease with the increase of depth several times. The increasing and decreasing trend of SPT values is dependent on some factors such as variation of particle sizes, density, natural moisture content capacity, saturation, pore pressure, and seepage as well as permeability.

Figure 6.8b shows that the natural moisture content decreases with the increases in depth. Considering particle sizes, it has been observed that the sand particles have increased gradually with the increases in depth as shown in Fig. 6.7. On the other hand, the clay particle is almost absent with the increases in depth. The sand-content soil has a low capacity to contain moisture in comparison to clay-content soil is recognized.

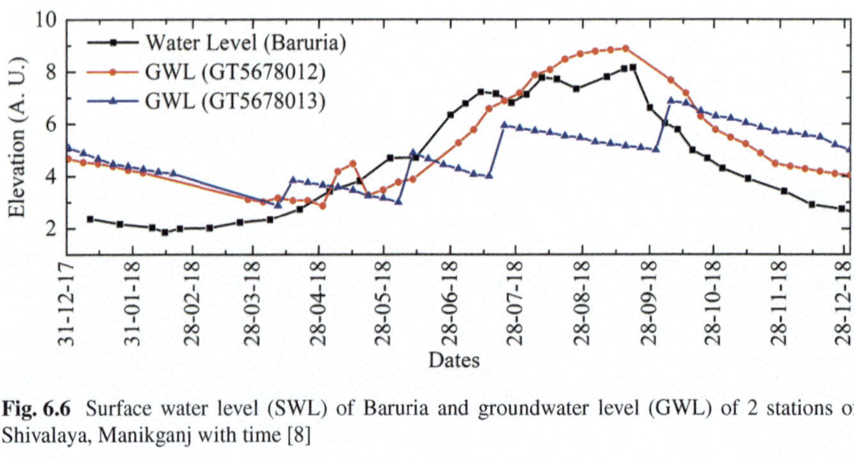

Fig. 6.6 Surface water level (SWL) of Baruria and groundwater level (GWL) of 2 stations of Shivalaya, Manikganj with time [8]

Table 6.7 The in-situ and laboratory investigation results with their corresponding depth of soil strata

Name of parameters	Range of values	
	Cohesive soils	Non-cohesive soils
Sand (%)	4–26	22–91
Silt (%)	64–84	9–77
Clay (%)	3–30	0–2
SPT-N values	3–14	4–55
Natural moisture content in (%)	31–43	21–33
Wet unit weight in kN/m^3	15–20	
Dry unit weight in kN/m^3	12–15	
Specific gravity, Gs	2.605–2.720	2.60–02.68
Compression index, Cc	0.129–0.215	
Cohesion c in kN/m^2	5–19	
The angle of internal friction ø in degree	18–20	
Saturation in (%)	93–100	
Pore pressure in kN/m^2	0–170	
Shear strength in kN/m^2	BH#1-(110–170)	BH#2-(0–21)
Permeability in mm/s	10^{-4}	10^{-3} to 10^{-2}
Color	Light brown to grey	

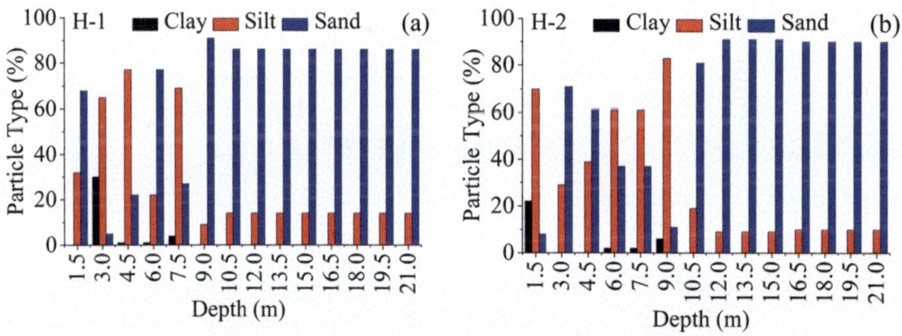

Fig. 6.7 Particle sizes versus depth graph of **a** Hole-1 (H-1) and **b** Hole-2 (H-2) of Baruria of Paturia side of Padma River

The effect of saturation on pore pressure and the effect of pore pressure on shear strength at Baruria of Paturia ferry ghat of Padma River have been shown in Figs. 6.9 and 6.10 respectively. From Fig. 6.9 it is observed that pore pressure increases with the increase of saturation, and the shear strength is observed to decrease with the increases in pore pressure as shown in Fig. 6.10.

From the table and graph, it has generally been observed that the soil is very soft up to 3.5 m following SPT values, and the soil after this depth is medium stiff and

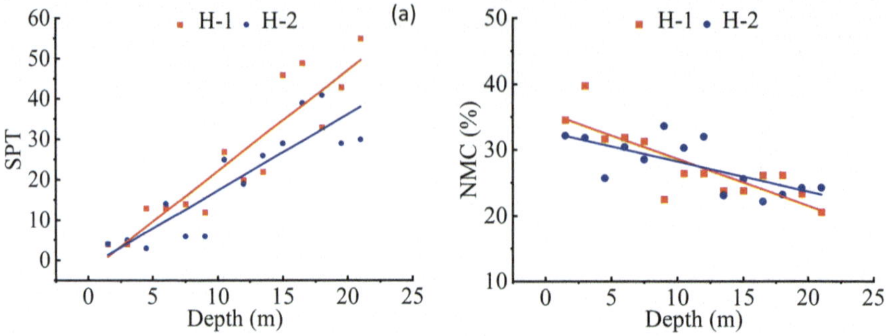

Fig. 6.8 Variation of **a** standard penetration resistance for the test (SPT) values and **b** natural moisture content (N.M.C) with the depth

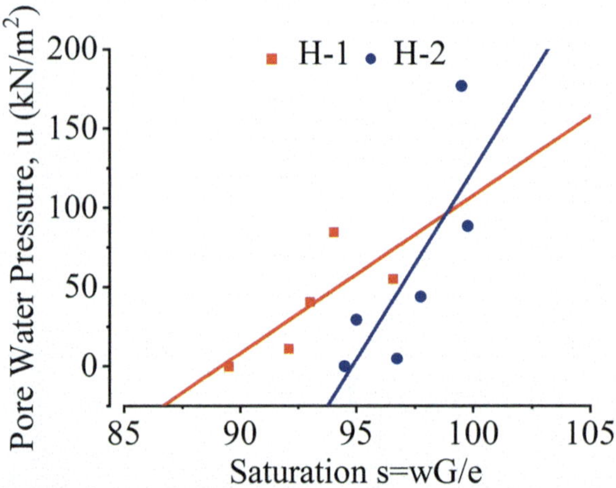

Fig. 6.9 Effect of saturation on pore pressure

medium dense up to 13.5 m, and the soil below, up to the depth of exploration (21 m) is dense [11]. The particle size distribution of the soil contains 31.81% silt and 68.19% sand at which there is no clay up to 1.5 m depth and the layer is non-cohesive. At 3.5 m of depth, the soil layer contains 30.55% clay, 64.82% silt, and 4.63% sand (USCS). Natural moisture content ranges are (39–43)% and the layers are very soft. The soils are soft and cohesive as the layers have 4 SPT values and contain above 30% clay. So, the bonding capacity of these layers is high. On the other hand, the soils are medium stiff cohesive up to 7.5 m. After that the soils are non-cohesive medium dense layers. From 15 m depth, the soils are observed non-cohesive dense layer. The permeability of the non-cohesive layers is 1.08×10^{-2} mm/s and cohesive soil layers are $(4.32–5.88) \times 10^{-4}$ mm/s (Fig. 6.11).

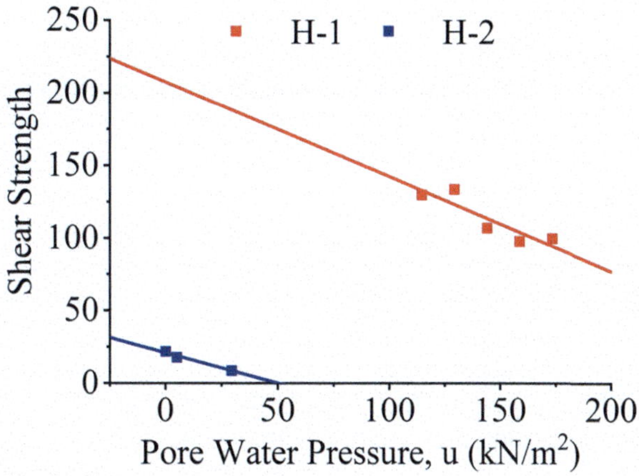

Fig. 6.10 Effect of pore pressure on shear strength

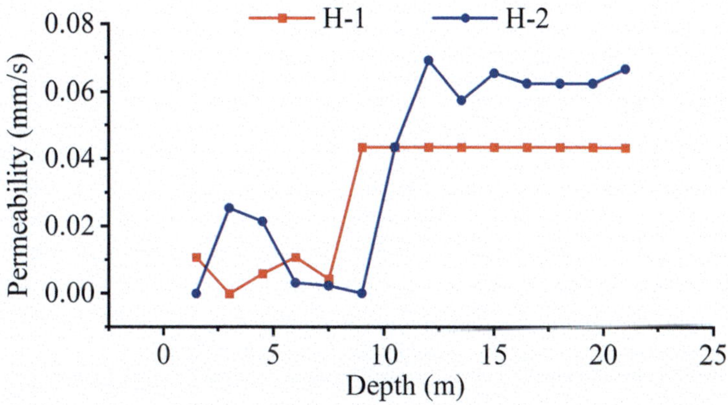

Fig. 6.11 Permeability phenomena below up to the depth of exploration at Baruria of Paturia side of Padma River

The non-cohesive soils having a coefficient of permeability in the range of 10^{-2} mm/s are pervious and the cohesive soils with a coefficient of permeability in the range of 10^{-4} mm/s are semi-pervious [2].

6.4 Conclusion

The low flood level at the Paturia side generally varies from 1.8 to 2.0 mPWD and the high flood level varies from 7 to 9 mPWD. The elevation at the Paturia ferry ghat adjacent to the bank line is 6.667 to 6.935 mPWD. It becomes submerged at monsoon. From the low flood level, the bank height is 4.867 to 4.935 mPWD. In this bank line, the layering of the bank material is fine sand mixed with some silt and trace mica. On the other hand, the stratification of the materials from the bank line to the ground surface consists of a cohesive layer wherever the high flood level varies from 7 to 9 mPWD. In this layer, about 30% of clay particles are presented over this layer which has been observed. The presence of such amounts of clay particles creates a higher level of bonding between the particles. The pore pressure has been developed due to increases in saturation. The increasing trend of pore pressure increases the soil mass which reduces the soil resistance as well as shear strength as soil-specific gravity is more than water-specific gravity.

Consequently, these soils are more resistant to surface erosion because they are less permeable. This reduces the effects of seepage and piping. However, because of low permeability, these soils are more susceptible to failure during rapid drawdown of water levels due to the increase in soil pore water pressures. However, it is noticed that waves are created by ferries and launch traffic near the bank line on the Paturia ferry ghat. As wave action has an impact directly on exposed soil as well as trafficking ferry and launch, so, it expedites the bank failure action in addition to the remaining pore water pressures of cohesive soil at the Paturia ferry ghat. The bank failure of the Paturia ferry ghat occurs due to a tension crack.

Recommendation

The design engineers may get an idea about the soil condition and bank failure situation in terms of bank materials. The study results indicate that the soil strata and its parameters with adjacent water levels permit the occurrence of bank failure due to tension cracks. It is recommended here that a complete geotechnical investigation is needed to accomplish the design of the protection work. Although the idea has been developed regarding the bank failure mechanism in the study area, it will be necessary to conduct a comprehensive geotechnical investigation.

References

1. Wazed A (1991). Bangladesher Nadimala [Rivers of Bangladesh]. Dhaka
2. Arora KR (2010) Soil mechanics and foundation engineering (reprint edition). Lomus Offset Press
3. Brammer H (1996) The geography of the soils of Bangladesh. University Press Limited
4. Food and Agriculture Organization (FAO) (1988) Land resources appraisal of Bangladesh for agricultural development, vol 2. Rome
5. Garg SK (2010) Soil mechanics and foundation engineering (first reprint). Khanna Publishers
6. Huq SMI, Shoaib JUM (2013) The soils of Bangladesh (World soils book series 1). https://doi.org/10.1007/978-94-007-1128-0
7. Inamul MH (2008) Water resources management in Bangladesh. Anushilon
8. River Research Institute (2020) Investigation of geotechnical reasons for bank failure on Daulatdia and Paturia side of Padma River of Bangladesh (No. 21)
9. Nasermoaddeli M, Pasche M (2013) Modelling of undercutting and failure of non-cohesive riverbanks

10. Talukdar B (2006) Riverbank erosion—a perspective. Assam Engineering College
11. Terzaghi K, Peck RB (1948) Soil mechanics in engineering practice. Wiley & Chapman & Hall
12. Thorne CR, Lewin J (1979) Bank processes, bed material movement, and planform development in a meandering river. In: Rhodes DD, Williams GP (eds) Adjustments of the fluvial system. Kendall/Hunt Publishing Co., pp 117–137
13. Watson AJ, Basher LR (2006) Stream bank erosion: a review of processes of bank failure, measurement and assessment techniques, and modeling approaches (ICM Report No. 2005-2006/01). Motueka Integrated Catchment Management Programme

Chapter 7
Estimation of Braided River Bathymetry by Data Fusion Method for Hydro-morphological Simulation: Applicability for the Riverbank Erosion

Mohammad Muddassir Islam, Shampa, and Israt Jahan Nejhum

Abstract Riverbank erosion by the Brahmaputra-Jamuna River has displaced thousands of people from their homes and pushed them into abject poverty in Bangladesh. Consequently, a better system for predicting riverbank erosion might be one tactic to mitigate the consequences of such a disaster. Numerical simulation of riverbed evolution can be a solution of that which necessitate high-resolution bathymetry data. The river bathymetry measurements taken by the Bangladesh Water Development Board (BWDB) at 4–6 km intervals are too coarse for numerical simulation. An appealing alternative to the costly and time-consuming classical bathymetric mapping of braided rivers is the data fusion of measured data with spectral reflectance obtained from satellite images. The data fusion method was used to construct 10 m resolution river bathymetry using sentinel 2 images and limited measured data. The estimated bathymetry has been evaluated for the adequacy of river erosion estimation using a 2D hydro-morphological model for the years 2017–2019. On average, between 2017 and 2019, the estimated bathymetry showed a depth of 10, 9, or 10 m, with a maximum depth of 17.5, 16, or 19.5 m. The simulated model's R2 value was close to 0.96 in relation to a discharge at Bahadurabad. The NSE, PBIAS, and RRMSE for the Mathura, Bahadurabad, Serajgani, and Chilmmari stations, respectively, show the level of satisfaction. Kappa statistics was used to measure the accuracy of erosion predictions, and the results for the three years 2017–2019 were around 67%, 78%, and 72%, respectively.

Keywords Passive bathymetry · Sentinel 2 · Numerical simulation · Brahmaputra-Jamuna · Erosion

M. M. Islam · Shampa (✉)
Institute of Water and Flood Management, Bangladesh University of Engineering and Technology, Dhaka, Bangladesh
e-mail: shampa_iwfm@iwfm.buet.ac.bd

I. J. Nejhum
Department of Urban and Regional Planning, Bangladesh University of Engineering and Technology, Dhaka, Bangladesh

G. M. Tarekul Islam et al. (eds.), *Sustainable Water Management and Environmental Resilience*, Springer Water, https://doi.org/10.1007/978-3-031-93350-9_7

7.1 Introduction

Typically, braided rivers have wide, unstable banks, a steep, shallow path, and several channel divides around braided islands. Changes in flow and sediment transport cause the channel to adjust and seek a stable equilibrium [1, 3, 13, 20, 34]. They change their geometry and planform so rapidly thereby tracking this modification and managing accordingly is one of the key management issues for such rivers [4, 33]. In the case of such river river-training works or environmental management, success depends on the proper management of bars and channels. For this purpose, river morphological data, i.e., river bathymetry with some resolutions, are crucial elements. However, in many countries, information on river bathymetry may be absent or maybe poor. As an example, Bangladesh Water Development Board (BWDB) measures the bathymetry of the large, braided river Brahmaputra-Jamuna with an interval of 4–6 km. But the confluence-bifurcation units (braiding unit varies from 500 to 30,000 m) are shorter than that [36, 38]. Air-born LiDAR measurement [40] can be an alternative but getting high-resolution bathymetry is very costly. Therefore, satellite-remote-sensing-based bathymetry with the accuracy level of erosion estimation can be a solution to this problem.

The theory of passive bathymetry measurement using satellite data was developed by Lyzenga [25, 26]) and expanded by Philpot [32]. Passive water depth measurement using satellite data, Lyzenga [25] assumed the lognormal relationship between water depth and surface reflection using one band, but in large areas, it has some limitations (e.g., these algorithms are limited by operational restrictions which reduce their applicability and utility, it is more complex and therefore somewhat more difficult to implement than the ratio methods. The subtraction and division operations required for the ratio methods can be implemented by either analog or digital processors, whereas the method defined here requires digital computation). Later, Lyzenga [26] introduced a system of linear equations to solve this problem. While using this method, the linear transform solution needs to be solved for five variables that must be determined empirically. Adjustment of five empirical variables for large areas is problematic where even with relatively small variations in water quality conditions. Furthermore, when the bottom albedo is low, as it is when thick macroalgae or seagrass are present, the bottom albedo is smaller than the water column reflectance for optically deep cells, which complicates depth estimation [39]. Later, Stumpf et al. [39] introduced 'Ratio' algorithm, where they scaled the ratio of two spectral bands. Although the ratio technique may not always discern fine morphology in relative depth shallow depth water, it is more robust than the linear transform. Later, researchers used the modified Lyzenga method or considered more bands and used the deep learning technique [14, 16, 18] which are location specific and usually recommended to apply in the low turbid zone.

Optical bathymetry mapping is a technique that uses the total amount of radiative energy reflected by the water column to determine water depth. The blue and green spectral bands of optical satellite imagery are capable of penetrating clear water, but the peak of the reflection shifts to longer wavelengths when there are higher concentrations of suspended sediments. And near infrared (NIR) could be used to discriminate for classified water and land [21]. When solar radiation passes through water, it is scattered and absorbed by water and in-water constituents, and then backscattered and recorded by multi-spectral sensors. This process enables the production of fine-resolution bathymetric information, but the range of detectable depth is reduced compared to microwave-based methods. Overall, optical bathymetry mapping provides a promising approach for generating accurate and detailed bathymetric data [5, 24, 27].

However, There are several categories of theories that attempt to explain why braiding occurs in alluvial river systems. The first is a practical explanation that relates the braiding to a combination of external forces, environmental factors such as discharge and sediment supply. The second step is to apply braid formation to theoretical stability analyses of channel bars in two-dimensional flow regimes. Finally, physical processes such as sedimentary and hydraulic conditions that initiate the braiding process have been investigated [2]. The Brahmaputra-Jamuna River is one of the largest rivers in the world and a dynamic transboundary braided river system [30]. Due to it's braided nature, continues measurement of bathymetry by tradtionally is very costly.

Given these circumstances, the primary goal of this study is to create high-resolution braided river bathymetry by data fusion method using satellite images and very little measured data. Then, using the generated bathymetry and numerical simulation estimation of the river erosion for a particular year and comparing it with the actual erosion.

The present study area is the Brahmaputra-Jamuna River in Bangladesh, as shown in Fig. 7.1. The erosion rate of the Brahmaputra-Jamuna is very alarming 17.05 km^2/y for the last four decades [7]. Bangladesh Water Development Board (BWDB) is continuously trying to improve the situation by building several bank protections structures. However, in monsoon, almost every year, new erosion locations are observed in unprotected areas as well as these hydraulic structures need regular maintenance.

7.2 Methodology

7.2.1 Data Collection

The study was conducted using a variety of data types, including satellite images, cross-profile data, water level measurements, discharge data and sediment data. These sources provided a diverse range of information that was necessary for the study's objectives. The satellite images were used to analyze water column reflectance. Cross-profile data allowed for the analysis of channel morphology and sediment transport. Water level measurements were used to assess hydrological conditions. Discharge data provided insights into the volume of water flowing through the river

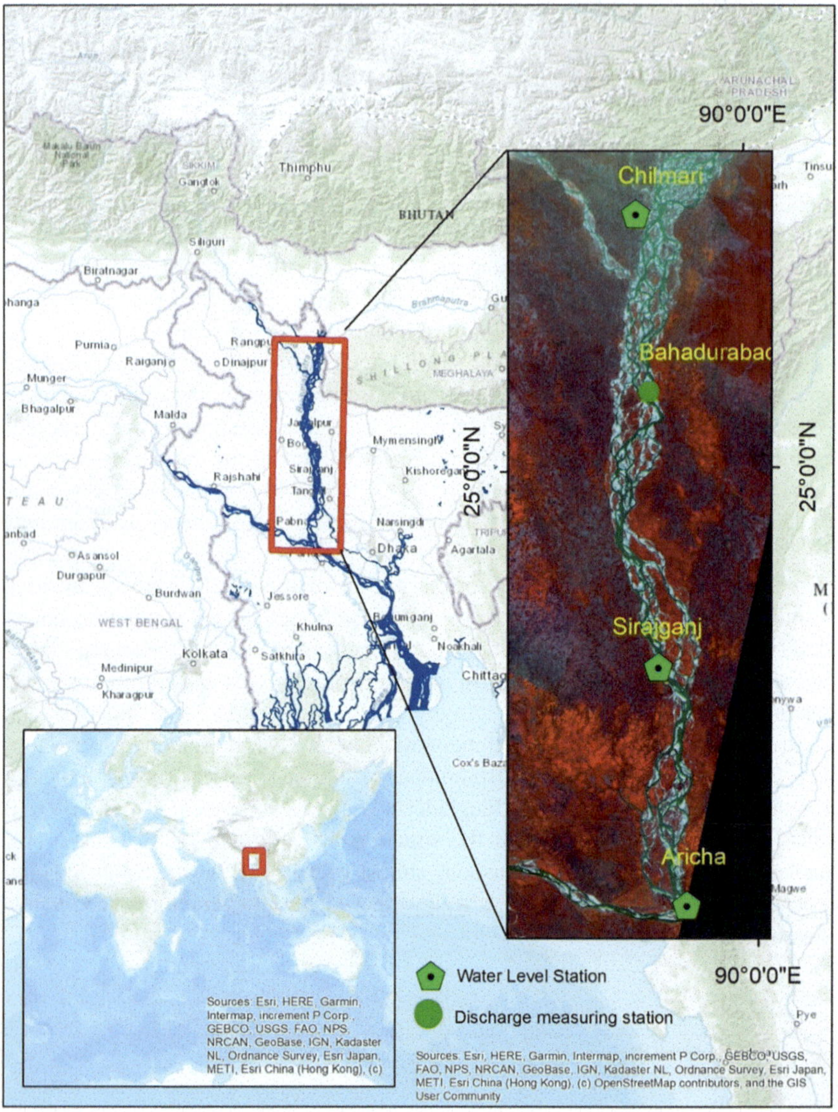

Fig. 7.1 Study area (Brahmaputra-Jamuna River)

system. The combination of these different data types allowed for a comprehensive analysis of the river system and its associated environmental processes. Table 7.1 shows the data list used in this study.

Table 7.1 List of data used for study

Data	Source	Period
Water level	BWDB	2017–2019
Discharge	BWDB	2017–2019
Sediment	BWDB	2017–2019
Cross-section	BWDB	2017–2019
Satellite images	Copernicus Open Access Hub	2017–2019

7.2.2 Methodology

This study combines satellite image data with measured data, which is subsequently employed as an input parameter in a hydro-morphological simulation. The following steps were followed (Fig. 7.2).

7.2.2.1 Generation of Automated River Bathymetry Using Optical Data

For passive bathymetry linear and ratio methods were used to derive the flow depth from the satellite images. In the linear method, it uses the albedo of the water body floor to calculate the water depth. The method requires prior knowledge of the river or sea floor type, which is used to estimate the bottom albedo [25, 26]. On the other hand,

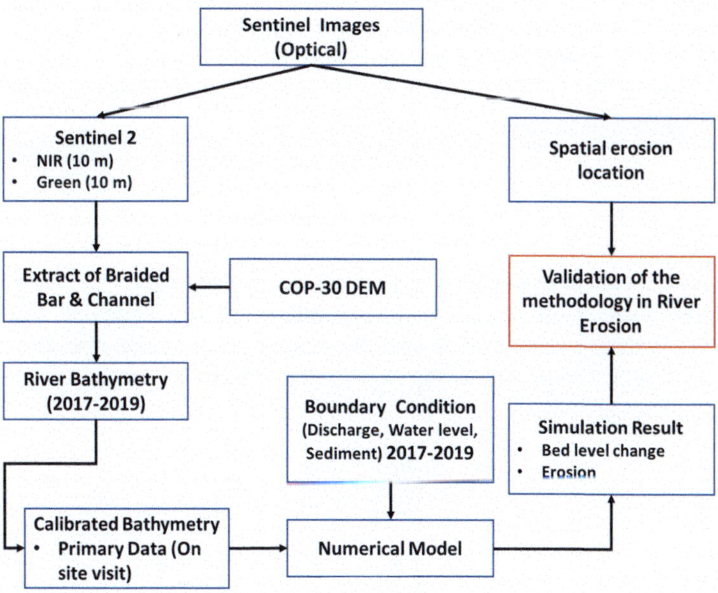

Fig. 7.2 Methodology framework of the study

the ratio method, is based on the principle that the ratio of reflectance values at two wavelengths is proportional to the depth of the water. This method is less dependent on knowledge of the river or sea floor type and can be more accurate than the linear method, particularly in turbid waters [39]. Stumpf et al. [39] introduced 'Ratio' algorithm, where they scaled the ratio of two spectral bands—band 'Blue' and band 'Green'. They use some fixed constant for their considered domain which needs to be chosen in such a way that the logarithm of the reflectance will be positive under any condition and that the ratio will produce a linear response with depth which is difficult in many cases. Addressing these issues, Shampa et al. [37] developed a method using the reflectance of 'Green' and 'Near Infrared' band are correlated with limited measured data. That methodology has been followed here. The spectral reflectance of the "Green" and "NIR" bands acquired from Sentinel 2 or optical images have correlated with insufficient measured data in a way that can be represented by Eqs. 7.1 and 7.2 for channel depth

$$z = h_{mL} + \frac{(e^k - e^k_{\min})(h_{mH} - h_{mL})}{e^k_{\max} - e^k_{\min}} \tag{7.1}$$

$$k = \frac{R(\lambda_g) - R(\lambda_{NIR})}{R(\lambda_g) + R(\lambda_{NIR})} \tag{7.2}$$

where the water depth (z) is expressed as the function of h_{mL} and e^k. The value of e^k is derived using Eq. (7.2). Here, $R(\lambda_g)$ and $R(\lambda_{NIR})$ the atmospherically corrected surface reflectance of 'Green' and 'Near Infrared' bands. h_{mL} which represents the 0.1st percentile of measured channel depth and h_{mH} represent the 99th percentile of measured channel depth. The 'Green' and 'Near Infrared' band from the Sentinel 2 satellite during 2017–2019 were used. As the measured data, the cross-section data measured by BWDB in 2017–2019 were used.

7.2.2.2 Prediction of the Monsoon Riverbank Erosion

The validated model of Brahmaputra-Jamuna developed Shampa et al. [36] was used here on open source Delft3D [22] platform for the numerical modelling, Delft3D is capable of predicting water flow in shallow seas, coasts, estuaries, lagoons, rivers, and lakes [10, 41]. The governing Eqs. 7.3–7.10 of the models are described shortly in the following sections.

$$\frac{\partial \boldsymbol{h}}{\partial \boldsymbol{t}} + \frac{\partial(\boldsymbol{hu})}{\partial x} + \frac{\partial(\boldsymbol{hv})}{\partial \boldsymbol{y}} = 0 \tag{7.3}$$

$$\frac{\partial \boldsymbol{u}}{\partial \boldsymbol{t}} + \boldsymbol{u}\frac{\partial \boldsymbol{u}}{\partial x} + \boldsymbol{v}\frac{\partial \boldsymbol{u}}{\partial \boldsymbol{y}} + \boldsymbol{g}\frac{\partial \zeta}{\partial x} + \frac{\boldsymbol{gn}^2}{\sqrt[3]{\boldsymbol{h}}}\left(\frac{\boldsymbol{u}(\boldsymbol{u}^2 + \boldsymbol{v}^2)}{\boldsymbol{h}}\right) - \boldsymbol{v_h}\left(\frac{\partial^2 \boldsymbol{u}}{\partial x^2} + \frac{\partial^2 \boldsymbol{u}}{\partial \boldsymbol{y}^2}\right) = 0 \tag{7.4}$$

$$\frac{\partial u}{\partial t}+u\frac{\partial u}{\partial x}+v\frac{\partial u}{\partial y}+g\frac{\partial \zeta}{\partial x}+\frac{gn^2}{\sqrt[3]{h}}\left(\frac{v\left(u^2+v^2\right)}{h}\right)-v_h\left(\frac{\partial^2 v}{\partial x^2}+\frac{\partial^2 v}{\partial y^2}\right)=0 \quad (7.5)$$

$$\frac{\partial(hc)}{\partial t}+\frac{\partial(huc)}{\partial x}+\frac{\partial(hvc)}{\partial y}=h\left[\frac{\partial}{\partial x}\left(D_H\frac{\partial c}{\partial x}\right)+\frac{\partial}{\partial x}\left(D_H\frac{\partial c}{\partial y}\right)\right]+hs \quad (7.6)$$

where ζ = is water level elevation with respect to a datum (here in m);
h = represents water depth (m);
u, v is depth average velocity in the x and y directions, respectively (m/s);
g = is the acceleration due to gravity (m/s^2);
v_h = denotes kinetic eddy viscosity (m^2/s);
n = represents the Manning's coefficient (sm-1/3).
c = mass sediment concentration (kg/m^3),
D_H = the horizontal diffusivity.
S = sediment source terms per unit area.
k = turbulent kinetic energy,
ε = dissipation is presented by. $k - \varepsilon$ turbulence model was used for turbulence closure.

The calculation of bed load transport is an essential aspect in understanding the dynamics of sediment transport in rivers and streams. While sediment transport formulas provide a useful framework for estimating bed load transport rates, they also have limitations. Many of these formulas were established based on limited flume and field data, and may not accurately reflect the complex processes involved in sediment transport in natural streams and rivers [11, 35].

In a study conducted by Kabir and Ahmed [17] they attempted to estimate the sediment load of the Brahmaputra-Jamuna by utilizing multiple sediment formulas. They found that among the formulas used, Van Rijn [35] model provided the most accurate prediction of the sediment load. As a result, the model was employed to calculate the bedload transport for their numerical model. The bedload transport rate $\left|\overline{S}_{bed}\right|$ was computed by Eq. (7.7)

$$\left|\overline{S}_{bed}\right|=\left\{\begin{array}{l} 0.053\sqrt{(s-1)gd_{50}^3}D_*^{-0.3}\left(\frac{\mu_c\tau-\tau_c}{\tau_c}\right)^{2.1} \; if\left(\frac{\mu_c\tau-\tau_c}{\tau_c}\right)<3.0 \\ 0.1\sqrt{(s-1)gd_{50}^3}D_*^{-0.3}\left(\frac{\mu_c\tau-\tau_c}{\tau_c}\right)^{1.5} \; if\left(\frac{\mu_c\tau-\tau_c}{\tau_c}\right)\geq 3.0 \end{array}\right\} \quad (7.7)$$

where s = the specific density of sediment particle;
$\left(\frac{\rho_s}{\rho_f}\right)$, d_{50} denotes particle size;
τ = bed shear stress;
τ_c = critical bed shear stress;
μ_c = the ratio between the total bed roughness and the grain-related bed roughness;
D_* = dimensionless particle parameter.

The bed elevation can be calculated by applying the mass-balance equation [12] using Eq. (7.8) and considering a size fraction of k in a mixed sediment transport

$$(1-\boldsymbol{\lambda})\frac{\partial \boldsymbol{\eta}_{bk}}{\partial t}+\boldsymbol{m}_f\left(\frac{\partial \boldsymbol{q}_{uk}}{\partial \boldsymbol{x}}+\frac{\partial \boldsymbol{q}_{vk}}{\partial \boldsymbol{y}}\right)+\boldsymbol{m}_f(\boldsymbol{E}_k-\boldsymbol{D}_k)=0 \tag{7.8}$$

$$\delta\eta_b=\sum_{k=1}^{M}\delta\eta_{bk}$$

The given Eq. 7.12 involves several variables, including λ, which represents porosity;

q_{uk}, q_{vk} = which denote the bedload transport vector for a specific size fraction k;

η_{bk} = the bed change due to size fraction k;

E_k, D_k are upward and downward suspended sediment transport flux near the bed for the size fraction k;

m_f = the morphological acceleration factor to reduce the computational time step to adapt the morphology.

The summation of the bed changes, $\delta\eta_b$ due to all the size fractions is the resulted in the total bed variation in one-time step.

Model Schematization

The numerical model of Brahmaputra-Jamuna was created with a curvilinear grid that spans 225 km in length, with an average width of 15 km. The model covers the stretch of the river starting from 5 km upstream of Noonkhawa water level measuring station and ending near Aricha water level measuring station (Fig. 7.3). The reach was discretized into 560 × 38 grid cells, with a grid cell size of 450 × 140 m^2. To account for the monsoon season when the entire braided plain may go underwater, a Manning's roughness coefficient of 0.027 was used. Additionally, the mean sediment size was assumed to be 200 μm [17, 31]. Estimated river bathymetry using satellite images was used. This technique is known as remote sensing bathymetry. The model was simulated for three years from 2017 to 2019. For the upstream boundary, the discharge data of Brahmaputra-Jamuna, Dharla, Teesta, Dudkumar river were also considered was considered (shown in Fig. 7.4) and the water level of Brahmaputra-Jamuna at Aricha station and the water level of Old Brahmaputra were chosen for the downstream boundary (Fig. 7.5). Table 7.2 shows the base parameters used in the model.

7.2.2.3 Accuracy Assessment

Statistical analyses were executed to evaluate the precision of the simulated outcomes for water levels and discharge. These analyses encompassed the application of well-established metrics, namely the Nash–Sutcliffe model efficiency coefficient (NSE), percent bias (PBIAS), relative root mean square error (RRMSE), and coefficient of determination (R^2) [29–40], as denoted by Eqs. (7.9)–(7.12) in the evaluation

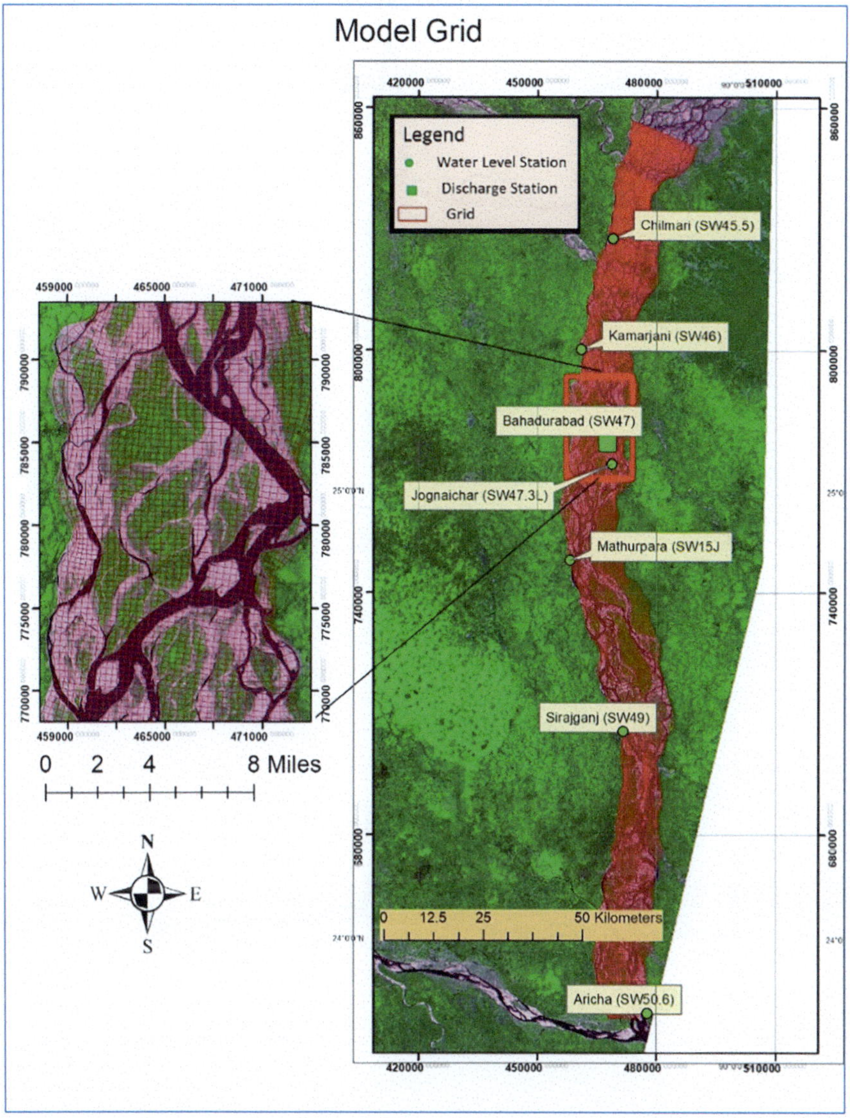

Fig. 7.3 Model grid for Brahmaputra-Jamuna

framework. These metrics collectively provided a comprehensive assessment of how effectively the simulation results aligned with observed data.

Here H_0^i represent the ith observed value of H parameter, H_s^i represent the ith simulated value of H parameter. n is the total number of observations and H_0^{avg} represent the average value of H parameter's observed data.

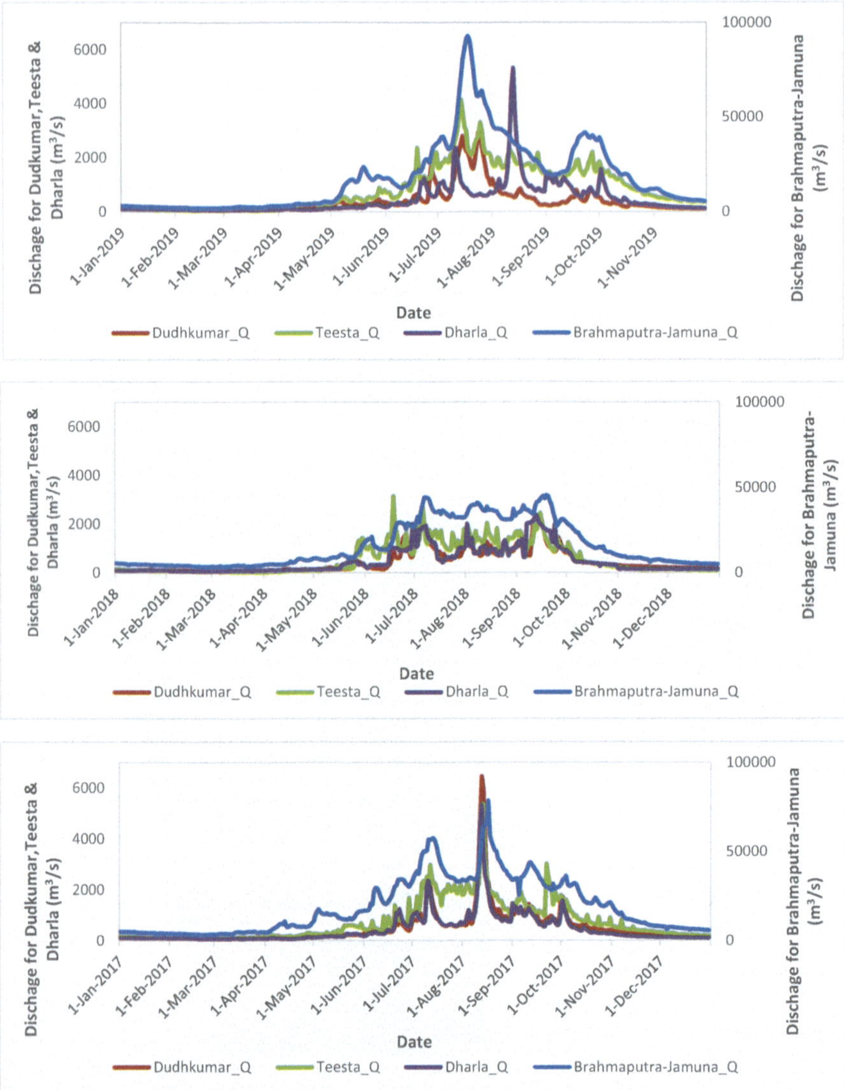

Fig. 7.4 The upstream boundary conditions (discharge) of the model Brahmaputra-Jamuna for the year 2019 (top), 2018 (middle), and 2017 (bottom)

$$NSE = 1 - \left(\frac{\sum_{i=1}^{n} \left(H_0^i - H_s^i\right)^2}{\sum_{i=1}^{n} \left(H_0^i - \overline{H}_0\right)^2} \right) \tag{7.9}$$

$$PBIAS = 100 * \left(\frac{\sum_{i=1}^{n} H_0^i - H_s^i}{\sum_{i=1}^{n} H_0^i} \right) \tag{7.10}$$

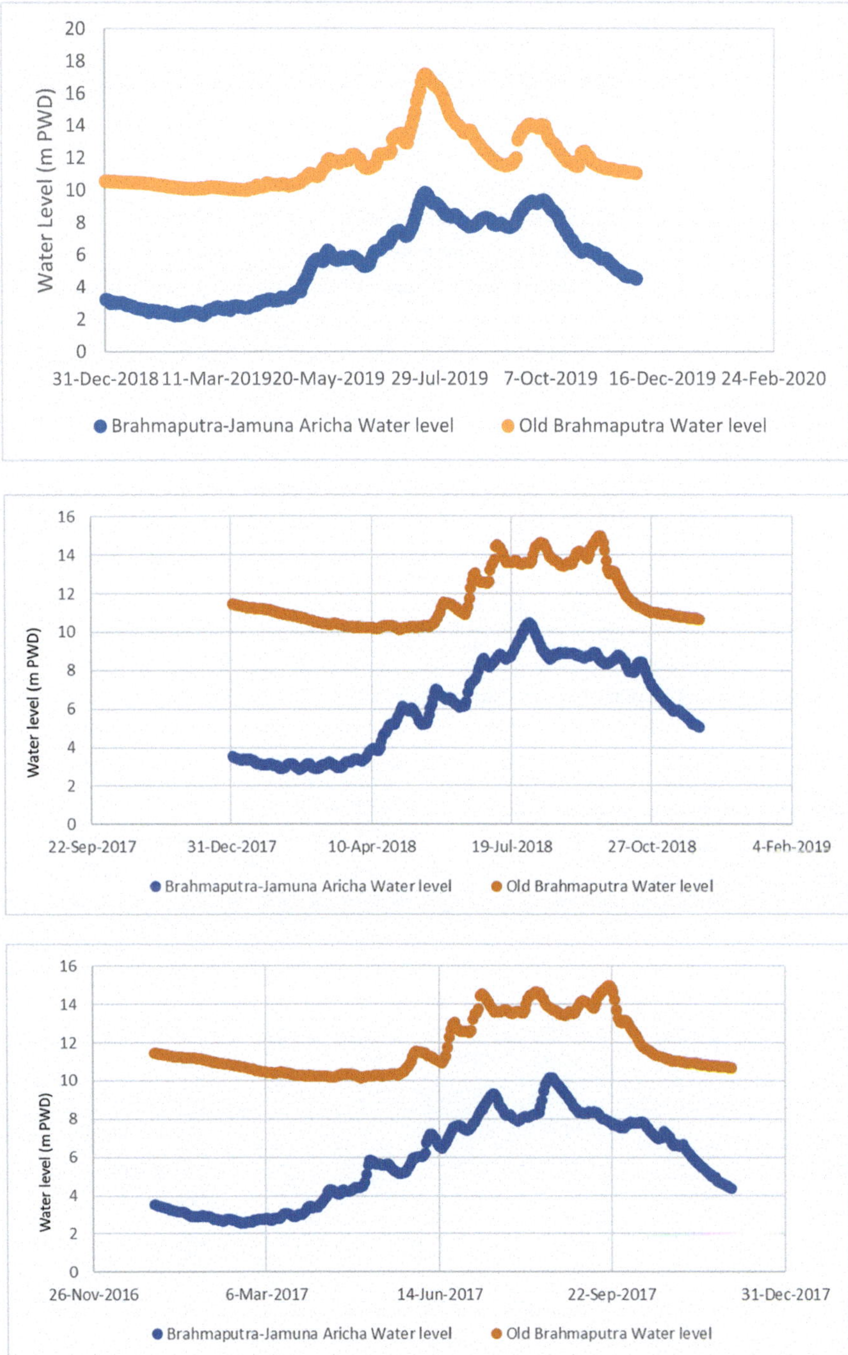

Fig. 7.5 The downstream boundary conditions (water level) of the model Brahmaputra-Jamuna for the year 2019 (top), 2018 (middle), and 2017 (bottom)

Table 7.2 The base parameters of the model

Parameter(s)	Unit	Value
Mean grain size	μm	200
Density of sediment	kg/m^3	2650
Density of water	kg/m^3	1000
Van Rijn's reference height factor	–	2
Horizontal eddy viscosity	m^2/s	1
Hydrodynamic time step	min	1
Roughness (Manning's)	sm$^{-1/3}$	0.027
Morphological acceleration factor, m	–	3
Threshold sediment thickness	m	0.005

$$RRMSE = \frac{\sqrt{\left(\frac{1}{n}\sum_{i=1}^{n}\left(H_o^i - H\right)^2\right)}}{\sum_{i=1}^{n} H_o^i} * 100 \tag{7.11}$$

$$R^2 = 1 - \frac{\sum_{i=1}^{n}\left(H_o^i - H_s^i\right)^2}{\sum_{i=1}^{n}\left(H_o^i - H_o^{avg}\right)^2} \tag{7.12}$$

Table 7.3 lists the performance ratings [23, 42]

Table 7.3 Performance ratings for NSE, PBIAS, RRMSE and R^2

Name of statistical analysis	Value range	Performance rating
NSE	$0.75 < NSE \leq 1$	Very good
	$0.65 < NSE \leq 75$	Good
	$0.50 < NSE \leq 65$	Satisfactory
	$NSE \leq 0.50$	Unsatisfactory
PBIAS	$PBIAS \leq \pm 10\%$	Very good
	$\pm 10\% \leq PBIAS \leq \pm 10\%$	Good
	$\pm 15\% \leq PBIAS \leq \pm 25\%$	Satisfactory
	$PBIAS \geq \pm 25\%$	Unsatisfactory
RRMSE	$RRMSE < 10\%$	Excellent
	$10\% < RRMSE < 20\%$	Good
	$20\% < RRMSE < 30\%$	Fair
	$RRMSE > 30\%$	Poor
R^2	Values of the coefficient Close to 1 indicate more efficient models	0–1

$$\text{Kappa Coefficient} = N\sum_{r=1}^{r} n_{ii} - \sum_{r=1}^{r}(n_{icol}n_{irow}/N^2) - \sum_{r=1}^{r} n_{icol}n_{irow} \quad (7.13)$$

$$\text{Producer's Accuracy} = n_{ii}/n_{irow} \quad (7.14)$$

$$\text{User's Accuracy} = n_{ii}/n_{icol} \quad (7.15)$$

$$\text{Overall Accuracy} = \frac{1}{N}\sum_{i=1}^{r} n_{ii} \quad (7.16)$$

Kappa coefficient calculation is described in Eq. 7.13. Where i is the class number, N is the total number of classified values compared to truth values, n_{ii} is the number of values belonging to the truth class i that have also been classified as class i, n_{icol} is the total number of predicted values belonging to class i, n_{irow} is the total number of truth values belonging to class i.

7.3 Results

7.3.1 River Bathymetry Generated from Optical Images:

River bathymetries estimated using Eq. (7.1) are illustrated in Figs. 7.6, 7.7 and 7.8 for the years 2019, 2018, and 2017. In 2019 (Fig. 7.6), the depth ranged from 0.5 to 19.5 m with an average depth of 10 m. Because of the higher braiding in the upper segment of the river (upstream of Chilmari, Bahadurabad), the depth was lower than in Sirajganj and Sariakandi upazilas. Upstream, the average depth was 4.5 m. The average depth downstream (near Aricha) was found to be 8 m. The depth of major channels ranged from 3.6 to 19.5 m. In minor channels (near Sirajganj), it ranged from 3.6 to 7.5 m. In 2018 (Fig. 7.7), the depth ranged from 0.8 to 16 m, with an average of 9 m. The average depth upstream was 6.7 m, whereas downstream it was 10.5 m. The depth of major channels varied between 12.5 and 16 m. In minor channels (near Sirajganj), it ranged from 3.5 to 9.6 m. Similarly, in 2017 (Fig. 7.8), the depth ranged from 0.5 to 17.5 m with an average depth of 10 m. The depths in major channels (downstream of Sarikandi upazila) ranged from 12 to 17 m. In minor waterways (near Sirajganj), it ranged from 3.6 to 9.5 m.

During the study (in 2019), the assessment depth from people's perceptions in major, secondary, and tertiary channels was collected at five locations in Jamalpur district: Guthail Bazar, Kulkandi, Muradabad Bazar, Muradabad Bazar, and Char Belghasa, as shown in Fig. 7.9. Table 7.4 shows a comparison of observed and estimated depths. It confirms the depth variations in different types of channels can be captured in estimated bathymetry as well. In Fig. 7.10 showed comparison of measured three cross section (RMJ 12) and estimated bathymetry for 2017 to

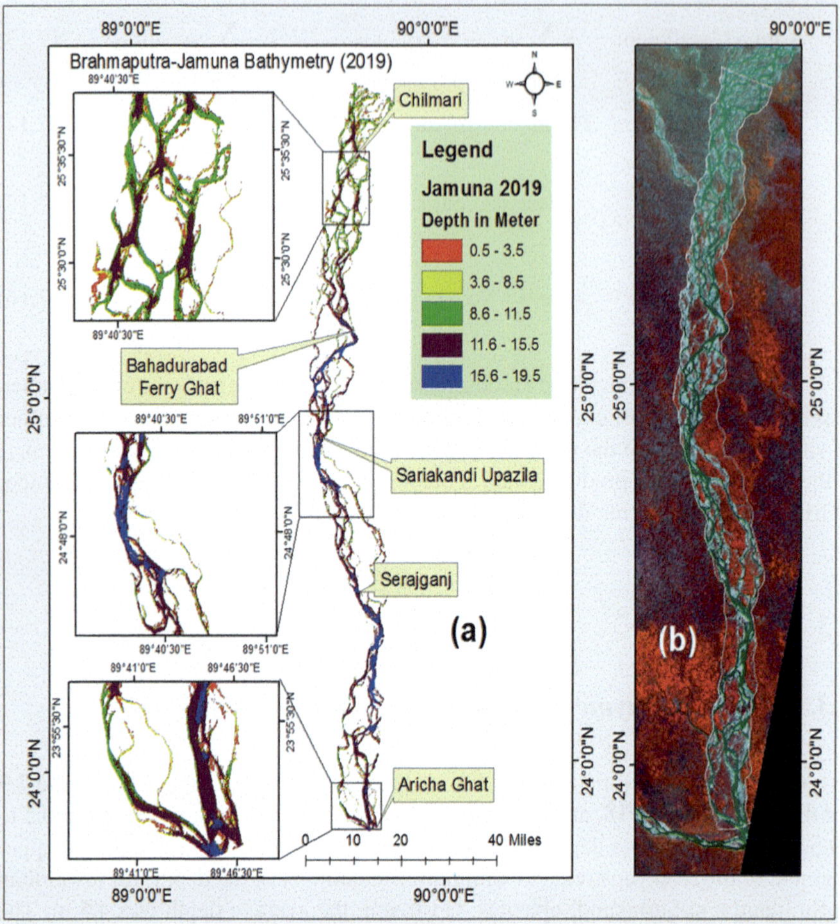

Fig. 7.6 **a** Brahmaputra-Jamuna generated Bathymetry and **b** Sentinel 2 image for the year 2019

2019 year. For 2017, measured elevation ranges from 22 m to -0.89 m and estimated bathymetry found 27.5 and − 2.64 m. For 2018, measured elevation ranges from 22.25 to 1.55 m and estimated bathymetry found 23.5 and − 2.5 m. For 2019, measured elevation ranges from 22.25 to 3.50 m and estimated bathymetry found 22 and 2.94 m. Figure 7.11 depicts the R^2 values of measured and estimated reduced levels, which were 0.39, 0.48, and 0.48 for the years 2017, 2018, and 2019, respectively.

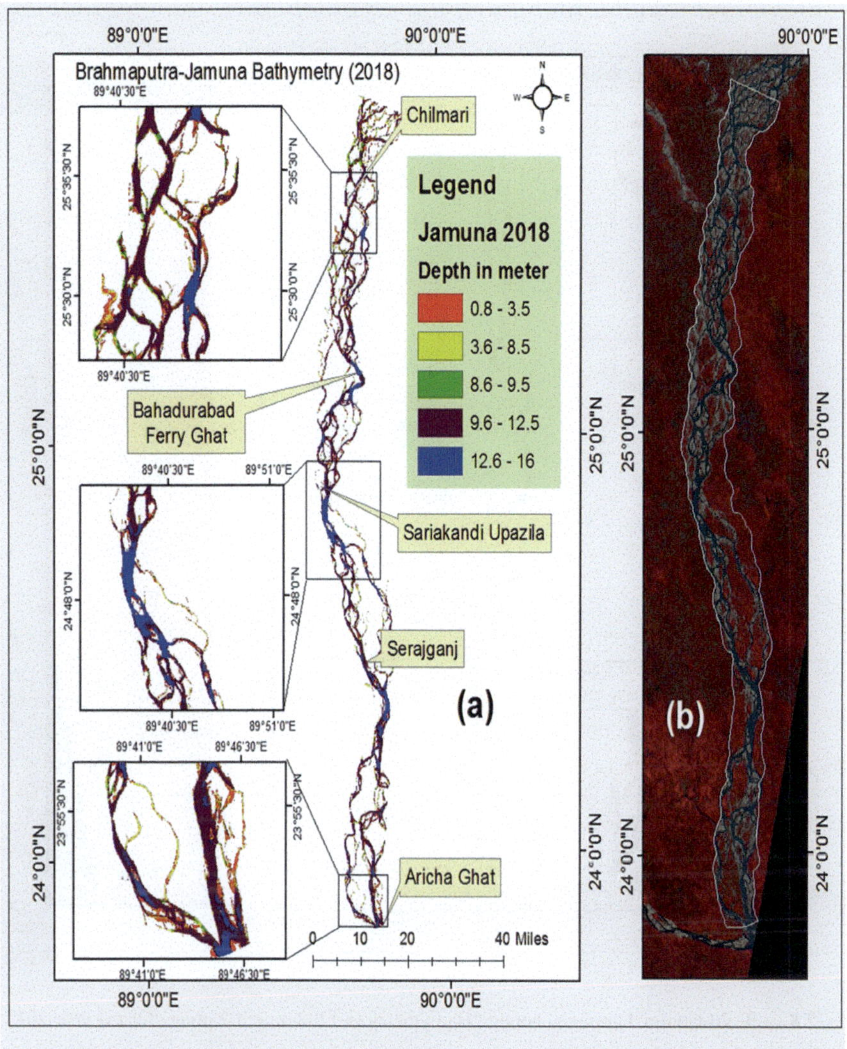

Fig. 7.7 **a** Brahmaputra-Jamuna generated Bathymetry and **b** Sentinel 2 image for the year 2018

7.3.2 *Estimation of River Erosion*

The critical bed shear stress is the minimum amount of stress required to initiate motion of sediment particles on a river bed. In the case of the Brahmaputra-Jamuna river, the critical bed shear stress for the sediment carried by the river is nearly 0.2 N/m^2 [17, 35]. It is important to note that the actual amount of bed shear stress required to initiate motion of sediment particles in a river can vary depending on a number of factors, including the size and shape of the sediment particles, the velocity of the water

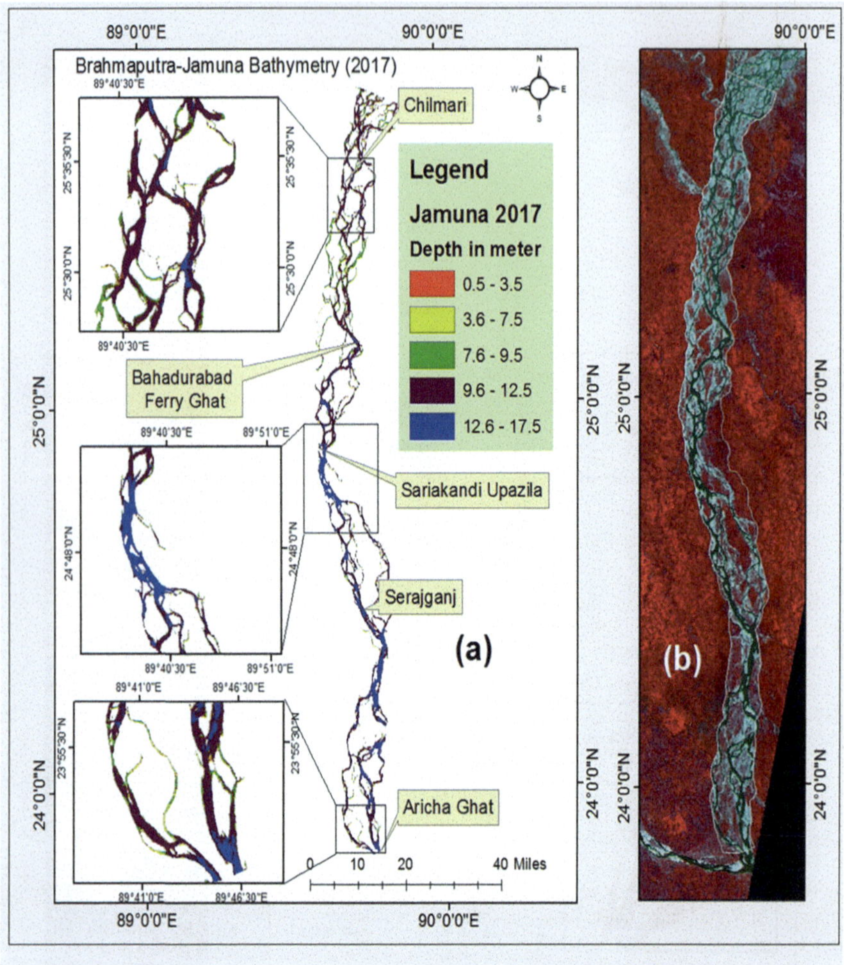

Fig. 7.8 **a** Brahmaputra-Jamuna generated Bathymetry and **b** Sentinel 2 image for the year 2017

flow, and the composition of the riverbed itself. Erosion-prone areas are locations in the river where the water flow is strong enough to cause sediment particles to move and be transported downstream.

Figures 7.12, 7.13, and 7.14 depicts a comparison of river erosion from Sentinel images in with model-generated river erosion for the year 2019, 2018 and 2017 respectively. In the case of 2019 the major channel degradation ranges up to 20 m, including local scour. The overall change was 0.04 m (including the chars). In 2018, the major channel deterioration ranged up to 19.5 m, including local scour but it was nearly 20 m in case of 2017. Tables 7.5, 7.6 and 7.7 shows the confusion matrix for erosion estimation in 2019, 2018 and 2017. After randomly selecting 80 points for accuracy testing (Eqs. 7.13–7.16), the user's accuracy (U_Accuracy) non-eroded

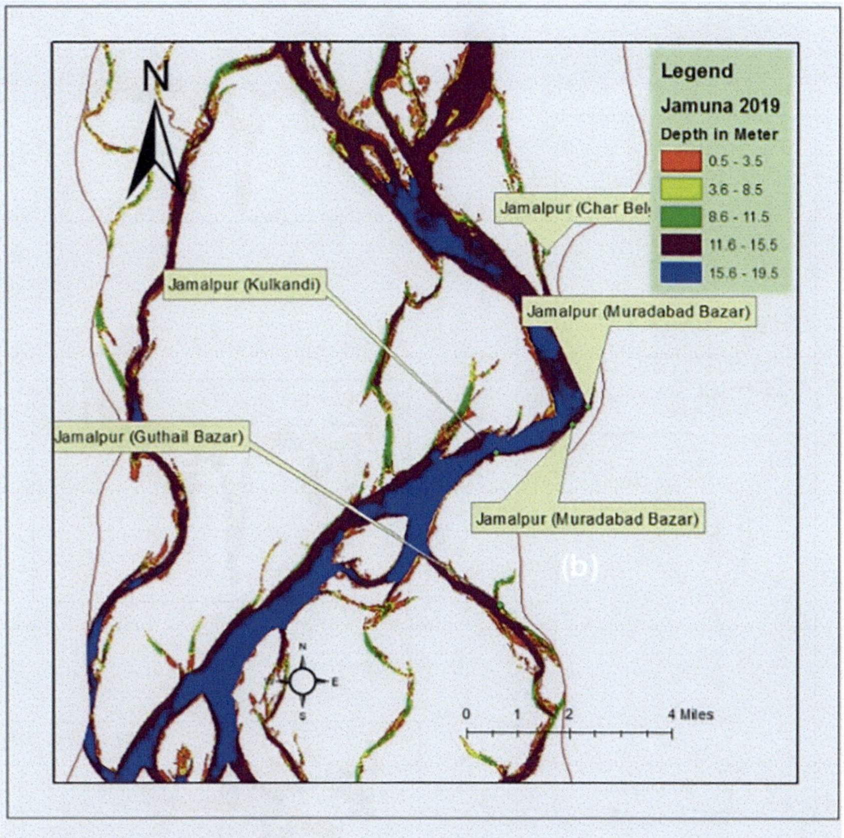

Fig. 7.9 Observed depth information location at Jamalpur Districts

Table 7.4 Assess the accuracy of the generated bathymetry using measured data and people's perceptions

Location	Observed depth (m) (2018–2019)	Estimated depth from satellite images (m) (2019)	Channel type
Guthail Bazar	15	12.8	Secondary channel
Kulkandi	20	15.6	Main channel
Muradabad Bazar	9	12.5	Main channel
Muradabad Bazar	9	12.3	Main channel
Char Belghasa	3	2.7	Tertiary channel

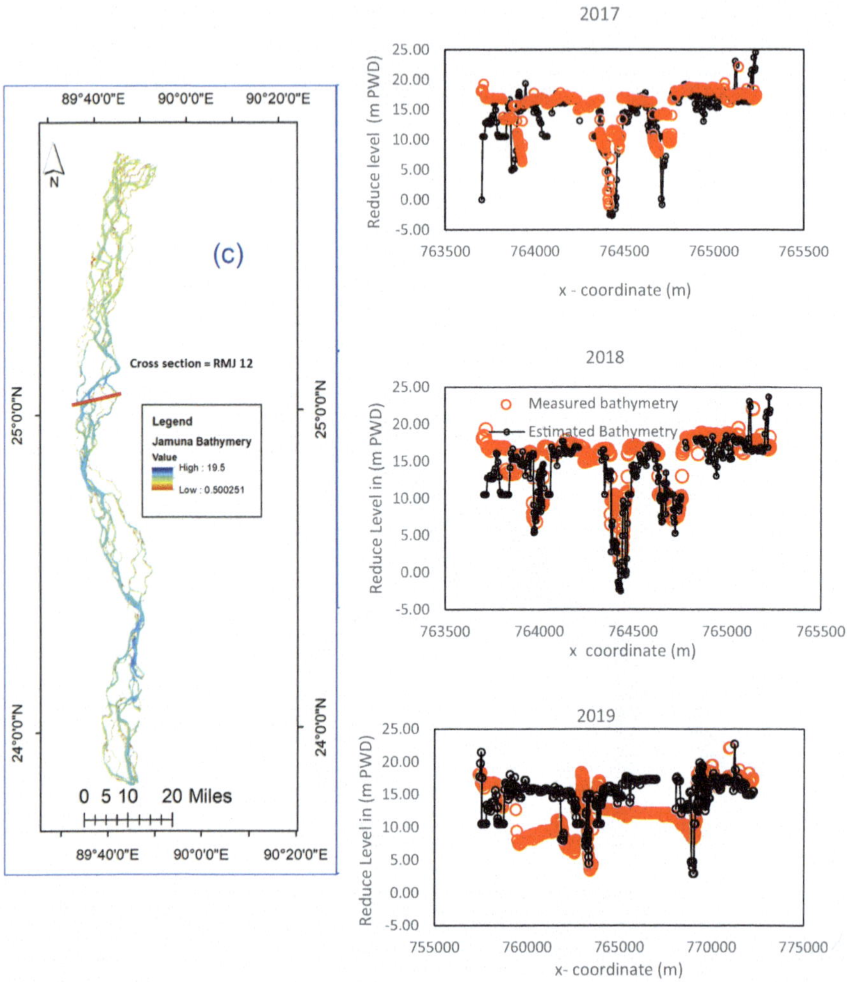

Fig. 7.10 Comparison one cross section (RMJ 12) Measured bathymetry and estimated bathymetry for 2017–2019

and eroded areas was 83% and 88%, respectively for 2019. The overall accuracy was found 88%, 86% and 84% for the year 2019, 2018 and 2017 respectively.

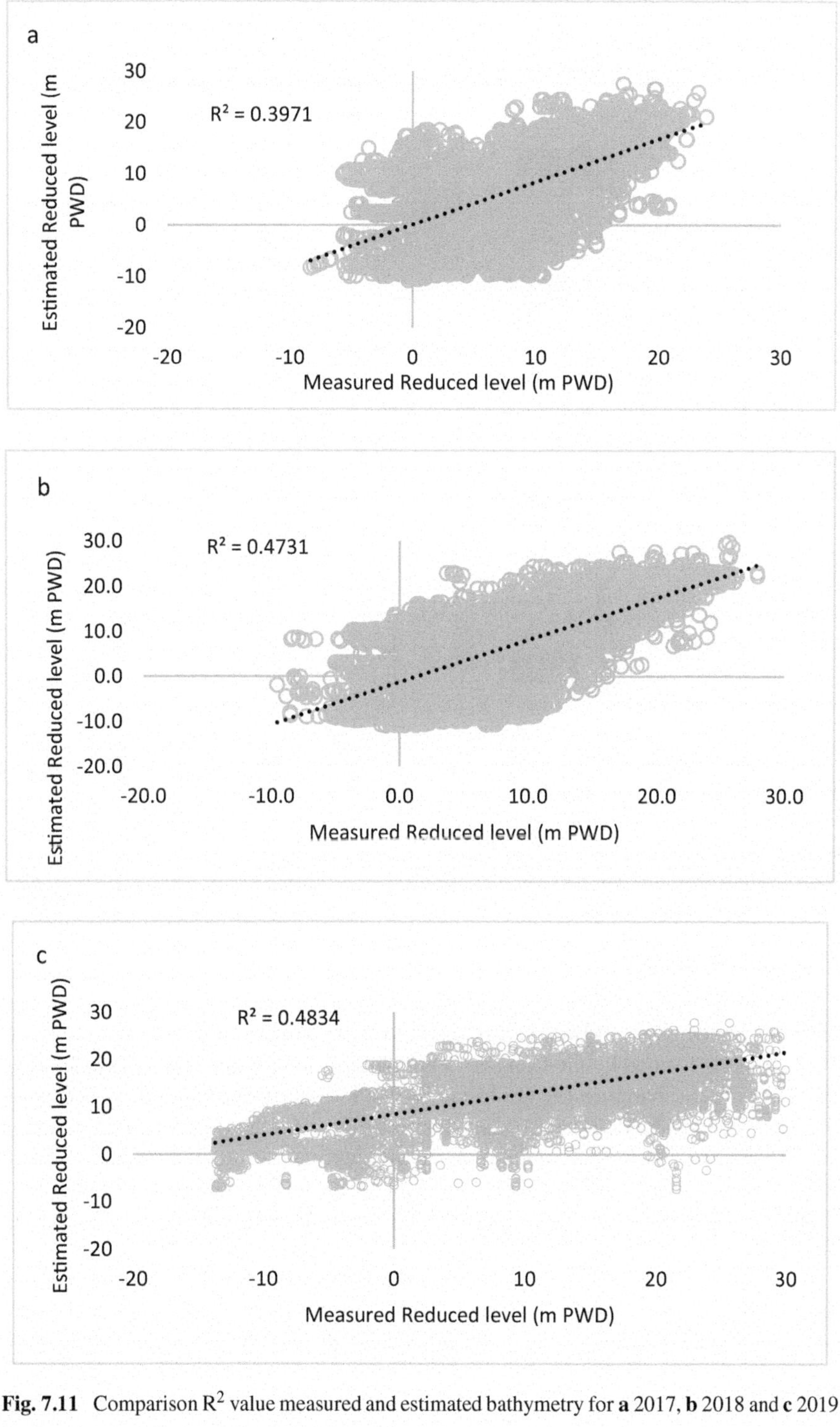

Fig. 7.11 Comparison R^2 value measured and estimated bathymetry for **a** 2017, **b** 2018 and **c** 2019

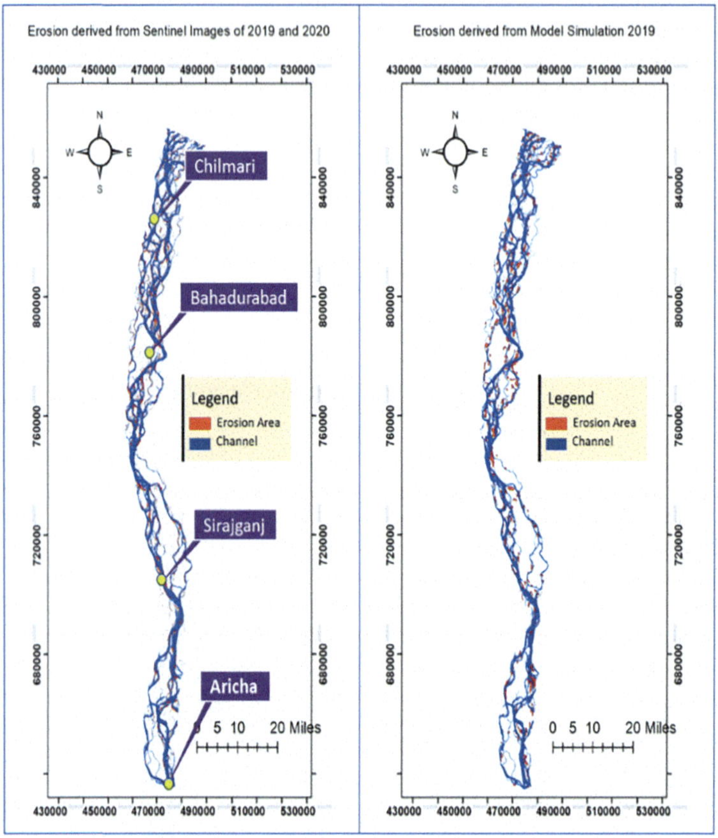

Fig. 7.12 Comparison of river erosion from the Sentinel images of 2019 and 2020 and model-generated river erosion from 2019

7.4 Discussion

The majority of rivers in Bangladesh pass through the unconsolidated sediments of the flood delta formed by the Ganges, Brahmaputra, and Meghna rivers [8]. Braided rivers are always changing, the traditional method of measuring bathymetry is expensive and needs to be done every year due to its dynamic nature. As an alternative, satellite-remote-sensing-based bathymetry with an accurate erosion estimation level could be a solution to this problem. It is a relatively new technique that has gained popularity as a cheaper and more spatially extensive alternative to traditional acoustic surveys.

This study found the estimated depth varies from 0.50 to 19.5 m, with an average depth of 9.5 m. In the previous literature of the related to Brahmaputra-Jamuna, the measured depth was found 3 to 20 m in some of the places. However, in the local

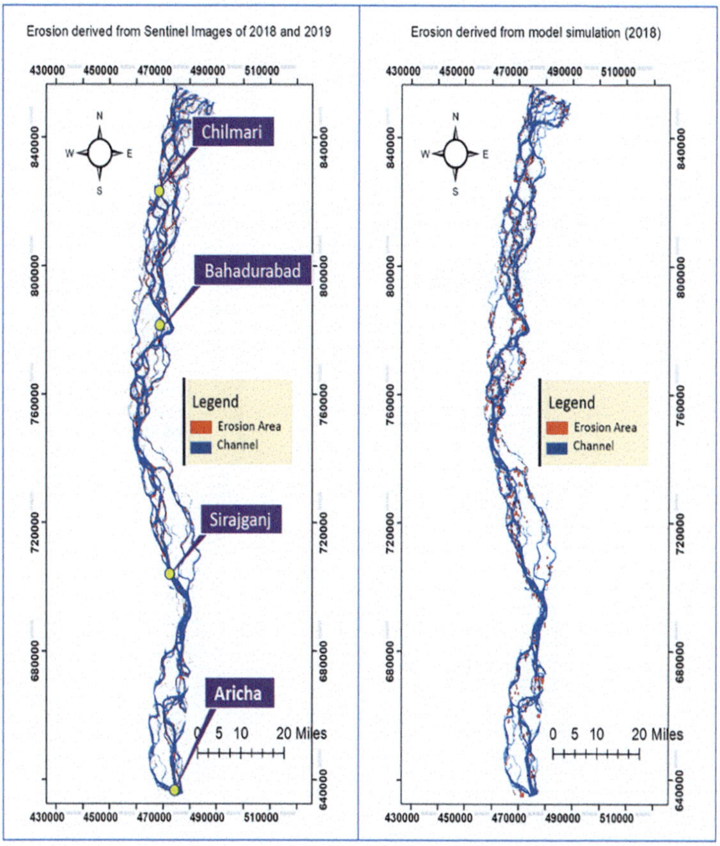

Fig. 7.13 Comparison of river erosion from the **a** Sentinel images of 2018 and 2019 and model-generated river erosion from 2018

scour due to any type of hydraulic structure the depth can be increased as up to 30 m [3, 6, 31].

In 2018 it shows erosion area was less at downstream of Bahadurabad and Sirajganj than compared to 2019. But in 2017 massive erosion occurred full Brahmaputra-Jamuna River derived from the satellite image analysis. From downstream of Bahadurabad to Sarishabari, the majority of the accretion has occurred. The effect of westward migration of Brahmaputra-Jamuna River as well as the construction of revetments, spurs, and other protective structures may play the key role for such accretion.

The confusion matrix statics, Kappa was found 72%, 78% and 67% for the year 2019, 2018, 2017 respectively. Kappa value matches with the 'Good' agreement (55–70%) and 'very good' agreement (70–85%) according to [28]. Nash Sutcliffe model efficiency coefficient (NSE) value for Chimari, Serajgani, Mathura and Bahadurabad station shows 0.9778, 0.985, 0.9278, 0.96531. Percent bias (PBIAS)

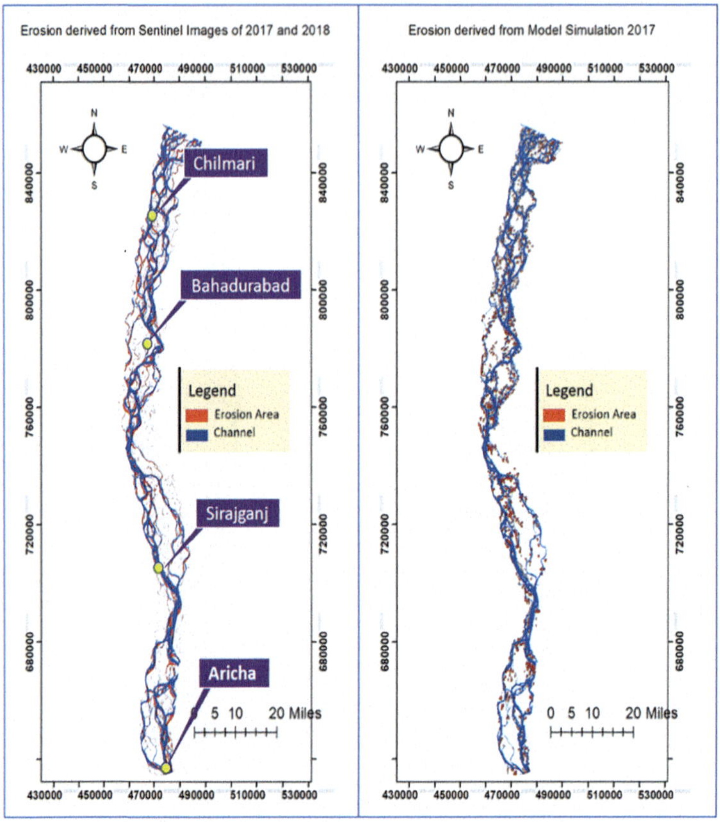

Fig. 7.14 Comparison of river erosion from the Sentinel images of 2017 and 2018 and model-generated river erosion from 2017

Table 7.5 Accuracy assessment of spatial erosion for 2019

Type	Non-eroded area nos.	Eroded area nos.	Total	U_Accuracy (%)	Kappa
Non-eroded area nos.	31	6	37	83	0
Eroded area nos.	5	38	43	88	0
Total	36	44	80	0	0
P_Accuracy (%)	86	86	0	86	0
Kappa (%)					72
Overall accuracy (%)					86

Table 7.6 Accuracy assessment of spatial erosion for 2018

Type	Eroded area nos.	Non-eroded area nos.	Total	U_Accuracy (%)	Kappa
Eroded area nos.	32	7	39	82	0
Non-eroded area nos.	2	40	42	95	0
Total	34	47	81	0	0
P_Accuracy (%)	94	85	0	89	0
Kappa (%)					78
Overall accuracy (%)					88

Table 7.7 Accuracy assessment of spatial erosion for 2017

Type	Eroded area nos.	Non-eroded area nos.	Total	U_Accuracy (%)	Kappa
Eroded area nos.	51	9	60	85	0
Non-eroded area nos.	6	30	36	83	0
Total	57	39	96	0	0
P_Accuracy (%)	89	77	0	84	0
Kappa (%)					67
Overall accuracy (%)					84

for Chimari, Serajgani, Mathura and Bahadurabad station shows 0.0351, 0.0314, 0.08112, 0.012268. Relative root means square error (RRMSE) for Chimari, Serajgani, Mathura and Bahadurabad station shows 3.43, 7.74, 7.45, 0.921. All of them showed satisfactory value.

There have some limitations of this study high turbidity levels result in more scattering of the light, causing it to be reflected by the sensor at different angles. This results in a blurred image, making it difficult to accurately determine the depth of the water. Conversely, low turbidity levels result in less scattering, allowing for clearer images and more accurate depth readings [15]. In this study two-dimensional model is used which gives bed level change from simulations. Therefore, the finest prediction depends on the numerical model's grid resolution.

7.5 Conclusion

This research presents a technique for generating high-resolution river bathymetry in data-poor regions. The resulting bathymetry serves as a valuable baseline for accurate estimation river erosion using numerical simulations, provided appropriate boundary conditions are applied. this study has been able to highlight some important issues which have been summarized below -

- Determining bathymetry from optical images was reliable to use in numerical model for river erosion estimation.
- Utilizing the 2D morphological model, accurate prediction of bed evolution in a compounded bar-dominated braided river is achievable.
- The model effectively identified the location of eroded areas, showcasing comparability with actual eroded locations. The kappa statistics shows accuracy average for three years was 72%, indicating reasonably good agreement.
- It proves beneficial in various other aspects, such as bed level change, sediment transport, bed shear stress, braided channels and flow patterns resulting from channel shifting, among other potential applications.

Acknowledgements The author wishes to thank the Bangladesh University of Engineering and Technology (BUET) for M.Sc. thesis and research funding.

References

1. Alam KF, Ahamed T (2023) Erosion vulnerable area assessment of Jamuna River system in Bangladesh using a multi-criteria-based geospatial fuzzy expert system and remote sensing. Asia-Pacific J Reg Sci 7(2):433–454
2. Ashmore PE (1991) How do gravel-bed rivers braid? Can J Earth Sci 28(3):326–341
3. Ashworth PJ, Best JL, Roden JE, Bristow CS, Klaassen GJ (2000) Morphological evolution and dynamics of a large, sand braid-bar, Jamuna River, Bangladesh. Sedimentology 47(3):533–555
4. Best JL, Ashworth PJ, Sarker MH, Roden JE (2007) The Brahmaputra-Jamuna River, Bangladesh. Large rivers: geomorphology and management, pp 395–430
5. Bramante JF, Raju DK, Sin TM (2013) Multispectral derivation of bathymetry in Singapore's shallow, turbid waters. Int J Remote Sens 34(6):2070–2088
6. Bristow CS, Best JL (1993) Braided rivers: perspectives and problems. Geol Soc London Spec Publ 75(1):1–11
7. Bryant S, Mosselman E (2017) Taming the Jamuna: effects of river training in Bangladesh. In: Hoitink AJF, de Ruijsscher TV, Geertsema TJ, Makaske B, Wallinga J, Candel JHJ, Poelman J (eds) NCR days 2017. Book of abstracts, 1st edn. NCR Publication 41-2017, p 114
8. Curtis S, Crawford T, Munshi K, Paul B (2017) Monsoon dynamics in the Ganges-Brahmaputra-Meghna basin. In: The 1st electronic conference on hydrological cycle (CHyCle-2017), vol 1, p 4865
9. Despotovic M, Nedic V, Despotovic D, Cvetanovic S (2016) Evaluation of empirical models for predicting monthly mean horizontal diffuse solar radiation. Renew Sustain Energy Rev 56:246–260
10. Duró G, Crosato A, Tassi P (2016) Numerical study on river bar response to spatial variations of channel width. Adv Water Resour 93:21–38
11. Engelund F, Skovgaard O (1973) On the origin of meandering and braiding in alluvial streams. J Fluid Mech 57(2):289–302
12. Exner FM (1925) On the interaction between water and sediment in rivers. Akad Wiss Wien Math Naturwiss Klasse 134(2a):165–204
13. Friend PF, Sinha R (1993) Braiding and meandering parameters. Geol Soc Spec Publ 75:105–111

14. Geyman EC, Maloof AC (2019) A simple method for extracting water depth from multispectral satellite imagery in regions of variable bottom type. Earth Space Sci 6(3):527–537
15. Islam A, Khondoker MS, Zahid Hasan Siddiquee M, Ashraful Islam M (2016) The challenges of river bathymetry survey using space borne remote sensing in Bangladesh. Atmos Ocean Sci 1(1):7–13
16. Jagalingam P, Akshaya BJ, Hegde AV (2015) Bathymetry mapping using Landsat 8 satellite imagery. Procedia Eng 116(1):560–566
17. Kabir MR, Ahmed N (1996) Bed shear stress for sediment transportation in the river Jamuna. J Civ Eng Inst Eng Bangladesh 24:55–68
18. Kanno A, Tanaka Y (2012) Modified lyzenga's method for estimating generalized coefficients of satellite-based predictor of shallow water depth. IEEE Geosci Remote Sens Lett 9(4):715–719
19. Knoben WJ, Freer JE, Woods RA (2019) Inherent benchmark or not? Comparing Nash–Sutcliffe and Kling–Gupta efficiency scores. Hydrol Earth Syst Sci 23(10):4323–4331
20. Lane EW (1957) A study of the shape of channels formed by natural streams flowing in erodible material (No. 9). US Army Engineer Division, Missouri River
21. Legleiter CJ, Overstreet BT (2012) Mapping gravel bed river bathymetry from space. J Geophys Res: Earth Surf 117(F4)
22. Lesser GR, Roelvink JA, van Kester JATM, Stelling GS (2004) Development and validation of a three-dimensional morphological model. Coast Eng 51(8–9):883–915
23. Li MF, Tang XP, Wu W, Liu HB (2013) General models for estimating daily global solar radiation for different solar radiation zones in mainland China. Energy Convers Manage 70:139–148
24. Lyons M, Phinn S, Roelfsema C (2011) Integrating quickbird multi-spectral satellite and field data: mapping bathymetry, seagrass cover, seagrass species and change in Moreton Bay, Australia in 2004 and 2007. Remote Sens 3(1):42–64
25. Lyzenga DR (1978) Passive remote sensing techniques for mapping water depth and bottom features. Appl Opt 17(3):379
26. Lyzenga DR (1981) Remote sensing of bottom reflectance and water attenuation parameters in shallow water using aircraft and Landsat data. Int J Remote Sens 2(1):71–82
27. Lyzenga DR, Malinas NP, Tanis FJ (2006) Multispectral bathymetry using a simple physically based algorithm. IEEE Trans Geosci Remote Sens 44(8):2251–2259
28. Monserud RA, Leemans R (1992) Comparing global vegetation maps with the Kappa statistic. Ecol Model 62(4):275–293
29. Moriasi DN, Arnold JG, Van Liew MW, Bingner RL, Harmel RD, Veith TL (2007) Model evaluation guidelines for systematic quantification of accuracy in watershed simulations. Trans ASABE 50(3):885–900
30. Mosselman E (2009) Bank protection and river training along the braided Brahmaputra–Jamuna River, Bangladesh. Braided Rivers 277–287
31. Nakagawa H, Zhang H, Baba Y, Kawaike K, Teraguchi H (2013) Hydraulic characteristics of typical bank-protection works along the Brahmaputra/Jamuna River, Bangladesh. J Flood Risk Manag 6(4):345–359
32. Philpot WD (1989) Bathymetric mapping with passive multispectral imagery. Appl Opt 28(8):1569–1578
33. Piégay H, Grant G, Nakamura F, Trustrum N (2006) Braided river management: from assessment of river behaviour to improved sustainable development. Braided Rivers 257–275.
34. Rahman MM, Hasan MS, Eusufzai MK, Rahman MM, Rahman MM, Hasan MS, Eusufzai MK, Rahman MM (2021) Impacts of dredging on fluvial geomorphology in the Jamuna River, Bangladesh. J Geosci Environ Prot 9(6):1–20
35. Van Rijn LC (1993) Principles of sediment transport in rivers, estuaries and coastal seas. Aqua Publications, Amsterdam
36. Shampa, Hasegawa Y, Nakagawa H, Takebayashi H, Kawaike K (2017) Dynamics of sand bars in braided river: a case study of brahmaputra-jamuna river. J Jpn Soc Nat Disaster Sci 36(Special):121–135

37. Shampa, Muktadir HM, Nijum IJ, Bhowmik A, Islam N, Fardin A, Biswas S, Islam MM (2023) Development of riverbank erosion warning system for braided river. In: Proceedings of the 4th IAHR young professionals congress, pp 130–131
38. Shampa (2019) Hydro-morphological study of braided river with permeable bank protection structure. Doctoral Thesis, Kyoto University, p 162
39. Stumpf RP, Holderied K, Sinclair M (2003) Determination of water depth with high-resolution satellite imagery over variable bottom types. Limnol Oceanogr 48(1 II):547–556
40. Tonina D, McKean JA, Benjankar RM, Wright CW, Goode JR, Chen Q, Reeder WJ, Carmichael RA, Edmondson MR (2019) Mapping river bathymetries: evaluating topobathymetric LiDAR survey. Earth Surf Proc Land 44(2):507–520
41. Van Der Wegen M, Roelvink JA (2008) Long-term morphodynamic evolution of a tidal embayment using a two-dimensional, process-based model. J Geophys Res: Oceans 113(3):1–23
42. Yorukoglu M, Celik AN (2006) A critical review on the estimation of daily global solar radiation from sunshine duration. Energy Convers Manage 47(15–16):2441–2450

Chapter 8
Experimental Investigation on Minimizing Erosion Near the First Groin in a Series Within a Channel

Md. Tofiquzzaman and Mohammed Alauddin

Abstract Groins are common hydraulic structures built in series to safeguard river banks. The stability of the groins and the protected zone depends on the stability of the first groin of a series. Despite the abundance of literature on groin design, groin effect, and so on, a few studies have focused on reducing thrust on the first groin due to upstream flow. This study explores a suitable approach to minimize the impact of upstream flow on the groins placed in a series focusing on the stability of the first groin. The study considers three different options: a set of five impermeable groins in a series (Case-1); the first one is a combined groin of Case-1 (Case-2); and an additional upstream minor combined groin in Case-1 (Case-3). All the laboratory experiments are conducted in this study under clear-water scour conditions. The results revealed that the first groin of combined nature in the series as well as the minor combined groin upstream of the main groins reduced the effect of upstream flow significantly. Compared to Case-1, Case-2 and Case-3 showed 50% and 47% reduction of scour, respectively. Therefore, Case-2 has been identified as an effective way of stabilizing groins and river banks in the field.

Keywords Bank erosion · Minor groin · Combined groin · Local scour · Groin stability

Md. Tofiquzzaman (✉)
River Research Institute, Faridpur, Bangladesh
e-mail: tofiqrri@gmail.com

M. Alauddin
Department of Civil Engineering, Dhaka University of Engineering & Technology, Gazipur, Bangladesh
e-mail: mauddin@duet.ac.bd

G. M. Tarekul Islam et al. (eds.), *Sustainable Water Management and Environmental Resilience*, Springer Water, https://doi.org/10.1007/978-3-031-93350-9_8

8.1 Introduction

Bank erosion is a serious problem in river channels, especially in alluvial deltaic plains and the lowest riparian countries like Bangladesh. Globally, groins are employed for river training that change the flow pattern significantly. Especially, in the first groin of a series, the effect of upstream flow is quite strong, which leads to the maximum scour around that [1], and the scour depth around the groin increases with an increase in velocity [8]. Local scour-induced instability threatens the safety of the groin structures. Thus, river engineers and researchers urgently need to improve groin stability by minimizing upstream flow effects or improving hydraulic conditions near groins.

Researchers have conducted many studies to minimize the effect of upstream flow and the local scour near the groin's tip, considering various parameters such as groin shapes [6, 7, 11], spacing of groins [3], change in groin permeability [4], groin orientation and spatial arrangement (big to small and vice versa) [2], installation of minor spur dike upstream of a series [5, 10], and so on. An impermeable groin produces a large scour hole due to rapid flow deviation and strong vortices near its tip, posing a major structural stability concern. Permeable groins slow down flow near the bank and deposit sediment there [9], but they can't divert the flow far enough away from the bank. Whereas, by improving flow field towards bank and reducing scour near groin tip, and producing a stagnant water zone near the bank line, combined groins function better over others [1]. However, the problem of groin instability still remains elusive. Therefore, a sustainable solution is essential to make the groin structures more stable, especially for the first one, by reducing the effect of upstream flow. Thus, this study investigates different groin conditions—altering the first groin's configuration and installing an additional minor combined groin upstream of impermeable groins. Analyzing the bed level variations in the groin area, the effectiveness of these groin arrangements and configurations is examined to improve their stability and functionality.

8.2 Methodology

8.2.1 Experimental Setup

The experiments are conducted in a 20.0 m long, 1.5 m wide, and 1.15 m deep straight rectangular flume at the Hydraulic Engineering Laboratory of Dhaka University of Engineering & Technology (DUET), Gazipur. The sidewalls and bottom of the channel are composed of reinforced cement concrete. The flume is filled with silt-free, 30.0 cm-thick fine sand with a median size (d_{50}) of 0.20 mm.

The groin models used in this study are made of a 1.0 cm thick transparent acrylic sheet. Besides, 5.0 mm diameter steel sticks with different spacing are also used only for the permeable portion of the combined groin. The groin models are 60 cm in height, with a projected length of 30 cm for the main groins and 15 cm for the minor groin. 30 cm height of these groin models are embedded in the sand bed, placing them perpendicular to the channel's right side. A spacing of 90 cm is maintained between two successive main groins, whereas the minor one is installed 60 cm apart, upstream of the first groin. The center of the first model of each set is installed at a distance of 10.0 m from the end of the inlet tank, which is theoretically sufficient to achieve a fully developed turbulent flow in the control area. A schematic representation of the experimental setup including top view of flume, and top view of groin models, is depicted in Fig. 8.1a and b, respectively.

In this study, all the tests are done under clear-water scour conditions. This scour happens when the bed materials of the channel are not in motion. This condition is developed by adjusting a control valve and a tail gate so that the flow velocity (u) does not exceed the critical velocity (u_c) of bed sediment upstream of the control section. To achieve sub-critical turbulent flow in the channel, the Reynolds number ($R_e = \frac{Uh}{\nu}$) is kept high enough and the Froude number ($F_r = \frac{U}{\sqrt{gh}}$) is kept low enough. The experimental conditions for all the test runs are provided in Table 8.1.

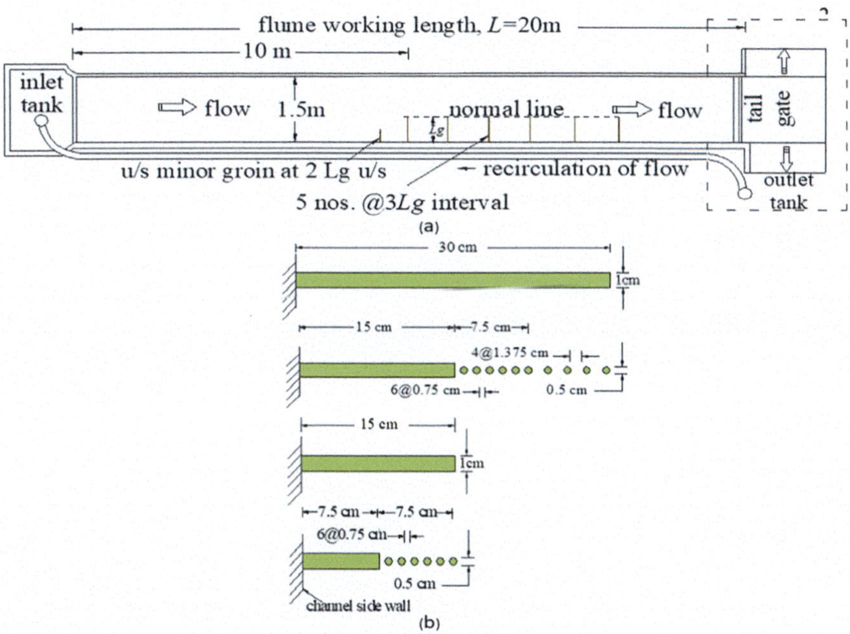

Fig. 8.1 Experimental setup; plan view of **a** model channel; **b** groin models

Table 8.1 Details of experimental conditions

Parameters	Value
Discharge (m^3/s)	0.03
Flow depth (cm)	10
Mean velocity (m/s)	0.2
$\frac{u}{u_c}$	0.88
Froude number, F_r	0.2
Reynolds number, R_e	20,000

8.2.2 Procedure

At first, the sand bed is carefully levelled with a wooden scraper after installing the model structures. Then, before starting the main test, water is gently allowed to enter into the flume and soak the sand bed at a low speed. After draining, the bed level is measured with a point gauge to determine the initial bed condition. Then, the flow is gradually allowed to enter into the channel with a controlled flow rate by a check valve and flow meter. The scour depth near the first structure is monitored by a point gauge to identify the equilibrium state and it is found to be maximum 6 h after running a preliminary test for 11.0 h. The final bed levels along some certain sections are measured after each test run. Figure 8.2 displays channel bed images for various test cases.

A precise measurement of bed level is done in the test area before and after each test run, respectively. To minimize bed disturbance, flow is gradually reduced before terminating the test run. The bed level is measured with a point gauge in the control region after draining and drying the bed, usually after one day of running. The point gauge is movable and is supported by a carriage that runs on steel frames on either side of the channel. In the groin area, the initial bed level for each case is measured at 10.0 cm grid points in both transverse and longitudinal directions. However, after ending of each test run, the final bed level in the entire groin area is measured at 5.0 cm longitudinal and 3 cm transverse intervals.

8.3 Results

Channel responses for the same flow condition are found different due to the groins of different configurations and arrangements. Figure 8.3a and b illustrates bed topographies developed in the experimental runs and bed level contours around the first groins.

The local scour around the first groin is more pronounced in all the cases as realized in Fig. 8.3a. Also, the local scour near the last groin can be recognized as higher over the other groins. Especially Case-1 exhibits a larger volume of scour formation around the first groin with a larger volume of sediment accumulation in

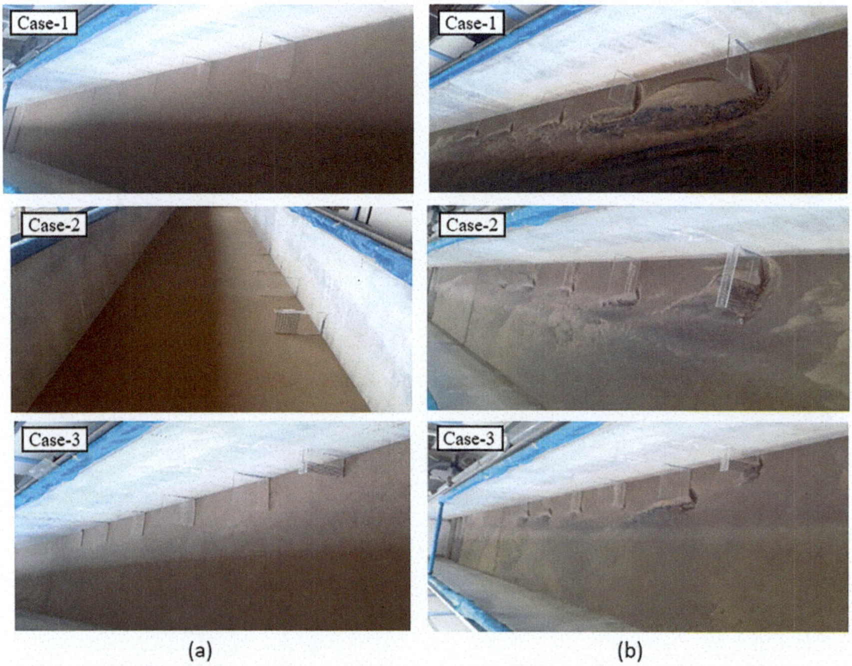

Fig. 8.2 Visual images of the channel bed for various test cases: **a** before run, and **b** after run

the downstream of it compared to other cases, and also, the bed formation is found irregular. A continuous bed erosion mark from the eroding zone around the first groin to the opposite side of the channel is observed more evidently in Case-1 than in Case-3. In Case-2, sediment accumulation is detected in the closer area of the channel side than in the other cases. From the expanded view of the first groin area, an extensive scour hole in a wider area can be recognized in Case-1 as in Fig. 8.3b, extending from the upstream to the downstream of the groin tip. In Case-2, the scour hole position is observed near the impermeable end of the groin. The scour zone does not greatly extend longitudinally along the channel as seen in the other cases. Compared to other cases, this case has the lowest scour depth and area. Similar to Case-1, in Case-3 a scour hole is observed extending from the upstream to the downstream of the groin tip, but the extent of the scour zone is significantly reduced.

Figure 8.4a and b depicts the bed level profiles in transverse and longitudinal directions near the groin head.

It is evident from Fig. 8.4a and b that the first groin in Case-1 gives the highest scour depth than the other cases. The scour zone presented with transverse and longitudinal sections in the figures is also evident higher in Case-1 than in the other cases. In Case-3, a greater scour reduction is observed than Case-1 for an upstream minor groin. However, the scour depth around the first groin in Case-2 is found lower than Case-3, despite the absence of a minor groin. In the channel's transverse

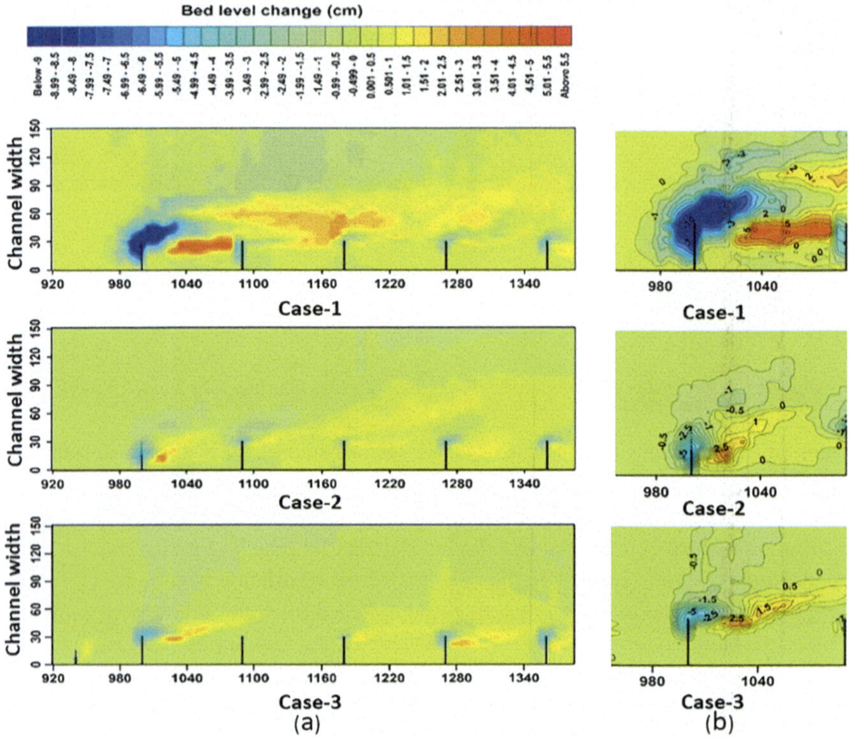

Fig. 8.3 **a** Channel bed topographies (all units are in cm) **b** bed level contours around the first groins

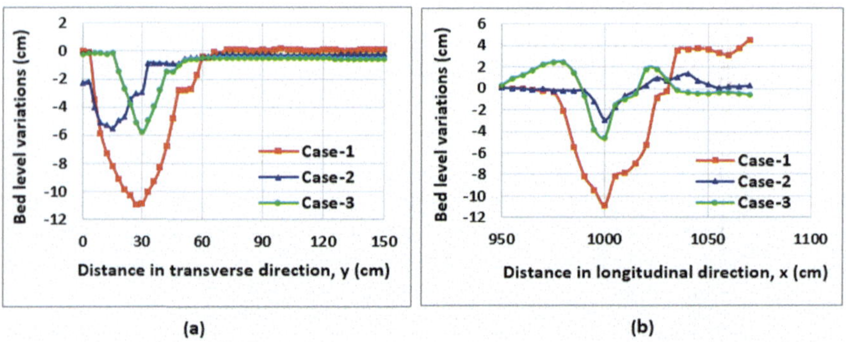

Fig. 8.4 **a** Transverse bed profile along the first groin (at 10.0 m from the u/s end) **b** longitudinal bed profile along the tip of the main groin

Table 8.2 Maximum scour depth around groin head and average deposition in first groin field

Test cases	Maximum scour depth (cm)						Sediment depos. (cm)
	Minor groin	1st groin	2nd groin	3rd groin	4th groin	5th groin	
Case-1	–	10.94	4.17	3.32	3.11	4.27	5.44
Case-2	–	5.49	3.87	2.36	4.05	4.16	3.7
Case-3	1.96	5.77	1.68	2.84	4.41	5.39	3.21

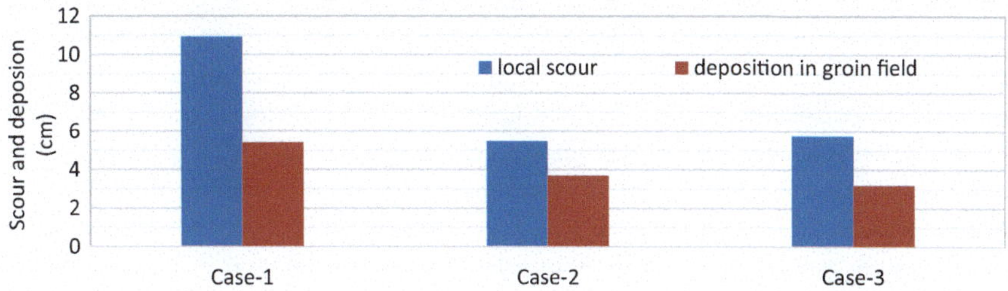

Fig. 8.5 Comparison of maximum scour depth near the first groin and deposition in the groin field

direction, the highest scour depth around the first groin is found 30 cm apart from the groin side for Case-1 and Case-3, and 15 cm apart for Case-2. It is also found from the analysis of bed level variations that Case-2 gives a shorter length of the scour-prone zone surrounding the first groin compared to all other cases, regardless of the transverse or longitudinal direction of the channel. Notably, the scouring at the tip of all groins exhibits a conical shape, implying a common scour pattern in the different cases. Table 8.2 and Fig. 8.5 present measured maximum scour depth and sediment deposition between the first and second main groin.

It is evident from Table 8.2 and Fig. 8.5 that, in Case-2, the measured maximum scour depth around the first and last groin is the lowest than in the other cases, and the sediment deposition in the groin field is found lower than Case-1, but higher than Case-3. The second-lowest scour depth around the first groin is observed in Case-3, with the highest scour depth observed around the last groin. Therefore, only replacing the impermeable conventional groin by a combined one (Case-2), the scour can be reduced by 50%, whereas 47% scour can be reduced by the minor combined groin in Case-3.

8.4 Discussion

There are three different cases considered in the study to explore the most suitable solution, where the first groin in all the cases gives the highest local scour. So, it can be explained by the fact that the first groin is exposed to the highest current by the upstream flow, which leads to a higher risk of groin failure. In each case, the local scour is also apparent for the last groin after the first groin in the series. This can be happened due to the return current by the reflected flow from the opposite side of the channel that can be significant to demolish downstream structures depending on its strength.

Based on the analysis of bed level variations, it is clear that different factors, such as groin permeability and the presence of minor groin, significantly influence the scour patterns. The modification of the configuration of the first groin in Case-2 and placing of an additional upstream minor groin of combined nature in Case-3 help attenuate the upstream flow impact and reduce the scour around the first groin. The minor combined groin in Case-3 can potentially reduce the scour around the first main groin, but the impact of upstream flow on scour formation still remains due to its impermeable nature. But the least effect on the first groin is observed in Case-2 without any minor upstream groin, and these can be attributed to the configuration of the groin, where a combination of permeable and impermeable portions is present. In a previous study, approximately 21% scour reduction around the first groin was observed in the case of a series of combined groins compared to a series of conventional impermeable groins [1]. But in this study, 50% scour reduction is resulted only by modifying the first groin of impermeable nature with a combined groin. Another previous study with an impermeable minor groin showed 46.2% scour reduction around the first groin [5], and a very close result is found in the present study, i.e., 47% scour reduction is identified with a combined minor groin.

Sediment deposition in the groin field varies with changing the groin configuration and arrangement in different test cases. It can be mentioned here that the experiments are carried out under clear water scour conditions, and even in the control area, the presence of suspended sediments is found to be insignificant. Therefore, the deposition pattern observed in this study may not properly reflect the natural flow environment, where siltation of fine suspended sediment particles typically occurs in the slow flow zones of the groin field causing huge deposition and hence land reclamation near the bank line.

8.5 Conclusion

This study is mainly aimed at stabilizing the groins in a series with minimization of the local scour around the first groin. The following conclusions have been drawn:

- The local scour depth around the first groin head is found higher in Case-1 with higher sediment deposition in the groin field, where all the groin models are of impermeable conventional type.
- The reduction of scour depth around the first groin is found higher in Case-2 compared to all other cases, with a higher sediment deposition than in Case-3, and the scour reduction is 50% compared to Case-1.

- Although Case-3 is still under the effect of upstream flow, there is a noticeable reduction in scour depth around the first groin due to the presence of an upstream minor groin, and this is 47% compared to Case-1.
- Local scour around the last groin is found more apparent than the other groins except the first one, possibly as a result of return currents and flow reflection from the other side of the channel.

Actually, the inclusion of a permeable portion in the first main groin in Case-2 significantly contributes to attenuating the impact of upstream flow and effectively reducing the local scour. Therefore, the configuration and arrangement of groins in Case-2 provide better efficacy in managing scour near the first groin in comparison to other cases and improve its stability.

Acknowledgements The authors would like to acknowledge the Department of Civil Engineering, Dhaka University of Engineering & Technology (DUET), Gazipur for providing all support during this research work. The first author would also be grateful to River Research Institute, Faridpur (Ministry of Water Resources, Bangladesh) for providing an opportunity to perform this research work.

References

1. Alauddin M, Tashiro T, Tsujimoto T (2011) Design of groynes modified with both alignment and permeability for lowland river problems. J Jpn Soc Civ Eng Ser A2 (Appl Mech) 67(2)
2. Choufu L, Abbasi S, Pourshahbaz H, Taghvaei P, Tfwala S (2019) Investigation of flow, erosion, and sedimentation pattern around varied groynes under different hydraulic and geometric conditions: a numerical study. Water 11(2):235
3. Kafle MR (2021) Numerical simulation of flow pattern in series of impermeable groynes in fixed bed. J Adv Coll Eng Manag 6:83–88
4. Kang J, Yeo H, Kim S, Ji U (2011) Experimental investigation on the local scour characteristics around groynes using a hydraulic model. Water Environ J 25(2):181–191
5. Karami H, Ardeshir A, Behzadian K, Ghodsian M (2011) Protective spur dike for scour mitigation of existing spur dikes. J Hydraul Res 49(6):809–813
6. Kumar T, Tyagi D, Aggarwal L, Kumar M (2018) Comparison of scour around different shapes of groynes in open channel. Int J Recent Trends Eng Res 4(3):382–392
7. Mohammed A, Pervin R, Hasan MZ (2022) Effect of various groins in a series on channel bed morphology: an experimental investigation. Stavební Obzor-Civ Eng J 31(1):211–221
8. Rashak BM, Khassaf SI (2021) Local scour around t-shape submerged groynes in clearwater conditions. GEOMATE J 20(77):163–172
9. Safie O, Tominaga A (2018) Effect of pile arrangement on flow characteristics around pile-group dike. J Jpn Soc Civ Eng Ser A2 (Appl Mech) 74(2)
10. Shafaie A, Ardeshir A, Sadat-Helbar SM, Saneie M (2008) The effect of minor spur dike on scouring at the first spur dike in the gravel bed. In: International conference on scour and erosion, 5–7
11. Uijttewaal WS (2005) Effects of groyne layout on the flow in groyne fields: laboratory experiments. J Hydraul Eng 131(9):782–791

Chapter 9
Assessing Water Insecurity for Bankline Communities Considering Riverbank Erosion: A Case Study Along the Padma River

Lamiya Sharmeen Jaren, Sabbir Ahmed, Syed Nazmus Sakib, Sara Nowreen, Ahmed Ishtiaque Amin Chowdhury, and Rabeya Sultana Leya

Abstract The concept of water security includes a diverse range of physical, socio-economic, environmental, and infrastructural water related challenges, traditionally measured by the Water Security Index (WSI). This study presents a disaster management inclusive approach for measuring WSI at the household level. The objective of this study is to assess water insecurity for bankline communities considering river erosion in selected unions in Harirampur, Manikganj, along the river Padma. By employing a mixed-method approach encompassing remote sensing, structured questionnaire survey of households, and participatory techniques, various indicators of water security were evaluated. WSI were calculated by using weighted additive function where component weights were assigned by pairwise ranking method through expert opinion. The results showed that the WSI value was 44.25, classifying the area as water insecure. The system is considered water secured when the WSI

L. S. Jaren (✉) · S. Ahmed · S. Nowreen · A. I. A. Chowdhury
Institute of Water and Flood Management, Bangladesh University of Engineering and Technology, Dhaka, Bangladesh
e-mail: lamiyasharmeenj@gmail.com

S. Ahmed
e-mail: sabbirahmed.j05@gmail.com

S. Nowreen
e-mail: snowreen@iwfm.buet.ac.bd

A. I. A. Chowdhury
e-mail: ishtiaquechowdhury@iwfm.buet.ac.bd

S. N. Sakib
Daffodil International University, Dhaka, Bangladesh
e-mail: sakib.esdm@diu.edu.bd

R. S. Leya
Urban and Rural Planning Discipline, Khulna University, Khulna, Bangladesh
e-mail: rabeya.leya@ku.ac.bd

G. M. Tarekul Islam et al. (eds.), *Sustainable Water Management and Environmental Resilience*, Springer Water, https://doi.org/10.1007/978-3-031-93350-9_9

value is greater than 56. The scores for the components of WSI, basic household needs, food production, environmental flows, risk management, and independence were 46.87, 69.1, 33.33, 39.37, and 34.27, respectively. Approximately 43.24 km^2 river eroded, and 17.00 km^2 river accreted in last two decades. Multiple frequencies of homestead shifting occurred among respondents which had a negative impact on water security. Moreover, the study identified risk factors contributing to water insecurity, such as inadequate disaster preparedness and erosion control measures. Based on these insights, a sustainable framework for addressing water insecurity in erosion-prone areas is proposed which provides actionable recommendations for enhancing resilience and sustainability in water management practices.

Keywords Water security index · Padma river · River erosion · Bankline community · Water insecurity

9.1 Introduction

"It's not about the water coming after us, but about us interfering with its path. Therefore it is not about fighting the river, but about adapting to it." -Salma Begum, 57 years old mother, living in Harirampur, Manikganj. Bangladesh, a predominantly riverine country, faces significant challenges from the highly dynamic and unstable banks of the Padma River, resulting in frequent and severe riverbank erosion [11]. Due to its placement in the Ganges–Brahmaputra-Meghna river system's alluvial floodplain and dynamic river patterns [9], there are several examples of frequent shifting of the bankline [5]. Around 20 of 64 districts of the country are vulnerable to riverbank erosion, which consumes around 8,700 hectares of land each year and affects over 200,000 people [3, 4]. Thus, riverbank erosion has become an increasingly serious problem in Bangladesh, with an annual erosion rate of approximately 1,200 km [18]. Riverbank erosion causes immediate losses of homes, assets, and income, and has long-term impacts on health, infrastructure, and education [13]. Severe erosion caused the river to encroach 12 km into the floodplain from 1963 to 2009, consuming 35,625.20 acres of land and displacing around 60,000 people, with half of this encroachment occurring in the last 15 years [26].

This chronic and recurring natural hazard threatens the bankline settlements along the left bank of the Padma in Harirampur Upazila of Manikganj District [9] and poses threat to overall water security in the region. The theory of water security has emerged from addressing merely water scarcity to overall human well-being [22]. Water security assessment considers political, hydrological, and social scales, ranging from municipal to country levels, basin boundaries, and distinctions between urban–rural, community, and household contexts [23]. River erosion and water insecurity are closely related, as erosion not only displaces communities and destroys infrastructure but also disrupts access to safe drinking water and sanitation services. This leads to increased vulnerability to waterborne diseases and long-term challenges in maintaining reliable water resources. While water security typically encompasses aspects such as water quality, quantity, reliability, equitable access, and environmental sustainability of water resources [17], it often overlooks the specific implications of natural disasters on water security. To address this limitation, it is necessary to redefine water security in the context of riverbank erosion.

Previous assessments of riverbank erosion predominantly focused on the geographical aspects [11, 15, 29]. Only few researches explored the effects of erosion on livelihood vulnerability [9, 18, 21], health [18], and food security [10]. Scholars such as Das et al. [13], Rana et al. [27], Das et al. [14], and Baishya [5] looked into some cases of riverbank erosion and their impacts. Billah et al. [10] and Rana et al. [27] tried to identify if people migrate permanently from one union to another union to ensure food and water security.

There is a dearth of studies examining the impacts of riverbank erosion on communities directly affected by this hazard. However, a substantial research gap exists regarding the understanding of water security impacts. Therefore, this study aims to fill this gap by assessing water insecurity resulting from river erosion among the communities residing along the banklines in Harirampur Upazila of Manikganj District. Current research is unique in the sense that it is the first to investigate the water security issues that evolve in erosion-prone areas in Bangladesh. The purpose of this article is to investigate the recent condition of the Padma River, as well as the long-term impact of riverbank erosion on the water security of the residents of the riverine areas of Harirampur. Apart from that, this article attempts to offer some insights for reducing the impact on residents and improving their living conditions. The objectives of the study are to assess the multidimensional aspects of water security in erosion-affected communities of Harirampur Upazila, to identify risk factors associated with water insecurity, and to develop a sustainable framework for addressing water insecurity in erosion-affected communities.

9.2 Methodology

9.2.1 Study Area

To assess river erosion-induced water insecurity, three different unions (the lowest tier of local administrative unit), namely Ramkhrishnapur, Kanchanpur, and Chala were selected from Harirampur Upazila of Manikganj District (Fig. 9.1). The selection was based on a detailed reconnaissance survey and the severity of river erosion as highlighted in existing literature and secondary data. These unions were chosen due to their geographical proximity to the Padma River, which significantly influences local water-related challenges. The Padma River is a major watercourse in the region, which was identified as a significant factor contributing to river erosion in Harirampur Upazila [11]. The river was eroded on both of its banks with the most extensive erosion occurring on the left bank in the Manikganj district, encompassing an area of 8,600 ha. Ramkrishnapur experienced significant erosion constituting 55% of the union and 3.59% of the total Upazila erosion whereas Kanchanpur faced more severe erosion with an eroded area of 85% of the union and 11.01% of the total Upazila erosion. In contrast, Chala did not experience any riverbank erosion [26]. Ramkrishnapur and Kanchanpur are directly situated along Padma's banklines, while Chala, although slightly further from the bank, was included due to the notable migration of bankline communities to this area. This strategic selection was made for a comparative analysis between communities directly impacted by river erosion and those that have been relocated.

Harirampur Upazila has an area of 244.30 km^2 with a population of 155,469 [8]. Agriculture serves as the predominant livelihood source, contributing 54.54% to the overall economic activities. In terms of water sources, tube-wells provide 96.7% of the drinking water, while tap water provides only 0.2%. Although 28% of shallow tube-wells were found to contain arsenic (Banglapedia, 2022). 57.8% of households use sanitary latrines, 39.2% use non-sanitary facilities, and 3.0% lack latrine access entirely.

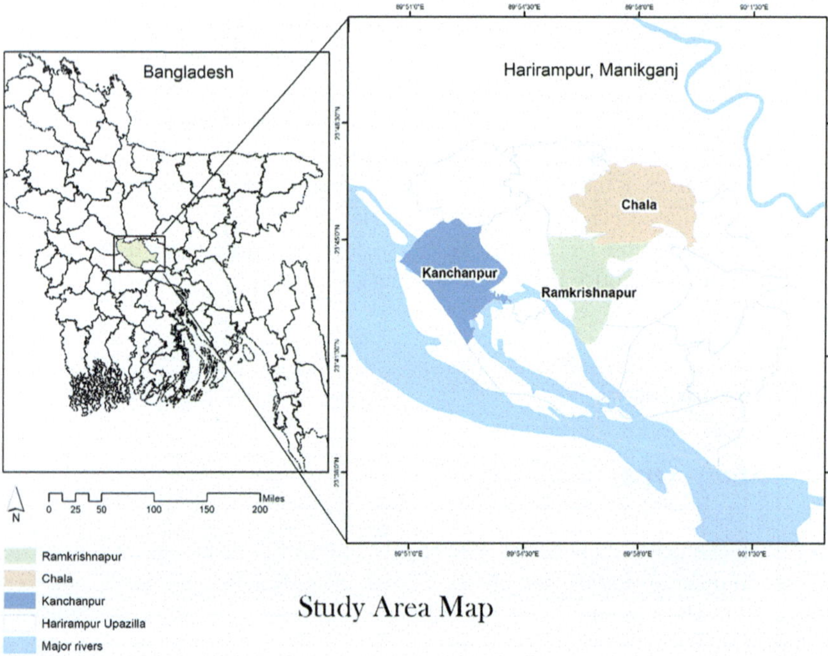

Fig. 9.1 Location of the study area: Manikganj District in Bangladesh (left) and selected unions in Harirampur Upazila (right)

9.2.2 Data

This study, conducted from March to August 2023, adopted a mixed-method approach, incorporating both primary and secondary data through remote sensing and field surveys. Primary data collection involved a structured questionnaire survey, prepared and pre-tested during reconnaissance visits, designed to comprehensively address water insecurity due to river erosion. Additionally, 3 Focus Group Discussions (FGDs) with bankline communities in 3 unions and 6 Key Informant Interviews (KIIs) with Upazila Nirbahi Officer, Upazila Agriculture Officer, Social Welfare Officer, Assistant Engineer of Local Government Engineering Department, local school teacher, and Executive Engineer of Bangladesh Water Development Board were conducted to understand the severity of riverbank erosion and its associated risk factors in the study area.

A purposive random sampling technique was used, with a sample size of 50 participants, including farmers, fishermen, and daily wagers. The target group was communities living along the bankline of Padma River. Concerning the breadth of the database, data samples need to be sufficiently extensive to yield statistically significant results [12]. Traditional research norms typically recommend a sample size exceeding 200 individuals to ensure a better representation of the overall population. Nonetheless, in behavioral studies, a minimum sample size of 50 participants per group is often adhered to (Barrett & Paul, 1981). Considering the limitations imposed by time and resources, and acknowledging the unique attributes of bankline communities with their inherent response variability, this study deliberately selected a sample size of 50 as a pragmatic and balanced approach.

Secondary data such as Landsat satellite images from 2002 and 2022, obtained via the United States Geological Survey EarthExplorer tool (https://earthexplorer.usgs.gov/, accessed on 01 March 2023) were analyzed to determine the Normalized Difference Water Index (NDWI), land use changes, and the extent of river erosion.

9.2.3 Materials and Methods

9.2.3.1 Identification of Indicators and Risk Factors of WSI

Water security was evaluated across five domains: basic needs, agricultural production, environment, risk management, and independence [19] (Table 9.1). Data for various indicators within each domain were collected through a household questionnaire survey and then normalized on a scale of 0 to 1 by dividing the actual value by its best value.

Table 9.1 WSI indicators and scoring

Components	Indicators	Scale
Basic household needs	Access to safe drinking water	0–100, where the numerical value represents the percentage of time during which the household has reliable access to safe drinking water
	Sanitation facilities	0–100, where no latrine: 0, open/katcha latrine: 10, semi-pucca latrine: 67, Pucca latrine: 100
	Electricity access[a]	0–100, where no access to a reliable electricity supply: 0, low likelihood of having a reliable electricity supply (less than 25% chance): score between 1 and 25, moderate likelihood (between 26 and 75% chance): score between 26 and 75 and high likelihood of having reliable electricity supply (more than 75% chance): score between 76 and 100
	Housing quality[a]	0–3, where traditional housing with high vulnerability to erosion is 0, improved housing with moderate vulnerability to erosion is 1, and resilient housing with low vulnerability to erosion is 2
Food production	Water availability for agricultural production	Rank 1–5, where low availability = 1 and high availability = 5
	Water use (withdrawal/person)	Rank 1–5, where low withdrawal = 1 and high withdrawal = 5
	Agricultural land[a]	Ratio scale, amount of arable land per household in decimal
	Crop diversity[a]	Ratio scale, number of different crops grown per household
	Average food-insufficient month[a]	0–12 (number of food-insufficient months in last year)
	Livestock ownership[a]	Ratio scale, number and type of livestock per household
Environmental flows	River ecosystem health	0–100, where poor: 25, fair: 50, good: 75, and excellent: 100 (rate water quality of the river)
	Wetland conservation	0–100, where 0–25: poor wetland conservation, 26–50: partial wetland conservation, 51–75: adequate wetland conservation, and 76–100: excellent wetland conservation
	Riparian buffer zones[a]	0–100, where 0–25: inadequate riparian buffer zones, 26–50: partially established riparian buffer zones, 51–75: well-established riparian buffer zones, and 76–100: extensive and well-maintained riparian buffer zones

(continued)

Table 9.1 (continued)

Components	Indicators	Scale
	Sustainable fishing practices[a]	0–100, where 0–25: unsustainable fishing practices, 26–50: partially sustainable fishing practices, 51–75: generally sustainable fishing practices, and 76–100: highly sustainable fishing practices
Risk management	Water storage capacity	0–100, where 33% (low water storage capacity), 66% (moderate water storage capacity) and 100% (high water storage capacity)
	Community-based disaster management (erosion control measures)	0–100, where 0% (limited erosion control measures), 33% (limited erosion control measures), 66% (some erosion control measures) and 100% (effective erosion control measures)
	Disaster preparedness[a]	0–100, where 0% (no shelters), 33% (low availability of shelters), 66% (moderate availability of shelters), and 100% (high availability of shelters)
	Nearest health center (minutes)[a]	Ratio scale, time in minutes
Independence	Diversification of income sources	0–100, where 33% (high dependence on agriculture), 66% (moderate dependence on agriculture), and 100% (other income sources)
	Social safety nets	0–100, where 33% (limited existence and coverage of government-led programs), 66% (partial existence and coverage of government-led programs), and 100% (reasonable existence and coverage of government-led programs)
	Support from neighbors during erosion	0–100, where 33% (low level of support received), 66% (some support received), and 100% (full support received)
	Access to credit and financial services[a]	0–100, where 33% (limited access to formal financial institutions), 66% (partial access to formal financial institutions), and 100% (reasonable access to formal financial institutions)
	Communication devices[a]	0–100, where 0 (no communication devices), 50 (mobile), and 100 (mobile, television, radio etc.)

[a] Risk factors (not incorporated in calculating index)

9.2.3.2 Calculation of Weights for the Indicators Through Pairwise Matrix Ranking

The data were normalized and weighted through the pairwise matrix ranking method, applying a scale of relative importance from 1 (equally important) to 9 (extremely more important), and their reciprocals [1, 28] to calculate the WSI. The matrix presented in Table 9.2 provides a visual representation of the pairwise ranking of indicators. In this context, each cell signifies the relative importance of one indicator in comparison to another, with values assigned according to Saaty's Analytic Hierarchy Process (AHP) scale. Relative importance ratings were informed by expert opinions gathered through KIIs. The subsequent steps involved calculating the sum and average for each row, followed by determining the total average by dividing the sum by the total number of indicators. The final normalization process was completed by dividing the average by the maximum average [25].

9.2.3.3 Construction of the Index

Simple additive weighted average method was adopted to calculate the final index value [16]. First, all the component values were calculated using the additive method, where each indicator value was multiplied by its respective weight, and the products were summed together. Subsequently, the same additive method was employed to derive the final WSI by combining the weighted values of individual components.

9.2.3.4 Changes in Water Bodies Over Time

Changes in water bodies were determined using the NDWI from Landsat satellite images, and land use land cover (LULC) changes were analyzed through supervised classification for agriculture, bare land, settlement, vegetation, and water bodies in 2002 and 2022. NDWI is derived from the Near-Infrared (NIR) and Green (G) channels. This formula highlights the amount of water in water bodies (Eq. 9.1). Additionally, the LULC raster image was converted to a polygon shapefile using the 'raster to polygon' tool. Then with the help of the 'erase' and 'intersect' tools, in the overlay layer in ArcGIS, the erosion-accretion and unchanged area were determined throughout the study area. The overall methodological diagram is presented in Fig. 9.2.

$$NDWI = \frac{G - NIR}{G + NIR} \tag{9.1}$$

Table 9.2 Pairwise ranking matrix for the indicators and components

Indicator	Pairwise ranking	Value/total	Total	Average	Total (average/max)	Normalized weight
A: Basic household needs						
A1: Access to safe drinking water	A1: 1, A2: 2	0.67, 0.67	1.33	0.67	1.00	0.67
A2: Sanitation facilities	A1: 1/2, A2: 1	0.33, 0.33	0.67	0.33	0.50	0.33
B. Food production						
B1. Water availability for agricultural production	B1: 1, B2: 3	0.75, 0.75	1.50	0.75	1.00	0.75
B2. Water use	B1: 1/3, B2: 1	0.25, 0.25	0.50	0.25	0.33	0.25
C. Environmental flows						
C1. River ecosystem health	C1: 1, C2: 2	0.67, 0.67	1.33	0.67	1.00	0.67
C2. Wetland conservation	C1: 1/2, C2: 1	0.33, 0.33	0.67	0.33	0.50	0.33
D. Risk management						
D1. Water storage capacity	D1: 1, D2: 5	0.83, 0.83	1.67	0.83	1.00	0.83
D2. Community-based disaster management	D1: 1/5, D2: 1	0.17, 0.17	0.33	0.17	0.20	0.17
E. Independence						
E1. Diversification of income sources	E1: 1, E2: 2, E3: 1/5	0.15, 0.6, 0.05	0.80	0.20	0.72	0.27
E2. Social safety nets	E1: 1/2, E2: 1, E3: 3	0.08, 0.3, 0.71	1.09	0.27	0.99	0.36
E3. Support from neighbors	E1: 5, E2: 1/3, E3: 1	0.77, 0.1, 0.24	1.11	0.28	1.00	0.37
WSI components						
A: Basic household needs	A: 1, B: 2, C: 3, D: 1/2, E: 4	0.24, 0.27, 0.27, 0.11, 0.62	1.52	0.30	1.00	0.31

(continued)

Table 9.2 (continued)

Indicator	Pairwise ranking	Value/ total	Total	Average	Total (average/ max)	Normalized weight
B. Food production	A: 1/2, B: 1, C: 3, D: 1/2, E: 1/2	0.12, 0.14, 0.27, 0.11, 0.08	0.72	0.14	0.47	0.14
C. Environmental flows	A: 1/3, B: 1/3, C: 1, D: 1/2, E: 1/2	0.08, 0.04, 0.09, 0.11, 0.08	0.41	0.08	0.27	0.08
D. Risk management	A: 2, B: 2, C: 2, D: 1, E: 1/2	0.48, 0.27, 0.18, 0.22, 0.08	1.24	0.25	0.82	0.25
E. Independence	A: 1/4, B: 2, C: 2, D: 2, E: 1	0.06, 0.27, 0.18, 0.44, 0.15	1.11	0.22	0.73	0.22

Note All values in the table are dimensionless, as they represent weightage derived from the pairwise ranking process and have no units

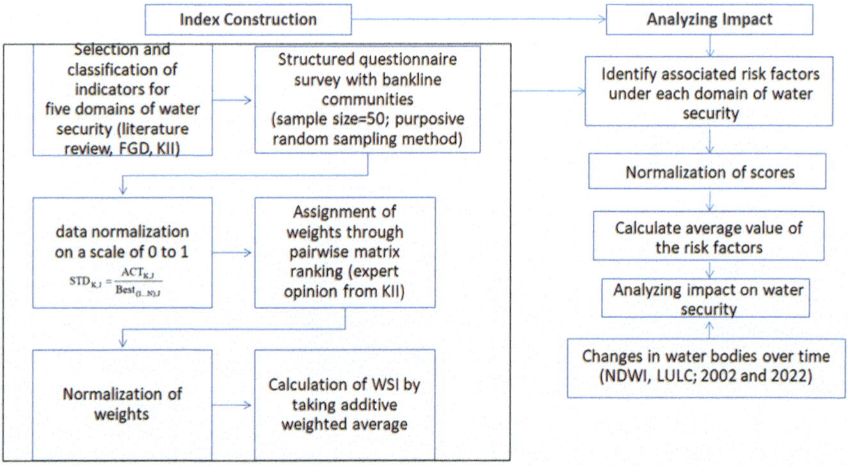

Fig. 9.2 Methodology of this study

9.3 Findings

9.3.1 *Bank Erosion*

The erosion, accretion, and unchanged area determined from Landsat satellite images for 2002 and 2022 showed that the river segment in the study area witnessed a notable change from 48.90 km^2 in 2002 to 75.08 km^2 in 2022. In these two decades, 43.24 km^2 of land was eroded by the river and 17.00 km^2 was accreted which means the portion of the river in the study area lost 26.24 km^2 land area. Additionally, 31.84 km^2 river area remained unchanged (Fig. 9.3). Field visits also revealed extensive bank erosion in the area. These on-the-ground observations confirmed the substantial and visible impact of river dynamics on the banks, with signs of erosion (Fig. 9.4).

9.3.2 *Changes in Land Use and Water Bodies*

Findings from both the NDWI and the supervised classification of LULC indicated a notable expansion of water bodies in the study area over the last two decades due to unplanned and illegal excavation of soil for brick kilns and other purposes (Fig. 9.5). Agricultural land, a crucial component of local livelihoods, experienced a moderate increase from 39.52 km^2 in 2002 to 43.54 km^2 in 2022 (Table 9.3). Conversely, barren land substantially reduced from 50.79 to 19.73 km^2, indicating potential transformations in areas vulnerable to erosion or abandonment. Barren land is associated with areas that are prone to erosion, devoid of vegetation cover, and susceptible to environmental changes. The most notable shift was observed in the built-up area, which has surged from 23.39 km^2 in 2002 to 76.18 km^2 in 2022. It suggested that human activities, driven by urbanization and infrastructure development, potentially brought settlements into closer proximity to river banks and, consequently, influenced their vulnerability to river-related processes. Vegetation, vital for ecological balance, encountered a significant decrease from 91.32 to 45.34 km^2. Waterbody areas increased from 58.22 to 78.56 km^2, which was attributed to the dynamic nature of river courses and the expansion of water bodies due to erosion-induced changes.

9.3.3 *River Erosion-Induced WSI*

The research findings indicated a WSI value of 44.25, falling within the range of water insecurity. The scores for the components of WSI, basic household needs, food production, environmental flows, risk management, and independence were 46.87, 69.1, 33.33, 39.37, and 34.27, respectively (Fig. 9.6). From the KIIs, the classification of WSI values was as follows: chronic water insecure (WSI < 35), water insecure

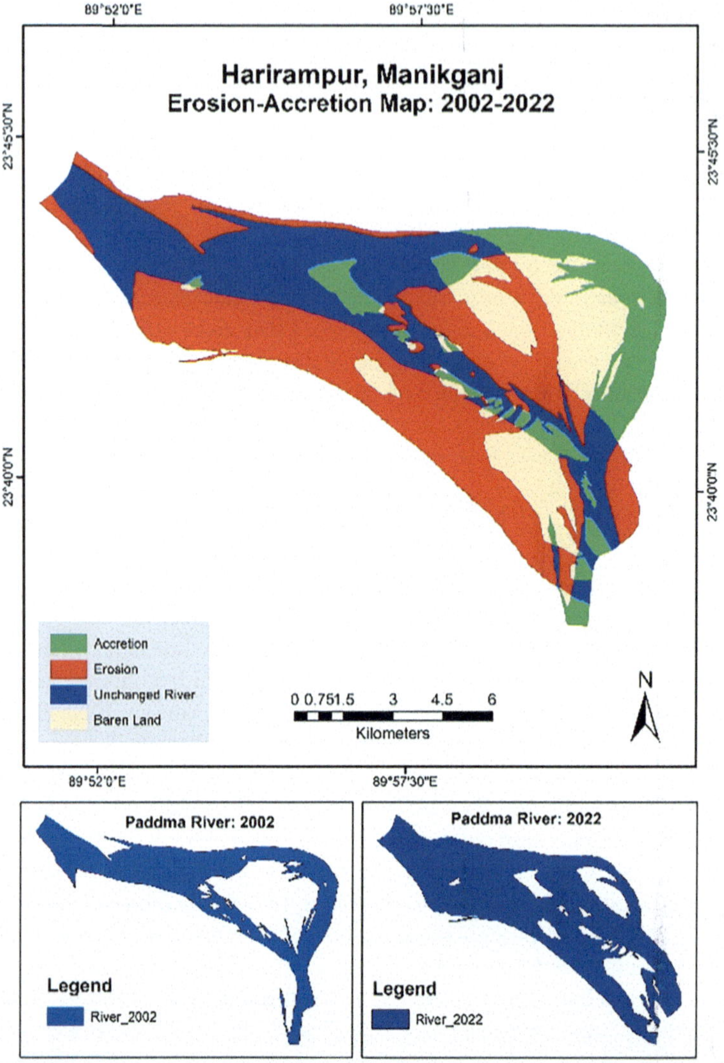

Fig. 9.3 Erosion, accretion, and unchanged area in Harirampur Upazila (2002–2022)

Fig. 9.4 River erosion in the study area (pictures taken on 20 March 2023)

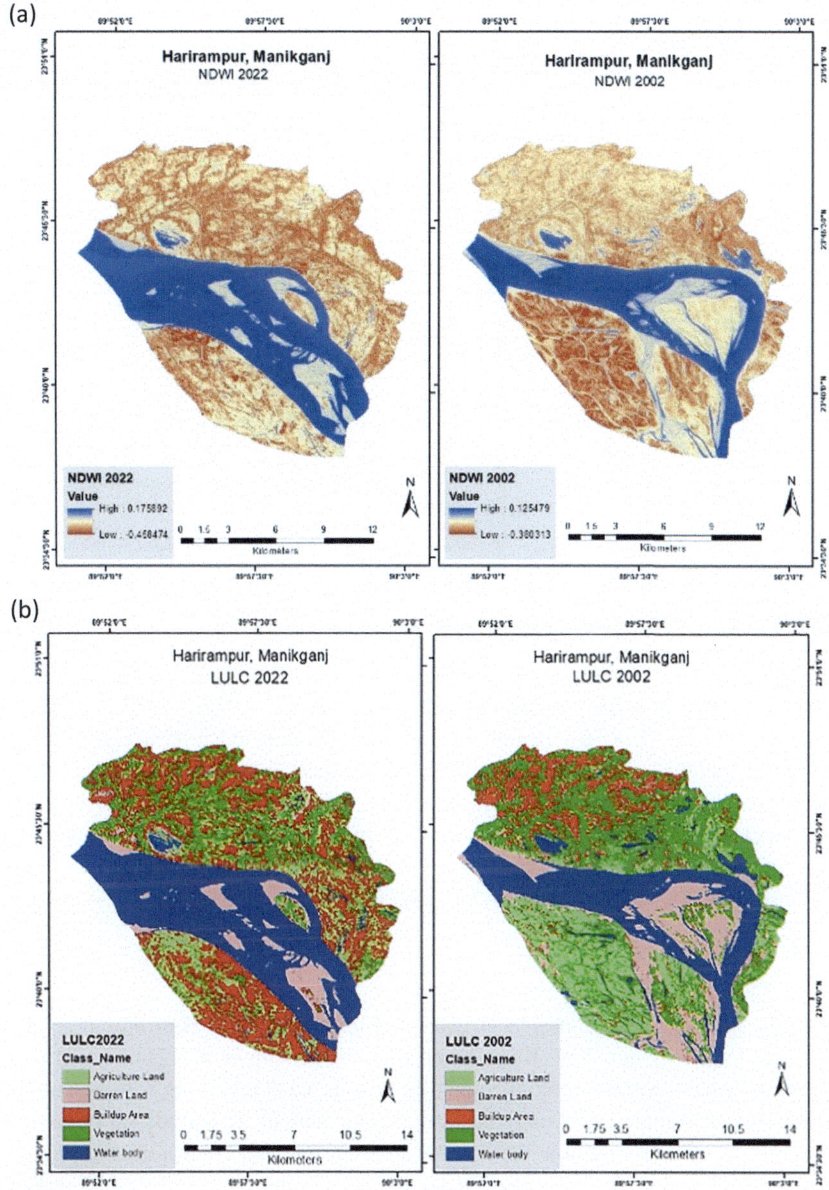

Fig. 9.5 Changes in **a** water bodies and **b** land use

Table 9.3 LULC classification (2002–2022)

Class name	Area-2002 (km^2)	Area-2022 (km^2)
Agricultural land	39.52	43.54
Barren land	50.79	19.73
Built-up area	23.39	76.18
Vegetation	91.32	45.34
Waterbody	58.22	78.56

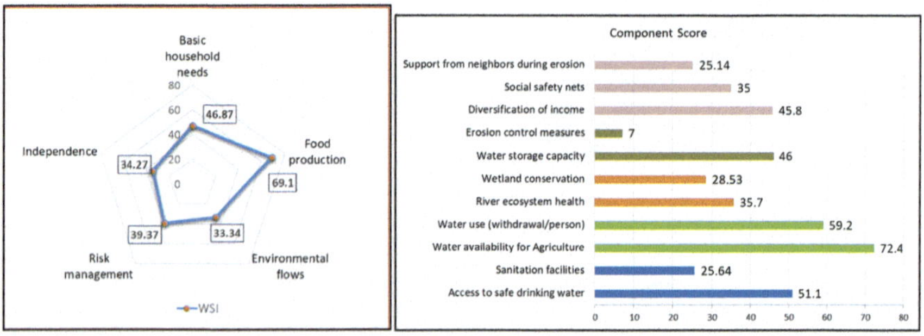

Fig. 9.6 Values of WSI components and component scores

or unsafe (35 < WSI < 48), seasonally water insecure (48 < WSI < 56), and water secured or safe (WSI > 56). Based on this classification, basic household needs fell into the category of seasonally water insecure class, food production fell under the water secured class, environmental flows and independence fell into the chronic water insecure class, and risk management fell under the water insecure category.

9.4 Discussion

This study indicated that massive erosion continued to affect Harirampur in the subsequent years, as noted by Rahman et al. [25]. Mishra et al. [21] argued that water security has several interconnected dimensions, each involving complex interactions between human society and the natural environment. This study supports this perspective, demonstrating how various socio-economic and environmental factors converge to influence water security in vulnerable communities. The AHP served as a valuable tool in this assessment, providing a structured method to prioritize attributes of water security based on expert opinion. According to [4], AHP collects information through paired comparison surveys using standardized decision matrices to identify priorities and obtain relative weights. This method's strength lies in its ability to capture expert insights on the importance of different attributes, reflecting a participatory approach that integrates experts and user opinions [25]. Additionally, the participatory approach involving expert opinion and community input is proved crucial for developing robust water security assessments. This approach not only aided in constructing a comprehensive WSI but also ensured that the index reflected the real-world complexities and stakeholder perspectives. Moreover, the fragmented distribution of data across various organizations and the lack of a universally applicable framework for measuring water security pose significant challenges [25]. The framework presented in this study addressed the need for a comprehensive approach to measure bank erosion-induced water insecurity. As Gain et al. [16] pointed out, country-level data aggregation fails to portray critical spatial and temporal variations, so this study utilized spatially detailed data for effective planning and policy-making.

This study considered access to safe drinking water and sanitation, denoted by the percentage of the population with access to these services. These are the crucial indicators for assessing water security [16] as the available freshwater is not always accessible due to various socio-economic and physical constraints such as water quality, lack of infrastructure, etc. The average values of access to safe drinking water and sanitation facilities are 51.1 and 25.64 respectively (Fig. 9.6). Only 14% households reported that they have access to safe water round the year. The findings on the water consumption patterns of the respondents showed that in Ramkrishnapur, households with an average of 4 members utilized 8 L of water daily for drinking, while Kanchanpur and Chala union residents used 7 L and 12 L, respectively. Data on water purification practices revealed that 90% of households in Ramkrishnapur and Kanchanpur do not engage in any purification practices. Others relied on methods such as boiling or cloth filtration. In Chala, a slightly lower percentage, 55%, also did not undertake any water purification methods. As the incidence rate of diseases is a crucial variable to measure water security [23], alarmingly, 85% of respondents in all three areas reported that their children suffer from water-borne diseases, particularly diarrhea, typhoid, and dysentery, annually. The health issues exacerbated during years of erosion and displacements. The component, water for food production exhibited a relatively higher value compared to other components. The indicators under this component, namely water availability for agriculture and water use (withdrawal/person), have values of 72.4 and 59.2, respectively (Fig. 9.6). The relatively high water use efficiency, as indicated by the withdrawal per person underscored a responsible approach to water utilization. These findings collectively contributed to a positive outlook on the region's ability to maintain agricultural productivity and, consequently, food security without significant compromise. Within the environmental flows component, the scores for river ecosystem health and wetland conservation stood at 35.7 and 28.53, respectively. These scores provided insights into the state of the river ecosystem and wetland conservation efforts, suggesting areas that require attention for improvement. Respondents were asked to assess the water quality of the river near their households and report any observed changes in biodiversity over the past 5 years. Findings revealed that 90% of the respondents rated the condition of wetland areas as poor, indicating a concerning perception of the environmental quality in those specific regions.

Under the risk management component, the indicators for water storage capacity and community-based disaster management (erosion control measures) scored 46 and 7, respectively. These values provided an assessment of the community's ability to manage and mitigate risks associated with water storage and erosion control measures. The findings indicated that the bankline communities have implemented limited erosion control measures, and there is a lack of awareness regarding the substantial risk reduction potential associated with these measures. Only 8% respondents reported that they understand the importance of planting and maintaining vegetation, such as grass, shrubs, or trees to reduce erosion nearby. Also, 95% of the respondents reported that they were excluded from decision-making processes regarding recovery planning and erosion control methods. Mamun et al. [20] emphasized the role of traditional erosion control measures over engineering erosion control approaches to lessen the impact of erosion on bankline communities. Under the independence component, the indicators for diversification of income sources, social safety nets, and support from neighbors during erosion scored 45.8, 35, and 25.14, respectively. These values represented the community's level of independence from external shock [20]. Only 4% of respondents experienced the existence and coverage of government-led programs in post-erosion situations. The level of support from neighbors was higher in Chala union than in Ramkrishnapur and Kanchanpur unions. The socio-economic dimensions of water security are equally vital. Billah et al. [10] highlighted that erosion-affected communities often adopt coping strategies such as migrating to urban areas or relying on less preferred food items to manage food insecurity. Our findings were aligned with this, revealing that only a few erosion-affected households were covered by government safety net initiatives like meal-for-work programs. The survey findings indicated varying frequencies of homestead shifting among respondents affected by river erosion. Specifically, 24% reported four shifts, 10% reported three shifts, 12% reported five shifts, 16% reported two shifts, 18% reported a single shift, and 20% never experienced homestead shifting due to river erosion. In the aftermath of erosion, 70% of respondents became reliant on their neighbors' tube wells for drinking water. Notably, 10% continued consuming untreated river water, 13% opted for boiled river water, 5% dismantled their tube wells, 1% purchased new parts for tube wells, and 1% took loans to reinstall their own tube wells. Regarding sanitation practices in the post-erosion scenario, 60% of the respondents utilized neighbors' latrines, 20% relied on temporary hanging latrines, 15% dismantled their own latrines, and 5% secured loans for latrine reinstallation. These findings underscored the significant disruptions in access to water and sanitation and strategies adopted by the community in response to the challenges posed by river erosion in the study area.

Table 9.4 outlines the associated risk factors for low water security across various components of the WSI. In terms of basic needs, a lack of electricity access and suboptimal housing quality contributed to a risk factor score of 6 and 40, respectively. The food production component was influenced by factors such as agricultural land availability (score of 24.3), crop diversity (score of 32), average food-insufficient months (score of 20.6), and livestock ownership (score of 34). Environmental flows, crucial for water security, are influenced by riparian buffer zones (score of 29.3) and sustainable fishing practices (score of 32.1) [20]. Risk management, an essential aspect of water security, was reflected in the disaster preparedness score of 11. Independence, assessed through access to financial services (score of 31.78) and communication devices (score of 46), underscored critical aspects that contribute to the overall water security. These risk factors collectively provided insights into potential vulnerabilities and areas for targeted interventions to enhance water security.

The correlation analysis between the WSI and associated risk factors revealed significant relationships that provided valuable insights into the determinants of water security. The positive correlation between WSI and electricity access ($r = 0.289$) suggested that improved electricity access is associated with lower water stress. A particularly strong positive correlation between WSI and housing quality ($r = 0.660$) indicated that suboptimal housing conditions contribute significantly to higher water stress levels. In the case of food production, a strong positive correlation with agricultural land availability ($r = 0.718$) and crop diversity ($r = 0.561$) underscored the pivotal role of these factors in determining water stress. Livestock ownership also exhibited a positive correlation ($r = 0.305$), emphasizing its influence on water stress levels. Environmental flows, measured by riparian buffer zones, showed a significant positive correlation ($r = 0.597$), suggesting that the health of river ecosystems and wetland conservation play a crucial role in mitigating water stress. Meanwhile, sustainable fishing practices exhibited a positive but moderate correlation ($r = 0.278$). Disaster preparedness demonstrated a moderate positive correlation ($r = 0.318$), indicating that communities with better disaster preparedness experience lower water stress levels. Independence factors, including access to financial services and communication devices, displayed strong positive correlations ($r = 0.659$ and $r = 0.539$, respectively), suggesting that financial inclusion and communication infrastructure contribute significantly to lower water stress.

There are critical strategic areas that demand attention in the aftermath of erosion, comparing the present scenario to the post-erosion scenario. In the present scenario, only 14% of households have full access to safe drinking water, a figure that sharply declines to a mere 2% in the post-erosion scenario. The availability of open or

Table 9.4 Risk factors for low water security

	Associated risk factors	Average normalized score (0–100)
Basic needs	Electricity access	6
	Housing quality	40
Food production	Agricultural land availability	24.3
	Crop diversity	32
	Average food-insufficient month	20.6
	Livestock ownership	34
Environmental flows	Riparian buffer zones	29.3
	Sustainable fishing practices	32.1
Risk management	Disaster preparedness	11
Independence	Access to financial services	31.78
	Communication devices	46

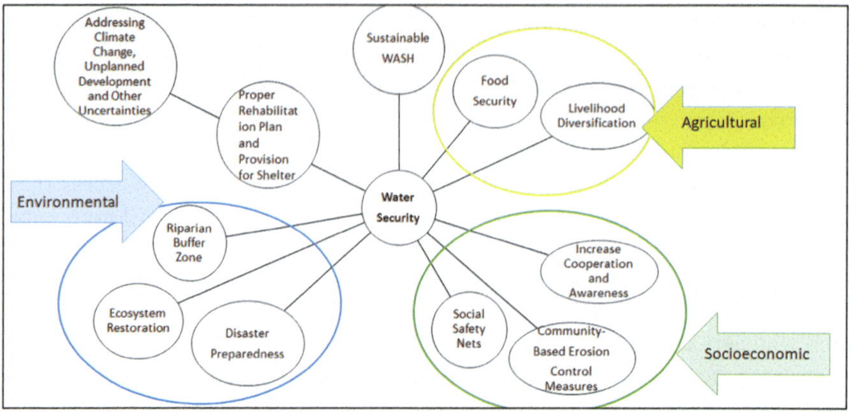

Fig. 9.7 Water security risk reduction framework

makeshift latrines also becomes a pressing concern, with 42% of households having such facilities currently, increasing to over 75% post-erosion. A noteworthy increase in the average number of food-insufficient months, from 4 to 5 months at present to 7–8 months post-erosion, signals heightened food insecurity. The river ecosystem health takes a considerable hit, deteriorating from fair to poor, with inadequately established riparian buffer zones. The availability of shelters and neighbors' support moderately increases post-erosion, but the water storage capacity diminishes, posing significant challenges. Additionally, erosion control measures drop significantly, with only 12% contributing in the present scenario compared to 0% in the post-erosion scenario. The existence of social safety nets remains limited in both scenarios, emphasizing the urgent need for strategic interventions to address these complex challenges and enhance the resilience of erosion-affected communities. Drawing on these findings, this study proposes a water security risk reduction framework in an erosion-prone area (Fig. 9.7).

9.5 Conclusion

The study reveals the significant influences of river ecosystem health, wetland conservation, water storage capacity, community-based erosion control measures, income diversification, social safety nets, and supportive neighbors in shaping water insecurity resulting from river erosion. Access to safe drinking water, sanitation, adequate agricultural water supply, and per capita water use play a significant role in ensuring better water security. This study presents a holistic approach that is vital in addressing these factors and tackling the challenges of water security in erosion-prone areas. However, one of the primary limitations of this study is the sample size. Although efforts were made to collect data from a diverse and representative sample of bankline households in Harirampur Upazila, the total number of respondents may not fully capture the variability within the entire population. The relatively small sample size could limit the generalizability of the findings to other regions or communities affected by similar issues. While the study employs robust methods, the sensitivity of these methods to changes in input data and weighting schemes must be considered in further studies. The findings highlight the interplay of various factors, signaling the need for comprehensive strategies and interventions to enhance water security in regions affected by river erosion. Policymakers, community leaders, and stakeholders need to consider these nuanced dynamics and implement context-specific measures to build resilience and ensure sustainable water management practices in the face of ongoing environmental changes.

Acknowledgements This study acknowledges the Interdisciplinary Field Research Methodology in Water Management course at IWFM, BUET for the field support.

References

1. Aboelnga HT, El-Naser H, Ribbe L, Frechen F-B (2020) Assessing water security in water-scarce cities: applying the integrated urban water security index (IUWSI) in Madaba, Jordan. Water 12(5):1299
2. Alam GMM (2017) Livelihood cycle and vulnerability of rural households to climate change and hazards in Bangladesh. Environ Manag 59(5):777–791
3. Ali MR, Ahmed Z, Islam AHMH, Rahman MM (2021) River bank erosion, induced population migration and adaptation strategies in the Sirajganj Sadar Upazila, Bangladesh. Eur J Environ Earth Sci 2(2):39–47
4. Arellanos E, Guzman W, García L (2022) How to prioritize the attributes of water ecosystem service for water security management: choice experiments versus analytic hierarchy process. Sustainability 14(23):15767
5. Baishya SJ (2013) A study on bank erosion by the River Baralia (Bhairatolajan) in Melkipara village of Hajo revenue circle, Kamrup district, Assam, India. Int J Sci Res Publ 13(9):1–10
6. Banglapedia (2024) Harirampur Upazila. Available at: https://en.banglapedia.org/index.php/Harirampur_Upazila. Accessed 8 Dec 2023
7. Barrett PT, Kline P (1981) The observation to variable ratio in factor analysis. Pers Study Group Behav 1:23–33
8. Bangladesh Bureau of Statistics (BBS) (2022) Statistical yearbook of Bangladesh Dhaka. Statistics Division, Ministry of planning, Government of the People's Republic of Bangladesh, Dhaka
9. Bhuiyan MAH, Islam SD-U, Azam G (2017) Exploring impacts and livelihood vulnerability of riverbank erosion hazard among rural household along the river Padma of Bangladesh. Environ Syst Res 6(25):1–15
10. Billah M, Majumdar A, Mohammad S, Rahman A, Hossain J (2023) Riverbank erosion and rural food security in Bangladesh. World 4(3):528–544
11. Billah MM (2018) Mapping and monitoring erosion-accretion in an alluvial river using satellite imagery—the river bank changes of the Padma river in Bangladesh. Quaestiones Geographicae 37(3):87–95
12. Booysen F (2002) An overview and evaluation of composite indices of development. Soc Indic Res 59:115–151

13. Das R, Samanta G (2023) Impact of floods and river-bank erosion on the riverine people in Manikchak Block of Malda District, West Bengal. Environ Dev Sustain 25(11):13595–13617
14. Das TK, Haldar SK, Das Gupta I, Sen S (2014) River bank erosion induced human displacement and its consequences. Living Rev Landsc Res 8(1):1–35
15. Dewan A, Corner R, Saleem A, Rahman MM, Haider MR, Rahman MM, Sarker MH (2017) Assessing channel changes of the Ganges-Padma River system in Bangladesh using Landsat and hydrological data. Geomorphology 276:257–279
16. Gain AK, Giupponi C, Wada Y (2016) Measuring global water security towards sustainable development goals. Environ Res Lett 11(12):124015
17. Gerlak AK, House-Peters L, Varady RG, Albrecht T, Zúñiga-Terán A, de Grenade RR, Cook C, Scott CA (2018) Water security: a review of place-based research. Environ Sci Policy 82:79–89
18. Kaiser ZA (2023) Analysis of the livelihood and health of internally displaced persons due to riverbank erosion in Bangladesh. J Migrat Health 7:100157
19. Lautze J, Manthrithilake H (2012) Water security: old concepts, new package, what value? Nat Resour Forum 36(2):76–87
20. Li P, Li D, Sun X, Chu Z, Xia T, Zheng B (2022) Application of ecological restoration technologies for the improvement of biodiversity and ecosystem in the river. Water 14(9):1402
21. Mamun AA, Islam ARMT, Alam E, Pal SC, Alam GMM (2022) Assessing riverbank erosion and livelihood resilience using traditional approaches in northern Bangladesh. Sustainability 14(4):2348
22. Mishra BK, Kumar P, Saraswat C, Chakraborty S, Gautam A (2021) Water security in a changing environment: concept, challenges and solutions. Water 13(4):490
23. Octavianti T, Staddon C (2021) A review of 80 assessment tools measuring water security. Wiley Interdiscip Rev Water 8(3):e1516
24. Paudel S, Kumar P, Dasgupta R, Johnson BA, Avtar R, Shaw R, Mishra BK, Kanbara S (2021) Nexus between water security framework and public health: a comprehensive scientific review. Water 13(10):1365
25. Rafaai NH, Lee KE (2024) Reconciling and contextualising multi-dimensional aspects for consolidated water security index: a synthesis. J Environ Manag 359:121067
26. Rahman MM, Islam MN, Islam MN (2016) Integrated approach of remote sensing and field survey data in assessment of bank erosion intensity of the Padma river in Bangladesh. Int J Geomat Geosci 7(2):285–297
27. Rana MS, Nessa AM (2017) Impact of riverbank erosion on population migration and resettlement of Bangladesh. Sci J Appl Math Stat 5(2):60–69
28. Saaty TL (2008) Decision making with the analytic hierarchy process. Int J Serv Sci 1(1):83–98
29. Thakur PK, Laha C, Aggarwal SP (2012) River bank erosion hazard study of river Ganga, upstream of Farakka barrage using remote sensing and GIS. Nat Hazards 61:967–987

Chapter 10
Assessing Spatial Thresholds of Indices-Based Water Mapping with Sentinel-2 for Ukhia and Teknaf of Bangladesh

Saifullah Sayed and Sara Nowreen

Abstract Efficient water planning crucially relies on monitoring surface water resources. In this context, threshold-based index methods are a less complex yet accurate approach to separate waterbodies from satellite imageries. Therefore, this study aims to identify efficient water mapping methods by comparing the performance of various water indices on Sentinel-2 scenes in the Eastern Hills (EH) of Bangladesh, using it as a case study. Here, the popular indices NDWI, MNDWI, NDMI, WRI, AWEInsh, AWEIsh, NWI, and SWI, are assessed. While finding the best panchromatic band, SWIR1 has been found to have a strong association ($R^2 = 0.512$) with the average of four bands: Red, Green, Blue, and NIR. Similarly, VNIR and SWIR2 are found to be strongly correlated with the Red Band with R^2 values of 0.68 and 0.43, respectively. High Pass Filter has the highest Universal Quality Index, lowest Spectral Angle Mapper, and lowest Root Mean Squared Error for pan-sharpening SWIR1, SWIR2, and VNIR bands. NDWI shows the highest Kappa (0.87) and F1-score (0.92). Following historical (2016–2023) analysis, Yen thresholding method shows an overall accuracy ranging from 82% in 2023 to 96% in 2021. Thus, the study suggests using Yen thresholding for NDWI in EH of Bangladesh.

Keywords Sentinel-2 · Water detection · Indices · Thresholding · Water mapping

10.1 Introduction

Since Rohingya influx in 2017, watershed degradation has been alarming in Cox's Bazar, particularly in Ukhiya, and Teknaf sub-districts [9]. Comprehensive basin-wide management planning is challenging due to lack of information on watershed and poor inventory of waterbodies. Accurate identification of changing waterbodies

S. Sayed · S. Nowreen (✉)
Institute of Water and Flood Management, Bangladesh University of Engineering and Technology, Dhaka, Bangladesh
e-mail: sara.sohel@gmail.com

G. M. Tarekul Islam et al. (eds.), *Sustainable Water Management and Environmental Resilience*, Springer Water, https://doi.org/10.1007/978-3-031-93350-9_10

or flow routes over time can yield useful information for resource management, flood monitoring, and other uses. Waterbody detection has been done either by ground measurements or by remote sensing. Traditional surveys can effectively monitor water but are expensive and time-consuming. In contrast, remote sensing imageries offer location accuracy, timely service, and abundant data with various spatio-temporal and spectral resolutions. While machine-learning approaches have shown better performances in extracting waterbodies from Sentinel-2 satellite imageries [23], there are limitations due to complexities of algorithms, lack of up-to-date referenced data, and necessities of professional expertise. Contrarily, index-based water detection (IWD) uses simple algorithms, is easy to deploy, and requires no prior high-level expertise. However, performance of different IWD methods varies depending on locations [5]. The primary function of IWD is to find optimal thresholds for waterbodies, which are difficult to determine. Examination of the performance of different IWDs is thus important, especially for Eastern Hill Tracts (EHT), as there are enormous unknown waterbodies which are hard-to-detect due to physical accessibility constraints. One problem of applying IWD for Sentinel-2 is inconsistent band resolutions [16]. Pan-sharpening is a commonly used technique to overcome this issue. However, performance of specific Pan-sharpening technique depends on bands and location characteristics [28], and hence the need for comparison among different techniques to determine an ideal IWD threshold for the study area. Also, most studies attempted watershed delineation using low-resolution DEMs (30–90 m) for hilly sites [1]. There are opportunities for using a 5 m Digital Terrain Model (DTM) in combination with 20 m Sentinel-2 bands to delineate waterbodies for EHT.

10.1.1 Study Area

EHT's Ukhia and Teknaf Upazilas have been selected as the study area. Based on data provided by the United Nations High Commissioner for Refugees (UNHCR), it is reported that a total of 967,842 Rohingya refugees now reside in Bangladesh, with 936,961 individuals specifically located inside the designated research region [22]. The inflow of Rohingya refugees has led to significant environmental consequences, including the damage of watersheds and a reduction in the recharging of groundwater.

With the exclusion of a few minor streams, the region under consideration has a restricted availability of surface water sources. The primary water sources, such as the Naf River and other major channels, are located at a considerable distance and exhibit salinity and brackishness, particularly in the downstream sections of these rivers [21]. Freshwater sources mostly consist of ponds and a limited number of minor streams that originate from nearby hills. The ponds and streams in the temporary camps are insufficient to supply the water demands of the people. However, they can be used for household purposes if they are maintained free from contamination caused by sewage. The availability of surface water is constrained, as the shallow water aquifer is experiencing depletion and may not be sufficiently recharged by precipitation. Moreover, the reliability of water supply from the deeper aquifer is unknown.

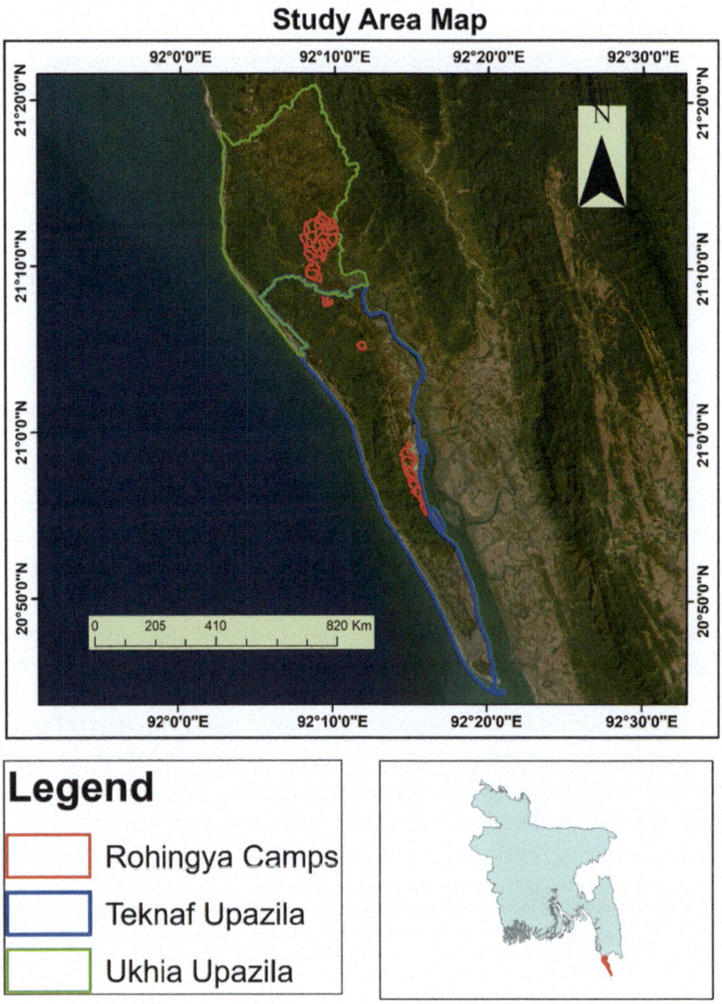

Fig. 10.1 Study area map

WBA quantification and characterization can assist government control degradation in a faster manner.

10.2 Methodology

Sentinel-2 satellite imageries with no cloud coverage during December–January for the years 2016–2023 were carefully selected to ensure the acquisition of high-quality data. The selection process involved leveraging the cloud cover percentage metadata provided by the Sentinel-2 dataset. Images with the least cloud cover were prioritized to ensure accurate and reliable information for subsequent analysis the images of Table 10.1 have been selected.

Sentinel-2 has thirteen bands, with Blue (B2), Green (B3), Red (B4), visible and Near Infrared (VNIR) (B5), Near Infrared (B8), short Wave Infrared 1 (B11), and short Wave Infrared 2 (B12) being employed in various water indices outlined in Table 10.1. However, a challenge arises due to differing spatial resolutions, with B2, B3, B4, and B8 at ten meters, while B5, B11, and B12 operate at twenty meters. To ensure uniform resolution, the study employs pan-sharpening methods to transform 20-m bands to 10 m, comparing five widely used methods IHS (Intensity, hue, and Saturation), Gram-Schmidt (GS), High Pass Filter (HPF), Wavelet Resolution Merge (WRM), and Sen2Res tool to identify the most effective approach for retaining image quality. Another problem of using pan-sharpening using sentinel-2 is the lack of panchromatic band which is solved by doing a linear regression.

Sentinel-2 imagery undergoes atmospheric correction using the Sentinel Application Platform (SNAP) software, specifically the Sen2Cor plugin. This crucial step eliminates atmospheric effects, ensuring accurate surface reflectance values in Level-2A products for subsequent analyses. The methodological framework (Fig. 10.2) illustrates the comprehensive process.

Post-resolution standardization, various indices-based methods are computed, addressing the thresholding challenge through a comparative analysis of seven global thresholding methods. Extensive field surveys validate the accuracy, with water and non-water maps cross-referenced against a 5m Digital Terrain Model (DTM)-derived stream. Furthermore, the optimal index and thresholding method are applied to higher-resolution Rapid Eye-3 and Planet Scope satellite imagery to assess potential enhancements in water detection. This meticulous approach enhances the study's reliability and precision, ensuring robust outcomes in surface water mapping.

Table 10.1 Date and cloud cover for selected Sentinel-2 images

Satellite	Date	Cloud cover (%)
Sentinel-2	05/01/2016	0
Sentinel-2	09/01/2017	0
Sentinel-2	04/01/2018	0
Sentinel-2	04/01/2019	0
Sentinel-2	09/01/2020	0
Sentinel-2	29/12/2021	0
Sentinel-2	14/12/2022	0
Sentinel-2	18/01/2023	0

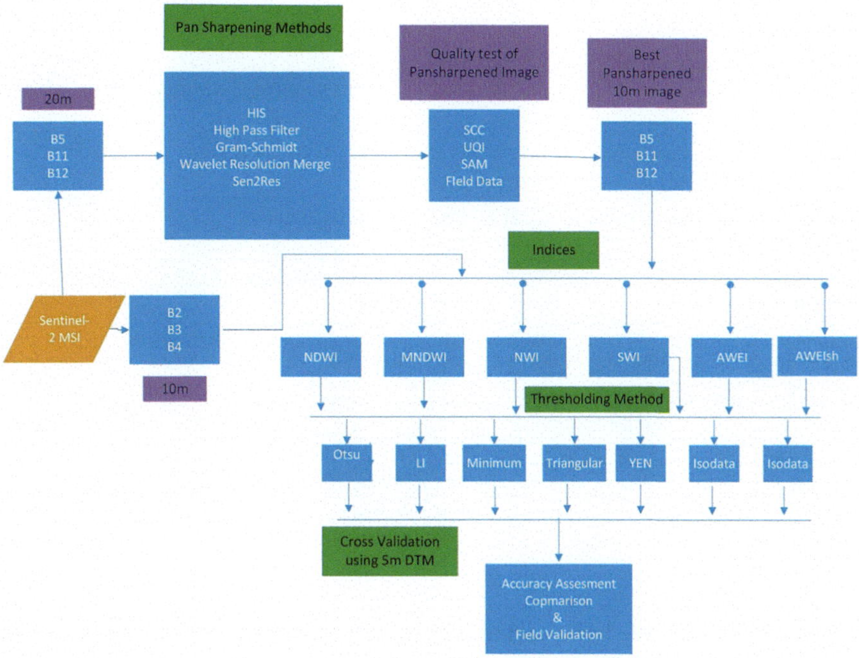

Fig. 10.2 Methodological framework of the study

10.2.1 Pan-Sharpening

As for the study pan sharpening of only 3 bands B5, B11, B12 is important 5 methods have been chosen to select the panchromatic band.

1. Producing the panchromatic band was by averaging all 10-m bands (Band 2–4 and Band 8)
2. Average of Band 4 and Band 8 (Red and NIR) as a panchromatic band for downscaling Band 5, and Band 8 as a panchromatic band for downscaling Band 11 and 12
3. Band 4 as a panchromatic band for Band 5, 11 and 12
4. Band 3 as a panchromatic band for Band 5, 11 and 12
5. Band 2 as a panchromatic band for Band 5, 11 and 12.

To find out the best fit panchromatic band the similarity between the 10 m and 20m bands have been evaluated with the help of linear regression. From the R^2 value most correlated band have been selected as a panchromatic band [6]. Before, doing the linear regression analysis Wald's protocol have been followed [17]. According to the protocol, a pan-sharpening method should have consistent property and two specific synthesis properties. By following the Wald's protocol before the linear regression 20 m bands are resampled into 40m and then again resampled into 20 m and the 10 m bands are resampled into 20 m and then the linear regression is performed.

Table 10.2 Water indices

Indices	Formula	References
Normalized difference water index (NDWI)	NDWI = (GREEN − NIR)/(GREEN + NIR)	[14]
Modified normalized difference water index (MNDWI)	MNDWI = (GREEN − SWIR2)/(GREEN + SWIR2)	[26]
Normalized difference moisture index (NDMI)	NDMI = (Red − NIR)/ (Red + NIR)	[27]
Automated water index (AWEInsh)	AWEInsh = 4 * (Green − SWIR2) − (0.25 * NIR + 2.75 * SWIR1)	[8]
Automated water index (AWEIsh)	AWEIsh = Blue + 2.5 * Green − 1.5 * (NIR + SWIR) − 0.25 * SWIR2	[8]
Water ratio index (WRI)	WRI = (Green + Red)/(NIR + SWIR2)	[19]
New water index (NWI)	$\mathrm{NWI} = \frac{\mathrm{Blue}-(\mathrm{NIR}+\mathrm{SWIR1}+\mathrm{SWIR2})}{\mathrm{Blue}+(\mathrm{NIR}+\mathrm{SWIR1}+\mathrm{SWIR2})}$	[7]
Sentinel water index (SWI)	$\mathrm{SWI} = (VRE1 - SWIR)/(VRE1 + SWIR)$	[11]
Land surface water index (LSWI)	LSWI = (NIR − SWIR1)/(NIR + SWIR1)	[25]

After finding out the best panchromatic band for each necessary band then the study moved forward to pan-sharpening. To perform the pan sharpening method ArcMap 10.8.2 Erdas Imagine and SNAP software have been used.

10.2.2 Indices

After finding out the best pan-sharpened image and all 10m high resolution band then the indices-based methods have been applied using the formulas of Table 10.2. To calculate all the indices ArcMap 10.8.2 and raster calculator have been used.

10.2.3 Thresholding

Global thresholding is a fundamental image segmentation method where a single threshold value is applied uniformly to the entire image. Pixels with intensity values above the threshold are classified as foreground, while those below are considered background. This method is straightforward and computationally efficient, making it suitable for images with well-defined intensity gaps between object and background. However, its effectiveness diminishes when images have uneven illumination, varying contrast, or complex structures, as it assumes a consistent threshold for the entire image. This thresholding method is widely used in water indices.

Yun Du have selected Otsu thresholding method to get the threshold of NDWI and MNDWI [6]. Feifei pan have also compared between H0 and Otsu methods and found out that Otsu performs better [15]. Aliihsan Sekertekin compared between 15 different thresholds and found out that minimum thresholding method was the best among the 15 thresholds [18]. Muhittin Karaman have compared between different thresholds Intermode, Isodata, IJ_Isodata, Minimum, Otsu, and Huang [12]. To find out the best thresholding method Otsu, Isodata, Minimum, Li, Triangular, Mean, and Yen these global thresholding methods have been compared in the study. To perform all the thresholding method skimage python library have been used.

10.2.4 Streamlines Generation from 5 m DTM

In this study, streamlines were generated from a high-resolution 5-m Digital Terrain Model (DTM) using ArcGIS. Initially, the DTM underwent preprocessing steps, including sink filling to enhance its accuracy. From Spatial Analyst Tool, Hydrology and then Fill tool is selected to fill the DTM. Subsequently, the flow direction and flow accumulation were computed to determine the directional flow of water and identify potential stream channels, respectively. From the same toolbox the flow direction and flow accumulation tools have been selected to calculate the potential channel. The latter involved specifying a threshold for flow accumulation based on landscape characteristics. Converting the thresholded flow accumulation raster into vector polylines was achieved using the Raster to Polyline tool. The resulting streamlines were subjected to additional analyses such as stream ordering. Visualization of the stream network, along with the original DTM, facilitated the refinement of the threshold and validation of results against existing hydrography data or field observations. The final stream network, represented as a feature class, was exported for further analysis or integration into the broader geographic information system (GIS) framework. This comprehensive methodology adheres to standard GIS procedures, providing a basis for accurate delineation and analysis of stream networks in diverse landscapes.

10.2.5 Sampling and Accuracy Assessment

The Indices maps of the study area after the thresholding generated a binary image. So, accuracy assessment of the indices needed collection of only two sample classes water and non-water. A combination of Purposive sampling and Simple random sampling have been selected in this study. In Simple random sampling each pixel in the raster has an equal chance of being selected [4]. To select the random sample, create random points the Purposive sampling method is selected to collect the streams in the study area. As the study area is has some very remote places selecting random sampling for the entire study area was not possible. So, some of the samples collected by going to the field, collecting GPS points, and calculating the width of the streams and other remote places have been randomly sampled and classified with the help of Google Earth Pro. In total 350 sample points have been collected and for accuracy assessment of each year a separate set of sample point have been taken into consideration. Among them 96 sample have been collected through field survey and the points of the remote places have been collected with the help of Google Earth Pro (GEP) (Fig. 10.3).

The classical confusion matrix approach is adopted for accuracy assessment of the indices-based water maps. Confusion matrices, the most often used technique of

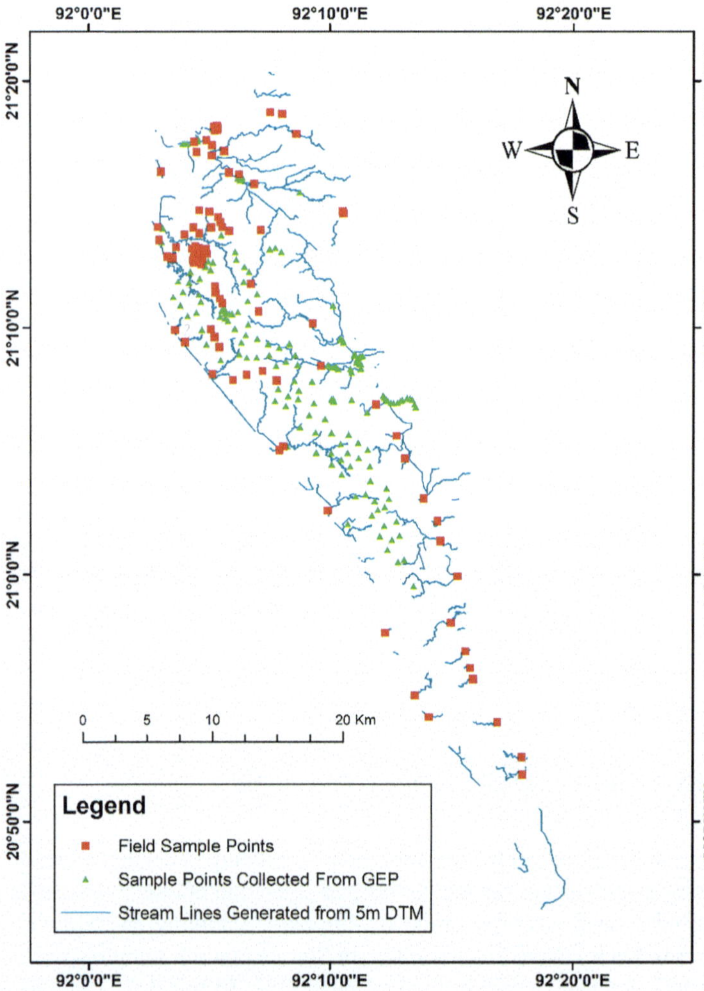

Fig. 10.3 Sample points for accuracy assessment

Table 10.3 The binary classification confusion matrix

	Water	Nonwater
Water	TP	FP
Nonwater	FN	TN

describing classification accuracy, are used to compare the reference data with the related classification outputs on a category-by-category basis [20].

The confusion matrix mentioned in Table 10.3. It is calculated for every indices-based method and the thresholding methods [13].

$$\text{OA} = \frac{TP + TN}{TP + FP + FN + TN} \tag{10.1}$$

$$\text{Sensitivity} = \frac{TP}{TP + FP} \tag{10.2}$$

The agreement Kappa coefficient is a multivariate statistical metric that can be used to verify classification accuracy. Sisay [20] based on the kappa values the most accurate water maps have been identified [2, 13].

$$\text{Specificity} = \frac{TN}{TN + FP} \tag{10.3}$$

$$\text{F1 Score} = 2 \times \frac{Precission \times Recall}{Precission + Recall} \tag{10.4}$$

$$\text{Kappa coefficient} = \frac{(TS \times TCS) - \sum(Column\,total \times Row\,total)}{TS^2 - \sum(Column\,total \times Row\,total)} \tag{10.5}$$

Here, TS = Total sample; TCS = Total corrected sample.

10.3 Results

10.3.1 Pan-Sharpening

To find the best pan-chromatic band for B11, B12 and B5 the correlation between these three bands and 6 different bands has been analysed. The highest correlation scenario is showed on Fig. 10.4. Best Pan-chromatic band for B11, B12 and B5.As average of 4 bands shows the highest R^2 value it has been chosen as a panchromatic band for B11. For both B12 and B5; B4 shows a high correlation in comparison to the other bands. After that, all the pan-sharpening algorithms have been run to calculate the pan sharpened images. To find out the best pan sharpened B11, B12 and B5 a qualitative and quantitative comparison have been done.

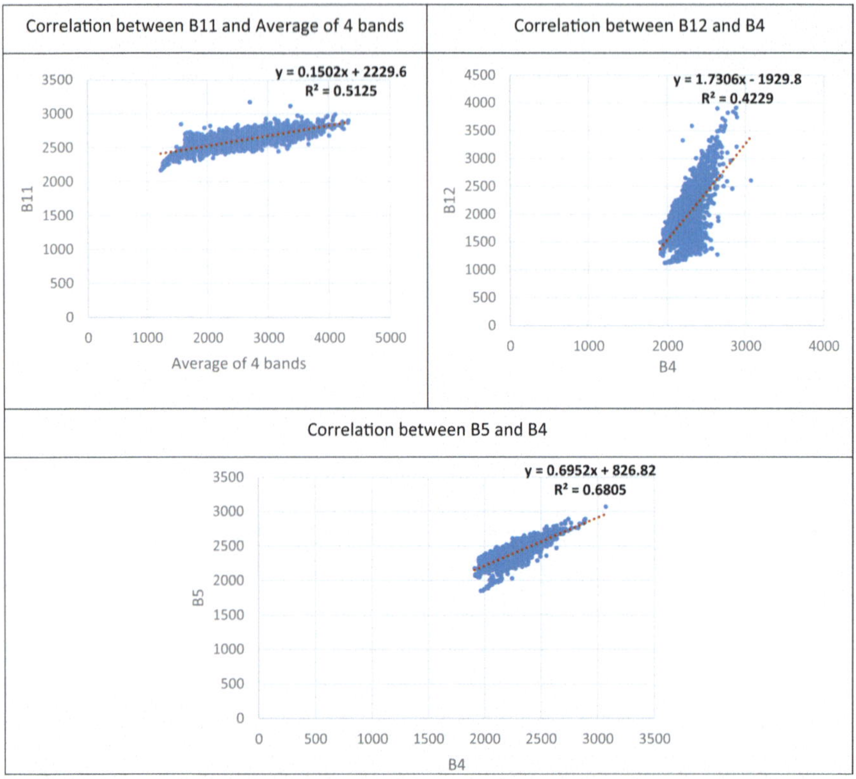

Fig. 10.4 Best pan-chromatic band for B11, B12 and B5

Table 10.4 represents the comparison between the pan sharpening methods and from the table it can be observed that HPF has the lowest SAM and RMSE value for all three bands. Though UQI value of B5; HPF is lower than GS, HPF has the highest UQI for B11 and B12. It represents that HPF generates the most detailed, high-quality pan sharpened image in comparison to other pan-sharpening methods.

From Fig. 10.5 the improved B11, B12 and B5 can be seen at the bottom right corner. On the top right corner, we can see the original B11, B12 and B5 which seems blurry, and the streams are almost unrecognizable but the pan sharpened B11 seems clearer, and the streams can be recognized more easily.

10.3.2 Accuracy of Indices Based Methods and Thresholding Methods

Figure 10.6 represents the best water maps of different Indices with the best thresholding method, with histograms on the bottom left corner and result of accuracy test

Table 10.4 Quality assessment of pan-sharpening algorithms for B11, B12 and B5

	SAM: spectral angle mapper	UQI: universal image quality index	RMSE: root mean squared error
B11			
Idel	0	1	Lower is better
IHS	0.03420	0.7079	29,167.8576
Brovey	0.10705	0.814839	6279.87389
GS	0.0275207	0.99429	1604.6464
HPF	0.02738	0.997189	1596.588
S2R	0.072258	0.840147	4225.5569
WRM	0.1475358	0.8055972	8691.26587
B12			
IHS	0.02794	0.98911	1629.038
Brovey	0.03082	0.863172	1797.217
GS	0.02857	0.99172	1665.6286
HPF	0.027414	0.9967	1598.2640
S2R	0.04652	0.86668	2714.9367
WRM	0.05280	0.84034	3082.8224
B5			
IHS	0.027668	0.99706	1613.2271
Brovey	0.03131	0.86457	1825.5309
GS	0.028353	0.997913	1652.957490
HPF	0.02170	0.8667	1265.2744
S2R	0.05844	0.84384	3413.8325
WRM	0.271578	0.79496	16,320.09286

above it. All the indices generate bell-shaped histogram. On the histograms horizontal axis represents the pixel values and the vertical axis represents count of the pixel. The thresholding methods are used to distinguish between two classes. Each dashed red line represents the threshold determined by the corresponding algorithm. If the threshold to the original image were applied, pixels with values to the right of the threshold line would be classified as water, and pixels to the left would be classified as non-water except for LSWI. In case of LSWI, pixels with values to the left of the threshold line would be classified as water, and pixels to the right would be classified as non-water.

To get the best thresholding method a detailed comparison has been done between various thresholding methods, including Otsu, Yen, Li, Minimum, Triangle, Isodata, Mean, and Ideal. With the help of confusion matrix, a detailed analysis of Specificity, Precision, F1 Score, Overall Accuracy (OA) and Kappa coefficient values has been analysed. Notably, these matric provide insights into the accuracy, reliability, and robustness of the thresholding techniques employed for water body detection,

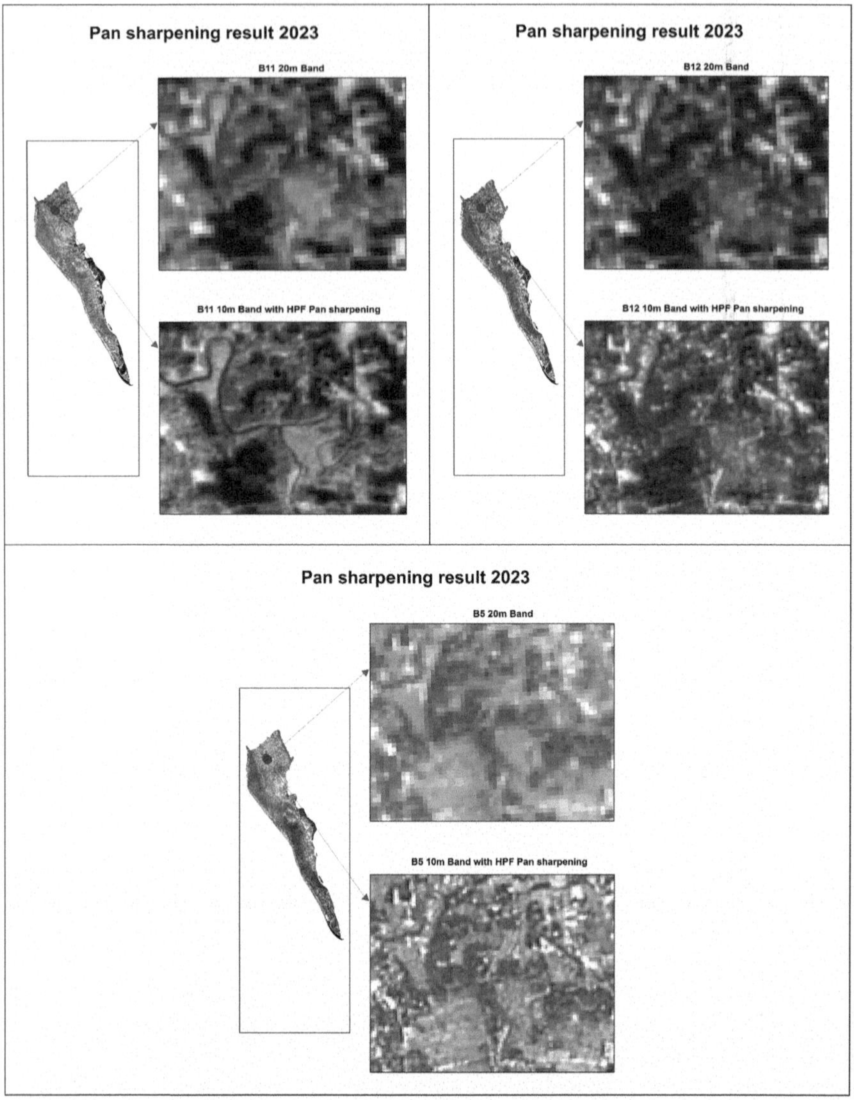

Fig. 10.5 Quality improvement of Sentinel-2 bands after going through pan-sharpening

with higher values indicating better performance. Figure 10.6 represents only the best thresholding methods that show the highest Kappa and F1 scores. The results highlight the varying effectiveness of different thresholding methods in delineating water features in the different indices-based method for the specified year.

From Fig. 10.6 it is apparent that for 2023, Triangle thresholding method performs the best with OA of 0.9130, Kappa value of 0.8134 and F1 score of 0.921. AWEInsh performs the worst OA of 0.44, Kappa value of 0.030 and F1 score of 0.60.

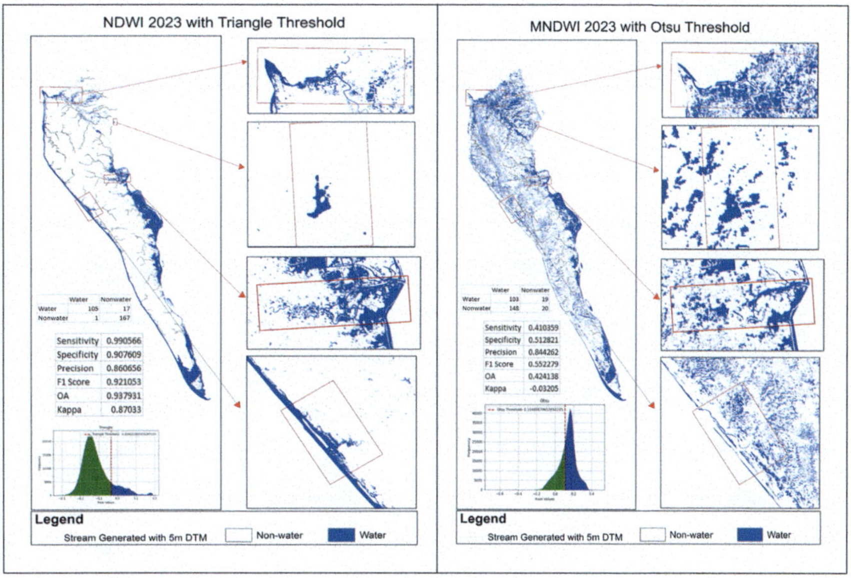

Fig. 10.6 Water maps of different indices with the best thresholding methods

10.3.3 Optimal Thresholding Method

For NDWI, a historical study of the thresholding methods Triangle, Isodata, and Yen was performed to determine the best thresholding approach. These three thresholding methods were chosen because they performed the best in the prior analysis (Fig. 10.7).

Considering the consistency of OA, Yen thresholding methods remain consistent for the images of all the years with never falling below 82% of overall accuracy. Isodata has the lowest OA among the three so this one can be ignored. Though the OA of Triangle thresholding method is lower than Yen, it also holds the OA around 80% and never falls below 78% so this method also can be used.

The selection of an appropriate thresholding method in the study is contingent upon specific objectives related to false positive values and the accurate identification of waterbodies. In instances where minimizing false positives takes precedence, the Yen thresholding method emerges as a viable option. Conversely, if the primary aim is to detect all waterbodies with a tolerance for a marginal occurrence of false positives, the Triangle thresholding method is recommended. This nuanced approach ensures alignment with the study's objectives, allowing for the tailored optimization of thresholding methods based on the desired balance between minimizing false positives and comprehensively identifying waterbodies.

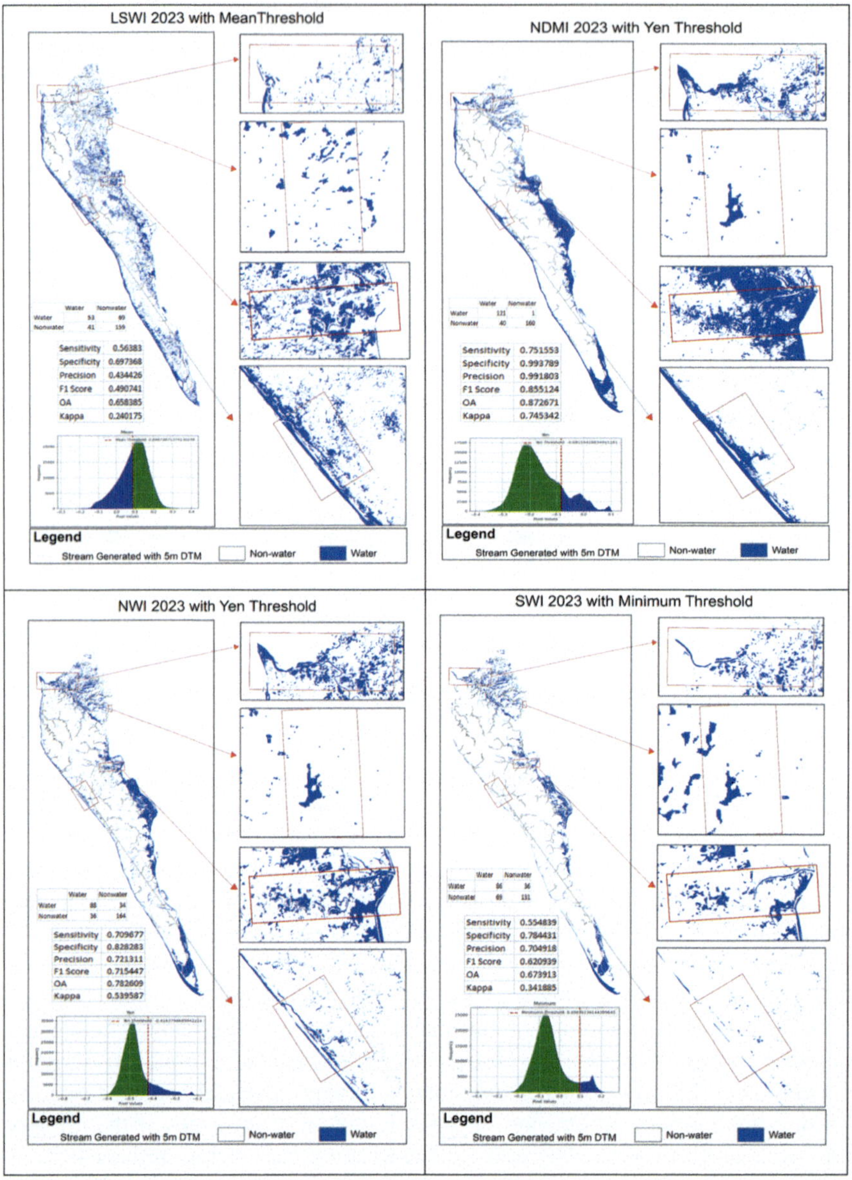

Fig. 10.6 (continued)

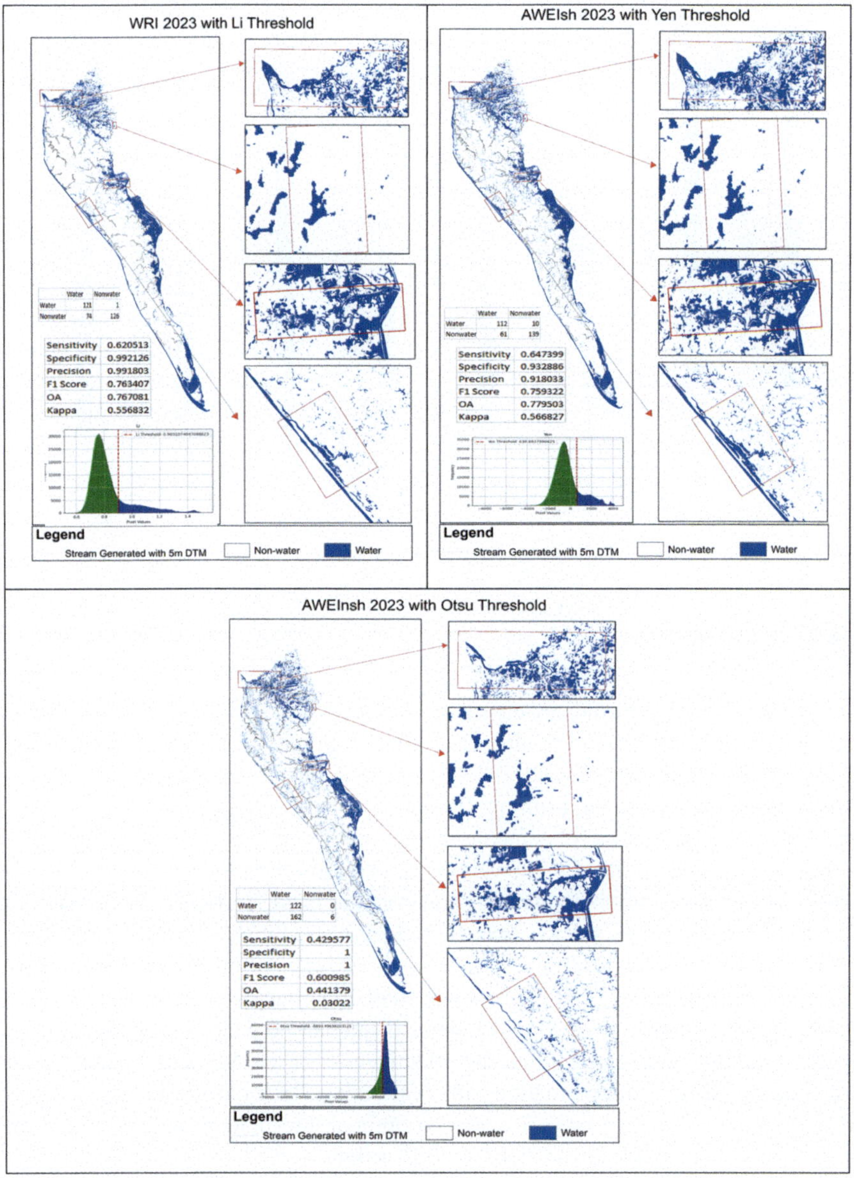

Fig. 10.6 (continued)

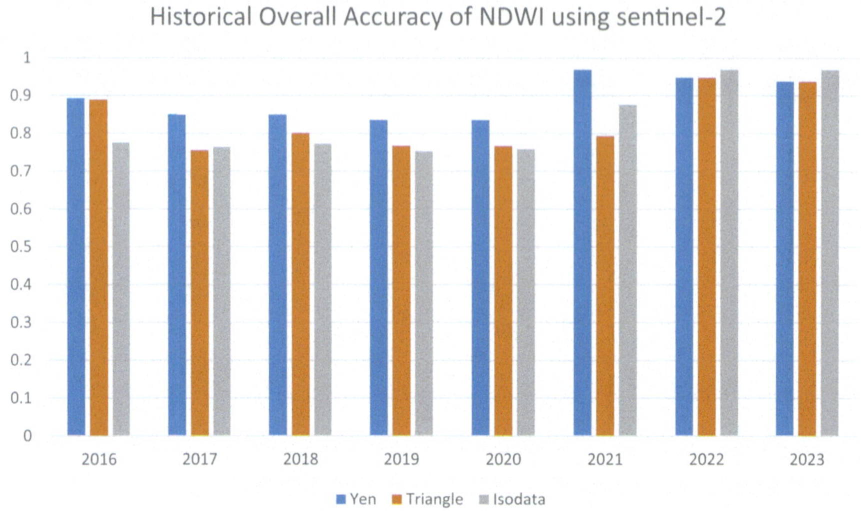

Fig. 10.7 Historical overall accuracy of NDWI using Sentinel 2

10.4 Conclusion

The primary function of indices-based water detection is to find optimal thresholds for waterbodies, which are difficult to determine from satellite imageries specifically where enormous unknown waterbodies are hard-to-detect due to physical accessibility constraints for ground truthing. But one problem with applying indices-based water mapping for Sentinel-2 is inconsistent band resolutions. Therefore, the study addressed challenges related to the varying spatial resolutions of Sentinel-2 imagery and the absence of a high-resolution panchromatic band. To tackle these issues, a simple linear regression analysis has been done to establish reliable relationships between different bands, helping to mitigate the absence of the Panchromatic band and it has been found out that the best pan sharpening method can help not only in water mapping but also in all kinds of analysis like land use land cover Mapping.

The findings have revealed that the effectiveness of the Normalized Difference Water Index (NDWI) in combination with Yen, Isodata, Triangle thresholding method works best for different years. To select an optimal thresholding method historical analysis using the images from 2016 to 2023 is done.

- Average of 4 bands (Red, Green, Blue and NIR) should be used as a Panchromatic band for SWIR1 and Red band as a Panchromatic band for SWIR2and VNIR.
- High Pass Filter (HPF) pan-sharpening method significantly improved image quality and clarity, contributing to better overall accuracy in identifying water bodies.

- The effectiveness of the Normalized Difference Water Index (NDWI) in combination with Yen, Isodata, Triangle thresholding method works best for years 2021, 2022 and 2023.
- Historical analysis of the images from 2016 to 2023 revealed that Isodata thresholding method generates too much false positive so that this method should be avoided.
- Yen threshold keeps a consistent Overall Accuracy (OA) over 82% reaching about 96% for 2021 so this thresholding method can be considered.
- Though the Triangle thresholding method generates some false positive results it also keeps an Overall Accuracy (OA) over 76% so this thresholding method can also be considered.

The implications of our research are significant and extend to resource management, environmental protection, and disaster preparedness in the Cox's Bazar area. These findings can serve as a valuable resource for similar regions worldwide facing the challenge of sustainable water resource management.

While the study represents progress in understanding surface water monitoring in complex landscapes, there are opportunities for future research. These may include automating the monitoring process, rigorously validating data, and fostering collaborations across disciplines and borders. By building on these insights, we can further enhance our approaches to surface water monitoring and contribute to the broader goal of securing water resources for a sustainable future.

References

1. Aher S, Kantamaneni K, Deshmukh P (2017) Detection and delineation of water bodies in hilly region using CartoDEM SRTM and ASTER GDEM data. Remote Sens Land 1(1):41–52
2. Arifeen HM, Chowdhury MS, Zhang H, Suepa T, Amin N, Techato K, Jutidamrongphan W (2021) Role of a mine in changing its surroundings—land use and land cover and impact on the natural environment in Barapukuria, Bangladesh. Sustainability 13(24)
3. Benedette Cuffari (2019) Can rainwater collection solve climate change? In: AZO Cleantech. https://www.azocleantech.com/article.aspx?ArticleID=979
4. Berndt AE (2020) Sampling methods. J Hum Lact 36(2):224–226
5. Bijeesh TV, Narasimhamurthy KN (2020) Surface water detection and delineation using remote sensing images: a review of methods and algorithms. Sustain Water Resour Manag 6(4):1–23
6. Du Y, Zhang Y, Ling F, Wang Q, Li W, Li X (2016) Water bodies' mapping from Sentinel-2 imagery with Modified Normalized Difference Water Index at 10-m spatial resolution produced by sharpening the SWIR band. Remote Sens 8(4)
7. Feng D (2009) Study on information extraction of water body with a new water index (NWI). Sci Surv Map
8. Feyisa GL, Meilby H, Fensholt R, Proud SR (2014) Automated Water Extraction Index: a new technique for surface water mapping using Landsat imagery. Remote Sens Environ 140:23–35
9. Hasan ME, Zhang L, Dewan A, Guo H, Mahmood R (2021) Spatiotemporal pattern of forest degradation and loss of ecosystem function associated with Rohingya influx: a geospatial approach. Land Degrad Dev 32(13):3666–3683
10. IUCN (2021) Groundwater Assessment Report of Ukhiya and Teknaf Upazila, vol 22. www.humanitarianresponse.info/files/documents/files/may_2021_ground_water_assessment_report_of_ukhiya_and_teknaf_upazilaiucn_fao_final_draft.pdf

11. Jiang W, Ni Y, Pang Z, Li X, Ju H, He G, Lv J, Yang K, Fu J, Qin X (2021) An effective water body extraction method with new water index for sentinel-2 imagery. Water (Switzerland) 13(12)
12. Karaman M (2021) Comparison of thresholding methods for shoreline extraction from Sentinel-2 and Landsat-8 imagery: Extreme Lake Salda, track of Mars on Earth. J Environ Manage 298:113481
13. Lee Y, Sim W, Park J, Lee J (2022) Evaluation of hyperparameter combinations of the U-Net model for land cover classification. Forests 13:1813
14. McFeeters SK (1996) The use of the Normalized Difference Water Index (NDWI) in the delineation of open water features. Int J Remote Sens 17(7):1425–1432
15. Pan F, Xi X, Wang C (2020) A comparative study of water indices and image classification algorithms for mapping inland surface water bodies using Landsat imagery. Remote Sens 12(10)
16. Park H, Choi J, Park N, Choi S (2017) Sharpening the VNIR and SWIR bands of Sentinel-2A imagery through modified selected and synthesized band schemes. Remote Sens 9(10)
17. Ranchin T, Wald L (2000) Fusion of high spatial and spectral resolution images: the ARSIS concept and its implementation. Photogramm Eng Remote Sens 66:49–61
18. Sekertekin A (2019) Potential of global thresholding methods for the identification of surface water resources using Sentinel-2 satellite imagery and normalized difference water index. J Appl Remote Sens 13(4):44507
19. Shen L, Li C (2010) Water body extraction from Landsat ETM+ imagery using adaboost algorithm. In: 2010 18th international conference on geoinformatics, geoinformatics 2010
20. Sisay A (2016) Remote sensing based water surface extraction and change detection in the Central Rift Valley region of Ethiopia. American J Geogr Inf Syst 5:33–39
21. UNDP (2018) Report on environmental impact of Rohingya Influx. https://www.undp.org/bangladesh/publications/report-environmental-impact-rohingya-influx
22. UNHCR (2023) Operational Data Portal Bangladesh. https://data.unhcr.org/en/country/bgd
23. Wieland M, Martinis S (2020) Large-scale surface water change observed by Sentinel-2 during the 2018 drought in Germany. Int J Remote Sens 41(12):4740–4754
24. Williams AT, Rangel-Buitrago N, Pranzini E, Anfuso G (2018) The management of coastal erosion. In: Ocean and coastal management, vol 156. Elsevier Ltd, pp 4–20
25. Xiao X, Boles S, Frolking S, Salas W, Moore III, Li C, He L, Zhao R (2002) Landscape-scale characterization of cropland in China using Vegetation and Landsat TM images. Int J Remote Sens 23(18):3579–3594
26. Xu H (2006) Modification of normalised difference water index (NDWI) to enhance open water features in remotely sensed imagery. Int J Remote Sens 27(14):3025–3033
27. Zhang K, Thapa B, Ross M, Gann D (2016) Remote sensing of seasonal changes and disturbances in mangrove forest: a case study from South Florida. Ecosphere 7(6):e01366
28. Zheng H, Du P, Chen J, Xia J, Li E, Xu Z, Li X, Yokoya N (2017) Performance evaluation of downscaling sentinel-2 imagery for Land Use and Land Cover classification by spectral-spatial features. Remote Sens 9(12):1274

Chapter 11
Assessing and Evaluating the Water Security: Water Quality, Accessibility, Availability, and Sanitation Practices Among the People of Sultanpur Village, Raozan, Chattogram

Sadia Salim and Sayed Mohammad Nazim Uddin

Abstract Groundwater is the only major source of potable water in Bangladesh. The concepts of quality, availability, accessibility, sanitation, and hygiene practices of drinking water all play a role in water security. This study evaluated the state of water security of Sultanpur village of Raozan Upazila, Chittagong through conducting a total of 210 household surveys, 4 Focus Group Discussions, 7 Key Informant Interviews, and lab tests to identify the issues at the grassroots level and contributing to the national development. The study's findings demonstrated that the water quality has deteriorated; people suffer from severe water scarcity, inaccessibility, sanitation, hygiene, and health issues. Other than pH, manganese (Mn), lead (Pb), and temperature, the parameters did not meet the drinking BD and WHO water standards. The survey and laboratory test results showed that excessive iron (highest = 21.08 mg/l) and low dissolved oxygen (2.3 mg/L) level posed health risks for the local population. This research emphasizes the urgent need for improved water resource management and alternative solutions to address the water security issues in Bangladesh.

Keywords Water security · Water quality · Availability · Accessibility · Health · Sanitation

S. Salim (✉)
Environmental Sciences and Public Health Programs, Asian University for Women, Chattogram 4000, Bangladesh
e-mail: sadia.salim@auw.edu.bd

S. M. N. Uddin
IMDEA Water Institute, 28805 Alcalá de Henares, Madrid, Spain
e-mail: sayed.uddin@auw.edu.bd

S. Salim · S. M. N. Uddin
Center for Climate Change and Environmental Health (3CEH) and Environmental Sciences Program, Asian University for Women, Chattogram 4000, Bangladesh

G. M. Tarekul Islam et al. (eds.), *Sustainable Water Management and Environmental Resilience*, Springer Water, https://doi.org/10.1007/978-3-031-93350-9_11

11.1 Introduction

Drinking water is one of the fundamental necessities of life and crucial for survival. Water security alludes to sustainable access to adequate and affordable water of suitable quality, enabling all individuals to live healthy and productive lives [7]. This is vital for healthy and dignified living, economic growth, social stability, and environmental sustainability. The Sustainable Development Goals specifically address water and sanitation issues, emphasizing the water management significance for overall development. Understanding the risks associated with water and society's capacity to ensure a consistent water supply is essential for sustaining aquatic and terrestrial ecosystems.

Several studies highlight the status of water insecurity worldwide. Based on some studies, approximately 780 million people worldwide, 8 million in Nigeria, 20% of Pakistan's total population, and 40% of Bangladesh's population stay without safe and clean water access [11]. Mexico City, Maharashtra, and Chennai are facing severe water scarcity, with nearly 65% of India's reservoirs empty by 2019 [10, 12]. During dry seasons, Bangladesh has experienced significant Fe contamination in groundwater [1]. Chittagong City has been suffering from inadequate water supply with poor quality [7]. Hence, poor water quality cause diarrhea, cancer, kidney, liver, and reproductive organ disorders and over-iron utilization may cause hemochromatosis, organ hurt, exhaustion, liver cirrhosis, joint torment, hepatocellular carcinomas, and hemosiderosis diseases [9, 11]. Besides, more than 2 billion people globally lack basic sanitation (Mekonnen and Hoekstra 2016). Because of water-borne and hygiene-related illnesses, Kenya represents a 70% morbidity and mortality rate whereas 40% of diseases and 60% of infant deaths are caused by diarrhea in Pakistan [4]. Globally, women who perform 64% of water collection tasks face the greatest risk of sexual assault, health issues, dropping out of school, and wasting time, all of which pose significant threats to their lives [2, 6].

While there are numerous studies on water-related issues in Bangladeshi communities, there is a lack of research on rural water security. This study focused on wards 4, 5, and 6 of Sultanpur village to understand the daily drinking water challenges and the villagers' methods of collection, transportation, and storage. The findings seek to improve the quality and accessibility of water sources and raise awareness about proper water handling and hygiene practices. The study aims to ensure water security for rural communities and assist planners, policymakers, and stakeholders in taking effective action.

11.2 Methodology

The conceptualization of water security in this study encompasses factors such as availability, accessibility, quality, sanitation, hygiene practices, and health risks which is an adaptation of the definitions suggested by Gain et al. [7]. 'Water security' refers to the conditions where sufficient water resources are easily obtainable in appropriate quality. Overall, it is a multifaceted context that depicts sustainable access to well-grounded and affordable quantities of water, of appropriate quality, to authorize all people to sustain healthy lives and ecosystems. This exploratory research consists of a mix of qualitative and quantitative surveys with FGD and KII, designed conveniently to fill each other's gaps, achieve the research objectives, and strengthen the data. Finally, the data and information were summarized and water tests were gathered for examination from different sources of wards no. 4, 5, and 6 to illustrate the water security situation of Sultanpur village. Due to limited time and budget, only a few water samples were taken, not covering all village water sources.

11.2.1 Study Area and Population

The study took place in wards 4, 5, and 6 of Sultanpur village in the Raozan Upazila, Chattogram. As per the Raozan Upazila statistics (2011), the total area of Raozan Upazila is 246.58 square kilometers. The research focused on the residents of Sultanpur village who rely on groundwater sources for drinking water, including women, children, adults, and the elderly.

11.2.2 Questionnaire Survey

There are published research articles on villages in Raozan [3] other than Sultanpur. To address the research gap in this area, Sultanpur village was chosen for this study based on piloting and observation. This is a non-probability convenience sampling, where all the participants were chosen based on their consent and convenience. A variety of socio-demographic questions were asked of the residents of the three wards of the Sultanpur village in the first section of the questionnaire, including their age, gender, educational level, and occupation. The information regarding their knowledge of water security is in the second section of the questionnaire: questions regarding the accessibility, availability, and drinking water sources. The third section details how they use water for drinking and household activities in a clean manner. The fourth part contains their well- being influences. Lastly, the fifth section shows their suggestions and expectations for the government to eliminate any risks associated with drinking direct groundwater. There were 29 questions in total in the survey for 210 individuals including 110 females (52.38%), and 100 males (47.61%). All data were collected through the questionnaire and face-to-face interviews.

11.2.3 Key Informant Interviews (KII)

Seven key informant interviews were conducted with individuals highly knowledgeable about water and sanitation issues in Sultanpur village. The interviews, involving government officials, local leaders, and other experts, focused on water quality, availability, accessibility, sanitation practices, and health issues. The discussions uncovered important details about the village's water sources, groundwater availability, sanitation, and past and present conditions.

11.2.4 Focus Group Discussion (FGD)

Four focus group discussions were conducted to identify the villagers' main concerns and brainstorm ideas and recommendations. FGD was conducted with water users (both surface and groundwater for drinking and sanitation) from various livelihood groups and age ranges, within a twenty-five-minute timeframe. Key topics included issues with tube wells and deep tube wells, access to safe drinking water, changes in groundwater availability over the past 10–20 years, water access conflicts, prevalent diseases, sanitation practices, and required facilities. The responses offered valuable insights into the community's attitudes, beliefs, perceptions, and ideas.

11.2.5 Water Quality Analysis

11.2.5.1 Water Sampling

The laboratory experiment collected six water samples from wards no. 4, 5, and 6, with two samples from each ward. The samples were taken from shallow tube wells and deep tube wells. 500-ml mineral bottles were used to collect the samples. House visits and observation were conducted to assess the water storage conditions.

11.2.5.2 Lab Test

The water quality was tested for various parameters by the Bangladesh Council of Scientific and Industrial Research (BCSIR) and Asian University for Women (AUW) labs. Heavy metals (Fe, Mn, and Pb) were tested at the BCSIR lab, while temperature, dissolved oxygen (DO), and pH were tested at the AUW lab. The pH was measured using the HI 8424 pH meter, DO with the Lutron YK 22dOA meter, and temperature with the Lutron YK 22dOA meter from the AUW Chemistry lab. Additionally, heavy metal testing was done with a Spectrophotometer at BCSIR. Due to suspected contamination, only one water sample was taken to BCSIR from a shallow tube well for E. coli testing.

11.3 Results and Discussion

11.3.1 Water Quality

The water quality of the area's sources was analyzed based on WHO and BD standards given in Table 11.1. Here, the pH levels were close to the acceptable range (6.5–8.5), with some samples slightly below the minimum limit. However, the dissolved oxygen (DO) levels in deep tube well samples were below the ideal range (6 mg/L), and iron concentrations exceeded safe limits for all wards, especially in Ward 4's deep tube well.

The presence of manganese, lead, and E. coli in the samples was within permissible levels. Approximately, 68.57% of people reported that their drinking water is yellow, and 4.76% said it is reddish, indicating a decline in water quality in terms of color.

Based on the lab test, survey, and FGD findings, the water quality in Sultanpur village has deteriorated, with excessive iron level (21.08 mg/L) and low dissolved oxygen level (2.3 mg/L) in most sources. The water is yellowish and reddish, with some sources having an unpleasant odor and smell. Deep tube well samples have the lowest DO level and highest iron level comparing to the shallow tube well samples, showing a correlation between the two. Although water temperatures were not excessively high, deep tube well samples had slightly higher temperatures, likely contributing to the lower DO levels in those sources (Figs. 11.1 and 11.2).

11.3.2 Water Availability

Water availability means water is available around the year and shortage means there is not sufficient water. The survey revealed that a significant majority of respondents in Sultanpur village face water shortages, with 68.57% reporting unavailability of water. The shortage occurs daily for 32.38% of respondents and monthly for 37.61%. Severe shortages are most prevalent in winter, affecting 55.23% of participants. Based on the FGD, KII, and observations, women, responsible for fetching water, face challenges due to groundwater levels falling in winter and summer. This impacts their education, safety, and quality time. The majority resort to neighboring sources when facing water deficits, some buy bottled water, and some are compelled to boil

Table 11.1 Laboratory test results (from BCSIR and AUW Lab)

Parameters	Standard	Range	Ward 4 deep tubewell	Ward 4 shallow tubewell	Ward 5 deep tubewell	Ward 5 shallow tubewell	Ward 6 deep tubewell	Ward 6 shallow tubewell
DO (mg/L)	BD Standard	6 mg/L	4.4	4.0	2.3	5.7	4.0	6.8
	WHO Standard	–						
Temperature (°C)	BD Standard	20–30 °C	24.1	23.9	24.5	24.0	25.1	24.7
	WHO Standard							
pH	BD Standard	–	6.39	6.41	6.36	6.38	6.31	6.61
	WHO Standard	6.5–8.5						
Iron (mg/L)	BD Standard	0.3–1.0 mg/L	21.08	2.5	7.83	2.87	9.29	5.0
	WHO Standard	0.3 mg/L						
Manganese (mg/L)	BD Standard	0.1 mg/L	0.21	0.31	0.28	0.20	0.20	0.25
	WHO Standard	–						
Lead	BD Standard	0.05 mg/L	BDL	BDL	BDL	BDL	BDL	BDL
	WHO Standard	0.01 mg/L						

(continued)

Table 11.1 (continued)

Parameters	Standard	Range	Ward 4 deep tubewell	Ward 4 shallow tubewell	Ward 5 deep tubewell	Ward 5 shallow tubewell	Ward 6 deep tubewell	Ward 6 shallow tubewell
E. coli	BD Standard	0 CFU (N/100 mL)	Not performed	Absent	Not performed	Not performed	Not performed	Not performed
	WHO Standard	–						

Note BDL = Below Detection Limit; The **BD Standard** and **WHO Standard** columns provide the acceptable ranges for each parameter according to the Bangladesh and WHO standards, respectively [14], *Water-quality-parameters* 2024)

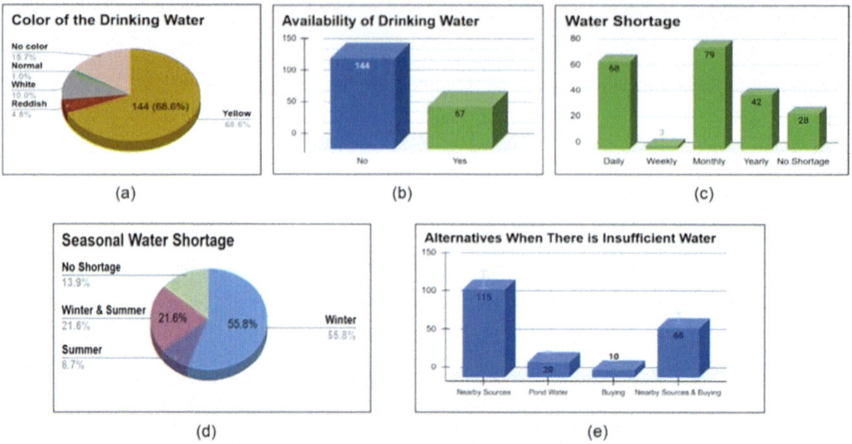

Fig. 11.1 **a** Color of the drinking water, **b** availability of the drinking water, **c** water shortage, **d** seasonal water shortage, **e** alternatives when there is insufficient water

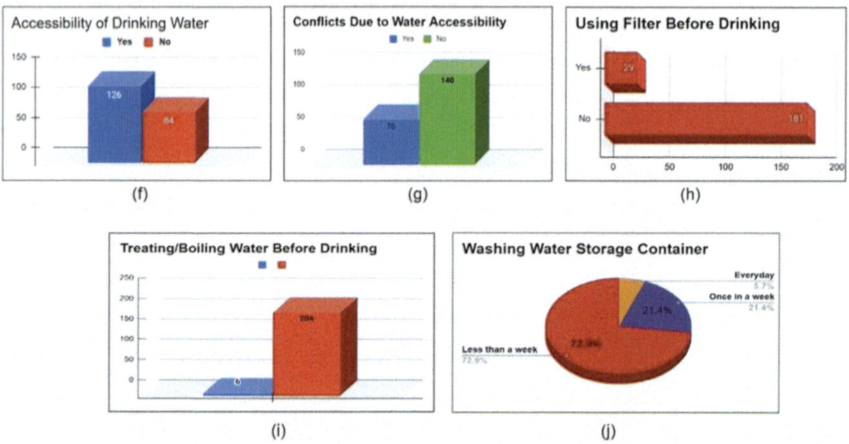

Fig. 11.2 **a** Accessibility of drinking water, **b** conflicts due to water accessibility, **c** using filter before drinking, **d** treating/boiling water before drinking, **e** washing water storage container

pond water, posing health risks. Overcrowding due to high population-to-tubewell ratio exacerbates the unavailability of water (Fig. 11.3).

11.3.3 Water Accessibility

Around 36.19% of people stated that they do not have access to water and they face difficulties obtaining it because of distance and time. Some of the participants

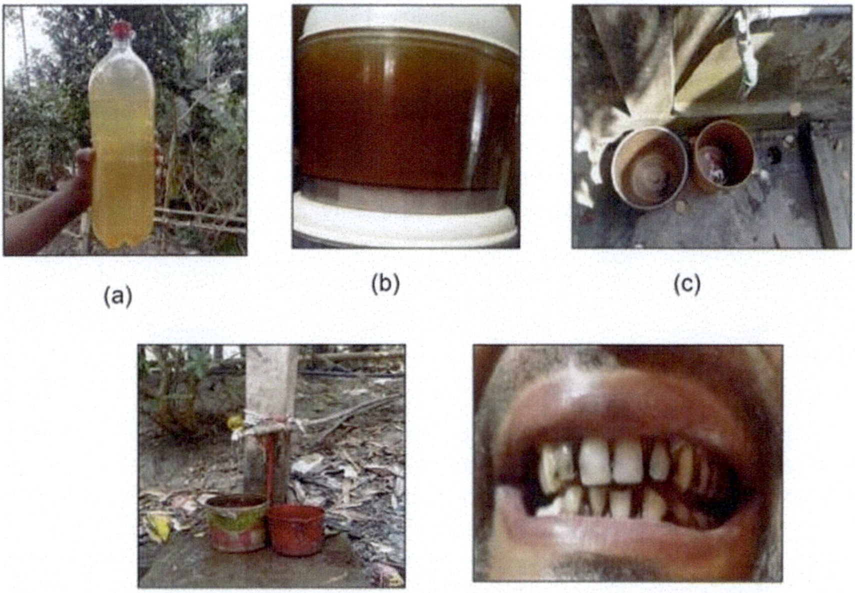

(a) (b) (c)

(d) (e)

Fig. 11.3 **a** Yellowish water, **b** yellow filtered water, **c** unavailable water during the summer season, **d** less water access, **e** black teeth

(33.34%) said that getting access to drinking water causes conflicts among them, most of which lead to arguments. Approximately 33.34% of them had conflicts with getting water access as Fig. 11.2b shows.

In addition, the FGD and KII findings revealed that the majority of respondents face daily water access challenges, mostly in the summer and winter. Many respondents of ward 6 have to walk 1–2 km daily which takes around 15–25 min to obtain fresh water, with a payment of 300 BDT for access to a pipeline connected to a deep tube well which is private property. The magistrate started water businesses and control the price and accessibility of water. However, the local people stated that they are not satisfied with the business model due to accessibility and reliability issues.

11.3.4 Sanitation and Hygiene Practices

The survey and other findings revealed that 80% of participants store their drinking water in plastic containers, with varying frequencies of cleaning and covering. About 72.85% people clean their containers once in a week, while 23.80% do not cover their water containers when transporting them. Most of the respondents (86.19%)

do not use any filter before consuming water (Fig. 11.2b). Additionally, a significant percentage do not have access to clean water for container cleaning. The study suggests that the majority of participants are unaware of health risks and fail to maintain proper hygiene and sanitation practices. Given the degraded water quality in the village, it is recommended that villagers treat the water or use filtration methods before consumption to avoid health risks. Studies show more bacteria in plastic containers, highlighting the importance of proper hygiene and sanitation to prevent contamination.

11.3.5 Health Risks

The study revealed that a majority of respondents (58.57%) attributed black teeth to high iron consumption in water, causing aesthetic concerns and health issues. One participant shared the negative impact of frequent teeth scaling on dental health. The degraded water quality and poor hygiene practices in Sultanpur village are posing health risks, as excessive iron intake may lead to serious conditions, for example, hemochromatosis, organ harm, weariness, liver cirrhosis, joint torment, hepatocellular carcinomas, and hemosiderosis. The study emphasizes the medical need to mitigate health risks in the community.

11.4 Conclusion

In conclusion, the study reveals water security and environmental health challenges in Sultanpur village. Locals struggle with inadequate water security, sanitation, and hygiene practices. By incorporating community perspectives and prioritizing grassroots-level challenges, policymakers can enhance the effectiveness of interventions aimed at improving water security and mitigating health risks associated with poor water quality. This approach is crucial for sustainable improvements in water quality, availability, and accessibility, aligning with community needs and ensuring equitable access to safe drinking water.

Acknowledgements The authors express their gratitude to Mr. Abu Bokor of BCSIR for his assistance with water sample laboratory analysis and to AUW, Bangladesh for funding support. Special thanks are extended to the participants of Sultanpur village for their active involvement in the survey, KII, and FGD interviews.

References

1. Ahmed I, Huq S, Mannan S, Mahmud MJ, Khan MR (2021) Ensuring water security for climate vulnerable communities in coastal Bangladesh. Towards a just climate change resilience: developing resilient, anticipatory and inclusive community response. Springer, Cham, pp 27–46
2. Akhter M, Uddin SMN, Rafa N, Hridi SM, Staddon C, Powell W (2020) Drinking water security challenges in Rohingya refugee camps of Cox's bazar, Bangladesh. Sustainability 12(18):7325
3. Ali MM, Chowdhury PA (2014) Assessment of water supply and sanitation in Dewanpur Village, Chittagong. Curr Urban Stud 2(4):380–384
4. Daud MK, Nafees M, Ali S, Rizwan M, Bajwa RA, Shakoor MB, Arshad MU, Chatha SAS, Deeba F, Murad W, Malook I, Zhu SJ (2017) Drinking water quality status and contamination in Pakistan. BioMed Res Int 2017(1):7908183
5. Department of Public Health Engineering. Government of the People's Republic of Bangladesh. Retrieved 25 Aug 2024, from: https://dphe.gov.bd/site/page/15fa0d7b-11f1-45c0-a684-10a543376873/Water-Quality
6. Eludoyin AO, Olanrewaju OE (2020) Water supply and quality in the Sub-Saharan Africa. Clean Water Sanitat 1–1
7. Gain AK, Giupponi C, Wada Y (2016) Measuring global water security towards sustainable development goals. Environ Res Lett 11(12):124015
8. Islam ARMT, Siddiqua MT, Zahid A, Tasnim SS, Rahman MM (2020) Drinking appraisal of coastal groundwater in Bangladesh: an approach of multi-hazards towards water security and health safety. Chemosphere 255:126933
9. Leya RS, Bala SK, Newton IH, Chowdhury MA, Haque SM (2022) Water security assessment of a peri-urban area: a study in Singair Upazila of Manikganj district of Bangladesh. Environ Dev Sustain 24(12):14106–14129
10. McCord H (2021) A sinking, thirsty city: the water crisis in Mexico City. Latin America Reports. Retrieved from: https://latinamericareports.com/a-sinking-thirsty-city-the-water-crisis-in-mexico-city/6075
11. Molla MH, Chowdhury MAT, Muhibbullah M, Ali KMB, Bhuiyan MHR, Das S, Morshed AJ, Das J, Islam S (2023) Suitability of drinking water quality in Chittagong Metropolitan City, Bangladesh: research on urban water bodies (UWBs) using multivariate analytic techniques. H2Open J 6(2):140–156
12. NASA Earth Observatory (2019) Water shortages in India
13. Water crisis in CHT (2018) The Daily Star, 26 Apr. https://www.thedailystar.net/editorial/water-crisis-cht-1567681
14. World Health Organization (2024) Drinking water quality guidelines. https://www.researchgate.net/figure/World-Health-Organization-WHO-Drinking-Water-Quality-Guidelines_tbl2_255700290

Chapter 12
Impact of Dasherkandi Sewage Treatment Plant on Balu River and Intake of Saidabad Water Treatment Plant

Rajib Ahmed, Sazia Afreen, and Md Mizanur Rahman

Abstract Megacity Dhaka spanning around 401 km^2 currently accommodates nearly 22 million people. Although the city has 5 peripheral rivers-Turag, Buriganga, Balu, Shitalakshya, and Dhaleswari; Dhaka WASA (DWASA), the authority providing water supply and sewerage facilities, is now forced to pump raw water from the Meghna and Padma rather than the aforementioned because of the deterioration of water quality which results in high operational costs. DWASA's Saidabad Water Treatment Plant (SWTP) (450 MLD) is pumping raw water from the Sarulia intake point of the Shitalakshya River. Meanwhile, Dasherkandi Sewage Treatment Plant (DSTP) (500 MLD) was established to reduce the pollution in the Balu River and eventually to improve the influent quality of SWTP. This study aims to assess the impact of DSTP on the Balu River and Sarulia intake. Water quality parameters such as DO, COD and NH_3–N were tested on samples collected from 11 potential sources. The study found that although the effluent quality of DSTP is within the Department of Environment (DoE) Standards, the influent quality of the Sarulia has not improved significantly due to untreated discharge from the industries along the Balu and Shitalakshya Rivers. The study suggests shifting SWTP intake to a nearby feasible upstream location.

Keywords Dasherkandi sewage treatment plant (DSTP) · Saidabad water treatment plant (SWTP) · Balu river · Shitalakshya river · Water quality

12.1 Introduction

Bangladesh's capital, Dhaka, has the third-highest population density in the world, placing a great deal of strain on utility services [20]. The per sq. km. population density in Dhaka city increased from 685 in 1974 to 2156 in 2022 [3]. Despite

R. Ahmed (✉) · S. Afreen · Md M. Rahman
Dhaka Water Supply and Sewerage Authority, Kawran Bazar, Dhaka, Bangladesh
e-mail: rajib.ahmed.dwasa@gmail.com

G. M. Tarekul Islam et al. (eds.), *Sustainable Water Management and Environmental Resilience*, Springer Water, https://doi.org/10.1007/978-3-031-93350-9_12

having different challenges Bangladesh is adamant to meet the targets of Sustainable Development Goals by 2030. To achieve SDG target 6.1 and 6.2 the major challenge is to ensure water quality and quantity for the huge population, at the same time safely managed sanitation and hygiene services need to be addressed [29]. One of the major challenges to achieve SDG target 6.3 is the proper application of laws and regulations to restore waterbodies. Uddin and Jeong [25] recommended that stringent laws and regulations, constant observation, and comprehensive research be implemented immediately to prevent river pollution in Bangladesh. Bangladesh is one of the most polluted nations, with 1176 factories and 0.4 million m^3 of untreated trash released into the rivers per day [23]. SDG 6.3 seeks to achieve improvement in ambient water quality by significantly reducing pollution flows into water bodies [29].

Dhaka city is blessed with 5 peripheral rivers-Turag, Buriganga, Balu, Shitalakshya and Dhaleswari [2]. DWASA has been using the Buriganga and the Shitalakshya as surface water sources for supplying potable water to Dhaka city dwellers [9]. However, because of significant pollution that has degraded the water quality of the aforementioned periphery rivers, DWASA is currently compelled to pump raw water from Meghna and Padma, which are distant from Dhaka metropolis [4, 8, 27]. The surrounding waterways of Dhaka city have dangerously high levels of toxicity due to a combination of untreated industrial, residential, urban, and agricultural waste [5, 27, 29]. Industries are the main sources of pollution, according to Islam et al. [16], since they use a lot of water and discharge untreated effluent during a product's production cycle. The high concentrations of pollutants have severe negative impact on livelihoods, public health, and ecosystem [2, 14, 27]. Food stuff growing with contaminated water irrigation or growing on the vicinity of contaminated river have higher concentrations of different toxic metals. Through the consumption of these foods, these metals come into the human body and cause toxic effects. Besides that, harmful microorganisms in surface water are responsible for serious health hazards [25]. Water contamination kills and inhibits the capacity of aquatic organisms to reproduce. For this reason, in September 2009, the four rivers Buriganga, Shitalakhaya, Turag and Balu have been declared by the Department of Environment (DoE) as Ecologically Critical Areas (ECA), for improving the condition of ecosystems [12, 25]. These rivers need to be restored to meet the target of SDG 6.3 and initiatives should be taken in an urgent basis to improve the water quality of these rivers [27, 29].

Water from the Balu and Dhaleshwari rivers flows into the Shitalakshya River. There were a number of industries located along the riverside through the channel. The quality of the water in the Shitalakshya River is greatly impacted by companies located along the riverside, including those that produce textiles, cement, sugar mills, fertilizer, pulp and paper, jute, etc. [25]. Additionally, the pollution loads were moved from the Dhaleshwari and Balu rivers to the Shitalakshya river. Industrial garbage is immediately dumped into the Begunbari and Narai canal, which transports it into the Balu River, according to Roy et al. [22] and Mottalib et al. [19]. The polluted Balu River ultimately flows on Shitalakshya River which is used in SWTP for meeting water demand of Dhaka city dwellers. As a result, the water of Shitalakshya River is deteriorating day by day. Numerous indicators of water quality, including turbidity, DO, BOD, and TDS, are in poor condition [7, 17]. In the Shitalakshya River, DO concentrations were found to be nearly nonexistent at certain locations during the dry season [25]. Ferdous et al. [11] calculated the Water Quality Index (WQI) to evaluate the DND Canal's water quality. After analysing the water quality parameters, it was observed that DO, BOD, phosphate, colour, turbidity, electric conductivity, hardness, chloride, free chlorine, salinity, TDS, TSS, NH_3, NO_2, NO_3, sulphate values exceeded the limits mentioned in Environment Conservation Rules [10] of Bangladesh.

SWTP with a capacity of 450 MLD is pumping raw water from the Sarulia intake point of the Shitalakshya River and it is one of the major surface water treatment plants under DWASA [4]. The main intake of SWTP is less than 1 km downstream of the meeting point of Balu and Shitalakshya rivers. Untreated water from the Shitalakshya River is pushed into the DND Canal, and from there, gravity directs the water through a closed tunnel to the SWTP's intake pump station. According to Atiq et al. [2], the contamination of the Balu River water poses a significant threat to both the SWTP's operation and the residential water supply in the Dhaka metropolitan area. For this reason, some researchers suggested that the current water intake point is not safe due to presence of toxic pollutants and microbial loads [25]. Talukdar [24] conducted research to determine the seasonal variations of the water quality index at the raw water collection point of the SWTP and he also analyzed its impact on plant operations. He pointed out that between 2017 and 2021, the Shitalakshya River's waterbody was in extremely bad shape. Additionally, he discovered that ammonia content significantly affects the plant's pre-treatment procedures. Talukdar [24] noted that SWTP requires certain pretreatment procedures when ammonia concentration is greater than 2 ppm. From 2017 to 2021, the pretreatment unit (Nitrification) was necessary for the pre-and post-monsoon periods, which significantly raised the SWTP's water treatment costs. According to Ferdous et al. [11], if the pollution rate of the DND canal is considerably lowered, the water treatment efficiency will increase, and the treatment cost will go down accordingly. He further suggested preventing discharge of untreated sewage and other industrial wastewater into canals to improve the water quality of DND canal.

Saidabad Water Treatment Plant (SWTP): The SWTP was planned to be constructed in three phases [4]. Phase-1 of the SWTP was constructed in 2002 with a capacity of 225MLD. Phase-2 was constructed in 2012 with the same capacity as Phase-1. Phase-3 of the SWTP, with the capacity of 450 MLD is under construction which will eventually raise the capacity of the SWTP to 900,000 m^3/d. Phase 2 included the inclusion of aeration and biological pretreatment (nitrification), as well as an increase in each treatment process's capacity [4]. The treatment process (Fig. 12.1) of the SWTP includes:

Dasherkandi Sewage Treatment Plant (DSTP): DSTP was established in April'2022 to improve the water quality of the Balu River and the influent of SWTP as influent water quality parameters like NH_3–N during the dry season reach up to 20 mg/L [11]. The government of Bangladesh undertook a project named "Hatirjheel Project" to mitigate severe pollution from surrounding areas. 11 Special Sewage

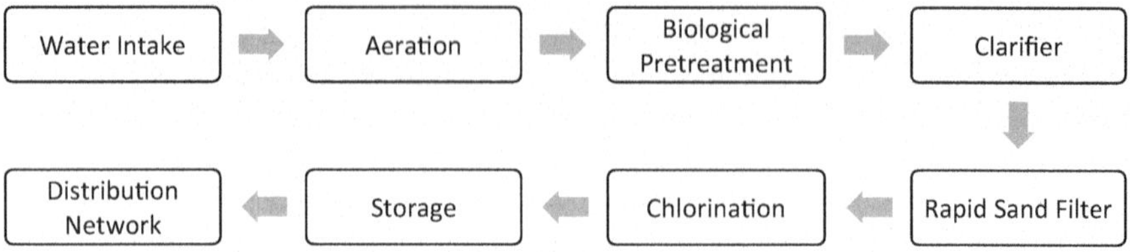

Fig. 12.1 Treatment process of the Saidabad water treatment plant

Diversion Structures (SSDSs) were constructed along the periphery under the said project and the SSDSs were connected by Main Diversion Sewers (Diameter 1200–1830 mm) [6]. It was planned to divert the incoming sewage to a Sewage Treatment Plant which was to be built in Dasherkandi village located 5 km south of Rampura Bridge and a permanent Sewage Lifting Station at Rampura and a 5 km long sewage conveyance pipeline were also included within the "Hatirjheel Project". DSTP is one of the largest sewage treatment plants in South Asia with a capacity of 500 MLD equipped with pretreatment, primary treatment, and secondary treatment facilities. The process incorporated for wastewater treatment is Activated Sludge Process which includes Anaerobic, Anoxic, and Aerobic chambers and is very efficient in removing Phosphate (P), Ammoniacal Nitrogen (NH_3–N), and Biochemical Oxygen Demand (BOD). There are 04 coarse screens (25 mm); 8 Nos of Fine Screens (5 mm) and 04 Nos of Vortex Grit Removal Chambers (0.15 mm) which comprise the pretreatment unit. The Primary treatment unit consists of 04 primary sedimentation tanks. The biological treatment unit consists of 4 A/A/A (or A/A/O) Tanks and 8 Nos of Secondary Sedimentation Tanks. The treatment plant also has a UV disinfection unit to kill the remaining pathogens and faecal coliform before discharging into the Gojaria khal, which eventually goes into the Balu River. The process flow diagram has been shown in Fig. 12.2. Table 12.1 shows the average influent and effluent quality from October'2022 to October'2023. Since DSTP went into operation, the water quality of the Hatirjheel has been improved significantly [21]. DSTP is also equipped with a sludge treatment facility which involves incineration of the sedimented and generated sludge during wastewater treatment. The flue gas that is generated during the incineration is purified using a dedicated flue gas purification unit before releasing into the atmosphere.

Despite the establishment of DSTP, industrial pollution is causing the water quality deterioration of the adjacent rivers. The degradation of surface water quality was identified by the National Sustainable Development Strategy (NSDS) of Bangladesh as being caused by unchecked industrial expansion, rural-to-city migration, encroachment of rivers and water bodies, overloaded infrastructure, uncertainty about institutional responsibility for the quality of urban water bodies, and inadequate enforcement of environmental regulations [18]. Most industries are often located around riverbanks. The river water is being exposed to hundreds of tons of waste items per day [14, 25]. Although there exist laws and guidelines, the industries do not appropriately abide by them. ETPs (effluent treatment plants) are owned by certain

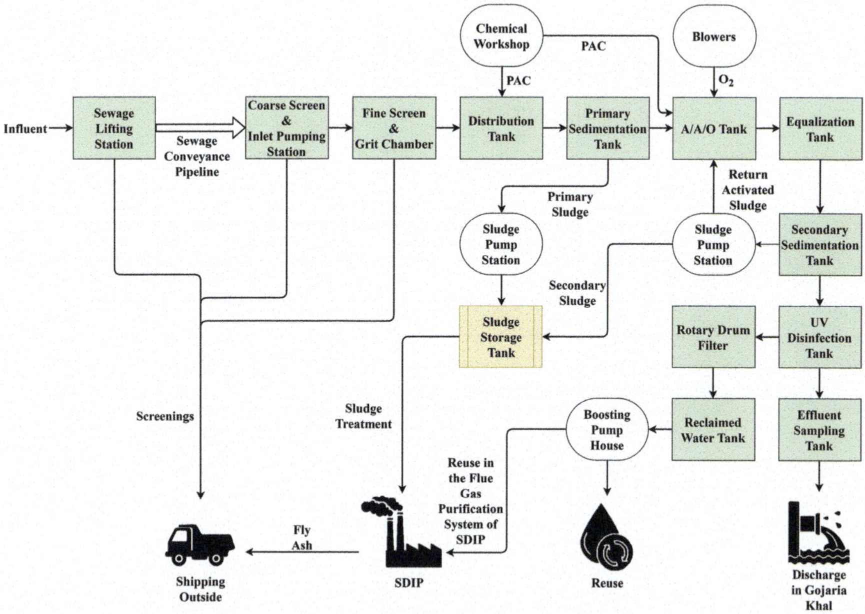

Fig. 12.2 Process flow diagram of Dasherkandi sewage treatment plant

businesses, but they are not used because of their large profit margins [15, 25]. River water thus turns harmful to living organisms. According to Ferdous et al. [11] and Uddin et al. [26], to improve water quality of DND canal existing laws and regulations should be implemented properly and suitable policy should be taken to raise public awareness. Sultana et al. [23] gave emphasis on proper enforcement mechanism for Balu River. An improved monitoring and regulatory system are required, and baseline data should be compared with the data over time [28]. The DoE should expand its capabilities to include real-time monitoring of effluent discharges and river water quality, as recommended by Whitehead et al. [29]. They recommended public dissemination of data on water quality as well.

12.2 Methodology: Study Areas and Sample Data Analysis

In order to evaluate the effects of Dasherkandi STP on the Balu River, the study region was selected with the least amount of susceptible sources of pollution in mind. Strategic ideas were taken from some previous research works (i.e. [13, 22]) for selecting the study area and data sampling locations. The study area is located from latitudes 23° 45′ 37.40″ N–23° 45′ 44.91″ N and longitudes 90° 27′ 55.44″ E to 90° 26′ 45.47″ along Rampura Khal. The study area also includes areas within

Table 12.1 Removal efficiency of Dasherkandi STP

Parameter	BOD_5			NH_3–N			Phosphate (As P)		
Month	Average value in influent (mg/L)	Average Value in effluent (mg/L)	Removal efficiency (%)	Average value in influent (mg/L)	Average value in effluent (mg/L)	Removal efficiency (%)	Average value in influent (mg/L)	Average value in effluent (mg/L)	Removal efficiency (%)
Oct-2022	75.86	1.65	97.82	21.83	0.18	99.18	3.42	1.37	59.94
Nov-2022	81.73	1.88	97.70	22.55	0.19	99.16	3.1	1.23	60.32
Dec-2022	89.64	1.91	97.87	24.59	0.22	99.11	3.48	1.1	68.39
Jan-2023	91.79	2.18	97.63	25.45	0.25	99.02	4.01	0.9	77.56
Feb-2023	98.58	2.59	97.37	25.08	0.3	98.80	4.94	0.58	88.26
Mar-2023	93.2	2.66	97.15	23.71	0.27	98.86	4.95	1.14	76.97
Apr-2023	101.77	2.82	97.23	19.93	0.34	98.29	5.62	2.16	61.57
May-2023	75.14	2.93	96.10	20.48	0.28	98.63	3.04	1.08	64.47
Jun-2023	73.43	3.01	95.90	20.11	0.3	98.51	2.93	1.03	64.85
Jul-2023	71.43	3.17	95.56	19.77	0.32	98.38	2.95	0.93	68.47
Aug-2023	74.29	3.23	95.65	19.31	0.33	98.29	2.91	0.9	69.07
Sept-2023	59.97	0.85	98.58	18.46	0.31	98.32	2.54	0.8	68.50
Oct-2023	66.47	0.87	98.69	16.58	0.23	98.61	2.16	0.62	71.30

latitudes 23° 43′ 2.72″ N–23° 47′ 52.29″ N and longitudes 90° 28′ 52.97″ E–90° 30′ 1.02″ E along Shitalakshya River. The map of the study area is shown in Fig. 12.3.

Samples were taken from 11 potential pollution sources along Rampura Khal, Balu River, Shitalakshya River and the vicinity of Sarulia Intake Point from November'2022 to April'2023. The sampling locations are given in Table 12.2.

Spot sampling technique was adopted for collecting the samples. The samples were taken from 1m depth from the water surface using a bucket and then the samples were filled in 1 L bottles [22]. Before sampling, the sampling vials were rinsed with the effluent to be sampled. After sampling, the bottles were promptly sealed. Quality data such as DO, COD, NH_3–N, Turbidity, TDS, pH have been tested in the SWTP laboratory. The amount of DO in water is one of the most used indicators of a canal's health [11, 27]. However, the study emphasized two specific parameters-DO and NH_3–N to analyze the pollutant loading as these two parameters are currently the major concerns of SWTP from technical and economic perspectives. To assess the industrial loading along Rampura Khal, Norai Khal, Balu River and Shitalakshya River, COD values were analysed [1], Standard method for determining DO was adopted. Ammoniacal Nitrogen (NH_3–N) (by Nessler Method) and COD (by Closed Reflux, Colorimetric Method) concentrations in water were obtained by using HACH UV Spectrophotometer DR/6000U [22]. Seasonal water quality data of the influent of SWTP from January'2021 to March'2023 were also collected from SWTP. Relevant

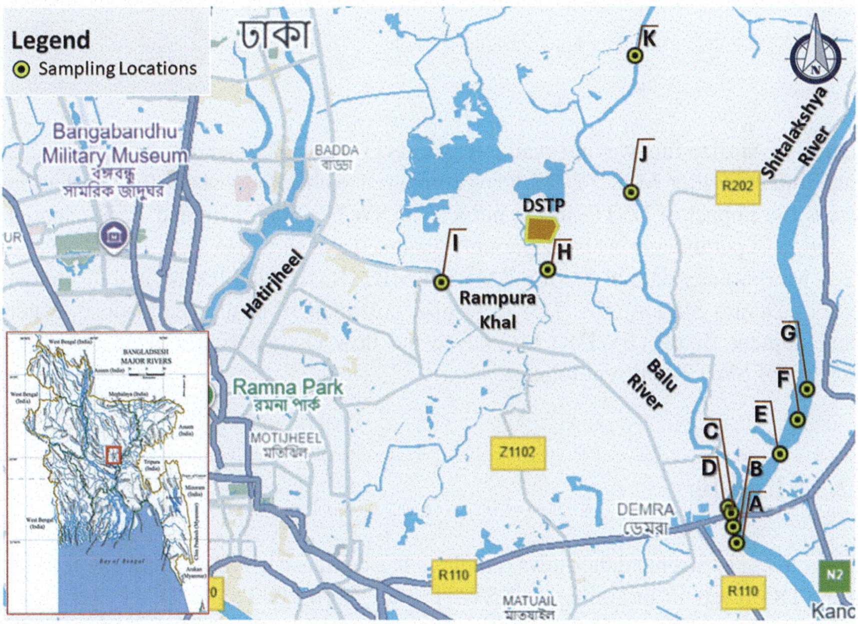

Fig. 12.3 Study area and spot sampling locations (latitude-longitudes have been shown in Table 12.2)

Table 12.2 Sampling locations

S. No.	Sampling location	Annotations in Fig. 12.3	Latitude	Longitude
1	Saidabad intake	A	23° 43′ 2.72″ N	90° 30′ 1.02″ E
2	DS RHD culvert	B	23° 43′ 12.29″ N	90° 29′ 58.78″ E
3	RHD culvert	C	23° 43′ 20.36″ N	90° 29′ 57.38″ E
4	US RHD culvert	D	23° 43′ 23.82″ N	90° 29′ 55.31″ E
5	DS AB textile	E	23° 43′ 55.75″ N	90° 30′ 29.53″ E
6	AB textile	F	23° 44′ 16.16″ N	90° 30′ 41.10″ E
7	US AB textile	G	23° 44′ 34.18″ N	90° 30′ 47.43″ E
8	Dasherkandi	H	23° 45′ 44.91″ N	90° 27′ 55.44″ E
9	Meradia	I	23° 45′ 37.40″ N	90° 26′ 45.47″ E
10	Eiderkandi	J	23° 46′ 30.87″ N	90° 28′ 50.54″ E
11	Beraid	K	23° 47′ 52.29″ N	90° 28′ 52.97″ E

Fig. 12.4 Comparison of seasonal average values of NH_3–N at Sarulia intake (data from April to July was not available)

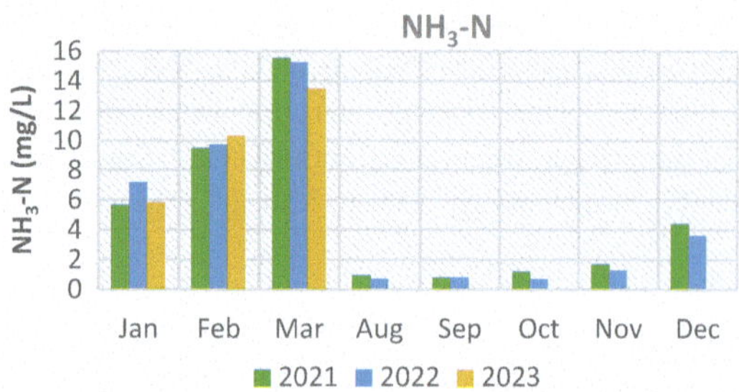

data from April to July was not available. As DSTP went into operation in April'2022, data collected after March'2022 were compared with previously collected data to assess the impact of DSTP on the influent of SWTP (Figs. 12.4 and 12.5). Spatial variation of parameters is also separately analysed and shown in Figs. 12.6, 12.7 and 12.8. Morphology data of Station RMB1, RMB2, RMB3 and RMB4 and discharge data of Demra Station (SW 7.5) were also collected from the Bangladesh Water Development Board (BWDB) to analyse the flow characteristics at an upstream location of SWTP's current intake.

12.3 Results

The collected seasonal data shows that there is a decrease in NH_3–N and an increase in the DO levels in the influent of SWTP after April'2022 since DSTP went into operation (Figs. 12.4 and 12.5).The average values of NH_3–N & DO are consecutively shown in Figs. 12.6 and 12.7. The mapped average values of DO and NH_3–N show that at the outlet of DSTP before meeting the Balu River, the DO level was 8.21mg/L

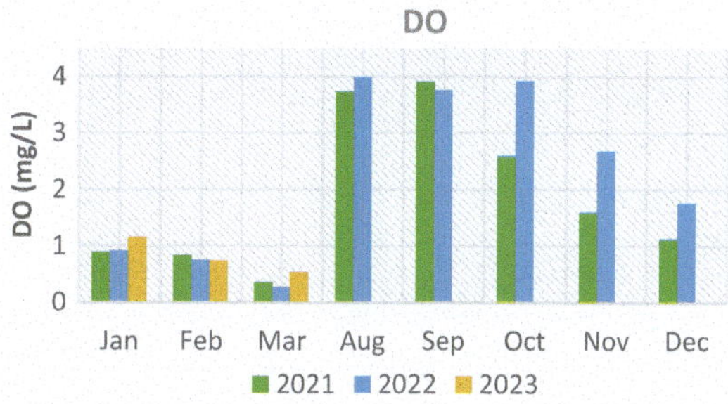

Fig. 12.5 Comparison of seasonal average values of DO at Sarulia intake (data from April to July was not available)

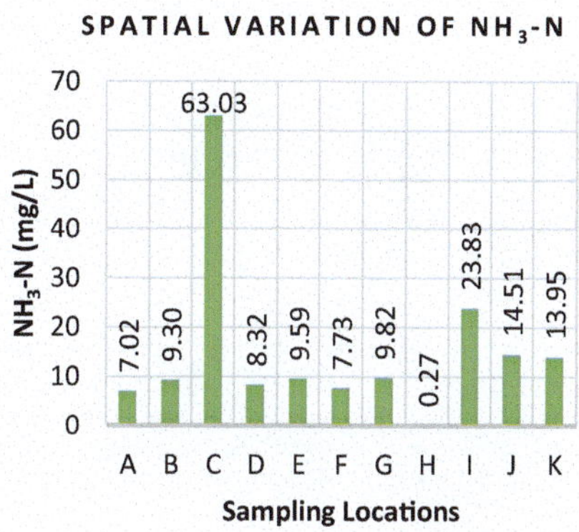

Fig. 12.6 Spatial variation of NH_3–N

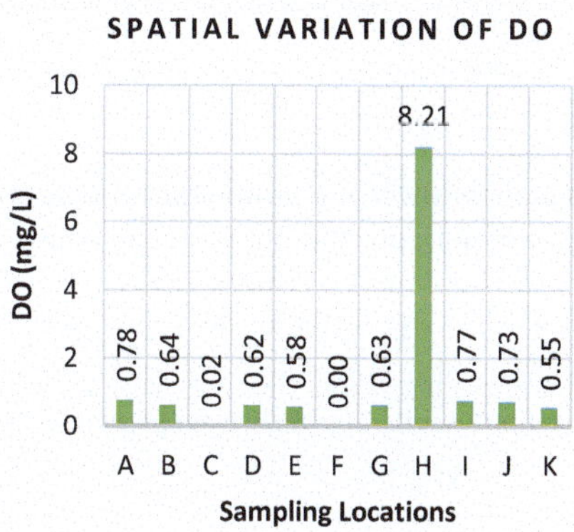

Fig. 12.7 Spatial variation of DO

and the NH_3–N was only 0.26 mg/L (Point-H, Fig. 12.3). However, huge pollutants from some jute mills and other industries are discharged into the contributing streams of the Balu River. During low flow season, the reduction of DO and spike in NH_3–N are exacerbated because of the Biswa Ijtema at upstream of Balu River [15]. The average DO value from 2017 to 2021 was 2.11 mg/l at the raw water collection station (Sarulia) of the SWTP and 2.2 mg/l at Demraghat. These values are not appropriate for the river's aquatic ecology. Between 2018 and 2020, the unit cost of treating

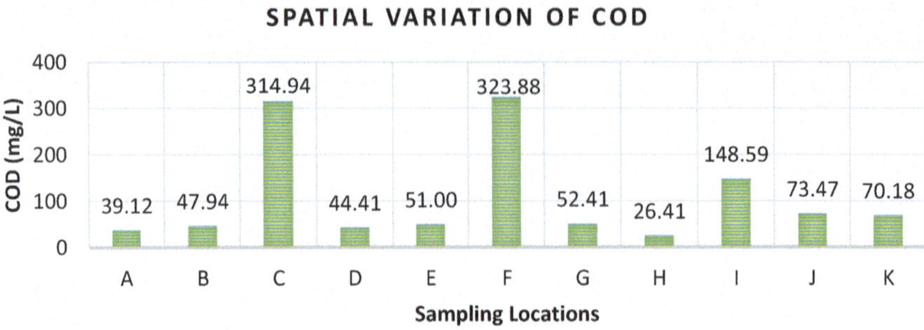

Fig. 12.8 Spatial variation of COD

Fig. 12.9 Fatal pollution source (denoted as "C" in Fig. 12.3)

water to lower the ammonia content grew from 19 to 151% [24]. The study found that major contributing pollutant sources are at Meradia DO = 0.77 mg/L and NH_3–N = 23.83 mg/L (Point I, Fig. 12.3) and at 525m upstream of the SWTP intake DO = 0.60 mg/L and NH_3–N = 63 mg/L (Point C, Fig. 12.3). The point is denoted as a fatal pollution source and is shown in Fig. 12.9. The spatial variations of NH_3–N, DO and COD are shown respectively in Figs. 12.6, 12.7 and 12.8.

12.4 Discussion

From the spatial variations (Figs. 12.6, 12.7 and 12.8) it can be assessed that sampling point "I" is a potential source of organic loading as the NH_3–N is significantly dominant and the fatal point source denoted as "C" in Fig. 12.3 is a source where mostly industrial waste is discharged as COD in this location is dominant. Due to the heavy NH_3–N loading just upstream, additional aeration and chemicals are necessary for bringing the NH_3–N within the Department of Environment (DoE)'s drinking water standard. This study found that the NH_3–N value at 670 m upstream is 8.31 mg/L (Point D, Fig. 12.3). The intake pumping station can be shifted to approximately 700 m upstream in the Shitalakshya River from the existing intake point (Point A, Fig. 12.3) the operation cost can be largely reduced. The discharge at station SW7.5 was 251.36 m^3/s during 2022, which complies with the SWTP design pumping rate (16.5 m^3/s). Similar research conducted on Shitalakshya River found that during dry season, the flow remained around 42.5% of its peak discharge [17].

12.5 Conclusion

Delivering drinking water to Dhaka city dwellers is one of the biggest challenges for DWASA. To reduce dependency on ground water and increase dependency on surface water; improvement of the river water quality has no other alternative. The study has found that although all the parameters of the discharge from Dasherkandi STP are within the DoE Standards, the water quality of Saidabad intake has not been improved on a significant level due to heavy pollutant loading in the near upstream location of the Sarulia Intake and resulting in high operational cost in SWTP. The shifting of SWTP intake to 700m upstream at the Shitalakshya River can be considered as one of the feasible options for improving the influent quality of SWTP and for this a buffer zone should be declared by the DoE. Strict regulations may be imposed for controlling the discharge of pollutants into the rivers from different industries and other potential sources.

References

1. APHA (2005) Standard methods for the examination of water and wastewater, 21st edn. American Public Health Association, American Water Works Association, Water Environment Federation, Washington DC
2. Atiq AR, Rahman MM, Sojib RM, Shil SC (2015) Water quality changes in Balu river during non-monsoon and monsoon period. J Water Resourc Eng Manag 2(3):38–49
3. BBS (2022) Bangladesh population & housing census 2022. Bangladesh Bureau of Statistics, Statistics and Informatics Division Ministry of Planning Population & Housing Census 2022 Preliminary Report Government of the People's Republic of Bangladesh
4. Bhari B, Kaamruzzaman AKM (2016) Saidabad water treatment plant, Dhaka, Bangladesh. Network on Water Technology in Asia and Pacific (NewTap), Japan Water Research Center (JWRC)
5. Bhuiyan NZ, Hossain MB, Ali MM, Habib A, Bhuyan MS, Arai T (2022) Seasonal variation of trace elements in water and sediment of the Turag and Balu rivers, Bangladesh. Egypt J Aquat Biol Fish 26(3):513–540
6. Biswas R, Ali MA (2014) Assessment of capacity reliability of newly constructed special sewage diversion system at Hatirjheel, Dhaka. In: 2nd international conference on advances in civil engineering. CUET, Chittagong, Bangladesh, 26–28 Dec 2014
7. Chowdhury RM, Ankon AA, Bhuiyan MK (2021) Water quality index (WQI) of Shitalakshya River near Haripur power station, Narayanganj, Bangladesh. J Eng Sci 12(3):45–55

8. DWASA (2014) Water supply master plan for Dhaka City. Dhaka water supply and sewerage authority (Dhaka WASA), Ministry of Local Government, Rural Development and Co-Operatives. https://dwasa.portal.gov.bd/sites/default/files/files/dwasa.portal.gov.bd/page/c0a3b947_9ad9_429a_8a3f_e320e33fea06/2021-01-17-16-51-f23ad05cc0f676fe25cca345e2def230.pdf
9. DWASA (2021) Annual report 2020–21. Dhaka water supply and sewerage authority (Dhaka WASA). https://dwasa.portal.gov.bd/sites/default/files/files/dwasa.portal.gov.bd/page/b5e42944_f9c0_430e_add7_b0888ba9e0d2/2022-08-11-09-09-60b874cbb0a32c32b0a65d7f60894703.pdf
10. ECR (1997) The environmental conversation rules, 1997. Ministry of Environment and Forest, Government of the People's Republic of Bangladesh
11. Ferdous R, Haque A, Adhikary KK, Huda N (2016) Water quality of DND canal from the perspective of Saidabad water treatment plant (SWTP). In: 3rd international conference on advances in civil engineering 2016 (ICACE 2016)
12. Hafizur RM, Nuralam HM, Rumainul IM (2017) Investigation of physicochemical parameter, heavy metal in Turag river water and adjacent industrial effluent in Bangladesh. J Sci Technol Environ Inf 05(01):347–360
13. Hasan MK, Happy MA, Nesha AK, Karim KHR (2014) Pollution status of Balu river due to industrial input at Dhaka, Bangladesh. Open J Water Pollut Treat 1(1)
14. Hossain MA, Uddin MK, Molla AH, Afrad MSI, Rahman MM, Rahman GKMM (2010) Impact of industrial effluents discharges on degradation of natural resources and threat to food security. Agriculturists 8(2):80–87
15. Hoque SF, Peters R, Whitehead P, Hope R, Hossain MA (2021) River pollution and social inequalities in Dhaka, Bangladesh. Environ Res Commun 3(9)
16. Islam JB, Sarkar M, Rahman AKML, Ahmed KS (2015) Quantitative assessment of toxicity in the Shitalakkhya River, Bangladesh. Egypt J Aquat Res 41(1):25–30
17. Islam MS, Uddin MK, Tareq SM, Shammi M, Kamal AKI, Sugano T, Kurasaki M, Saito T, Tanaka S, Kuramitz H (2015) Alteration of water pollution level with the seasonal changes in mean daily discharge in three main rivers around Dhaka City, Bangladesh. Environments 2(3):280–294
18. Mohiuddin AK (2019) Chemical contaminants and pollutants in the measurable life of Dhaka City. J Environ Sci Publ Health 3(2018):57–73
19. Mottalib MA, Roy S, Ahmed MS, Al-Razee ANM (2017) Comparative study of water quality of Buriganga and Balu river, Dhaka, Bangladesh. Bangladesh Int J Curr Res 9(10):59132–59137
20. Nur SMS, Mitra C (2023) Urban growth and multi-seasonal land cover classification of Dhaka, Bangladesh: an approach using google earth engine. In: Mookherjee D, Pomeroy GM, Huong LTT (eds) Advances in 21st century human settlements
21. Rahmat NS, Islam NB, Badruzzaman DABM (2023) Assessing the effect of Dasherkandi sewer treatment plant on the water quality index of Hatirjheel. In: 9th international conference on water management—ICWFM 2023
22. Roy S, Banna L, Hossain M, Rahman H (2014) Water quality of Narai Canal and Balu river of Dhaka City: an impact of industrialization. J Bangladesh Agric Univ 12(2):285–290
23. Sultana MN, Hossain MS, Latifa GA (2019) Water quality assessment of Balu River, Dhaka Bangladesh. Water Conserv Manag 3(2):8–10
24. Talukdar S (2022) Seasonal variation of water quality index at raw water collection point of Saidabad water treatment plant and its impact on plant operations. Master of Engineering thesis, Department of Water Resources Engineering, Bangladesh University of Engineering and Technology
25. Uddin MJ, Jeong YK (2021) Urban river pollution in Bangladesh during last 40 years: potential public health and ecological risk, present policy, and future prospects toward smart water management. Heliyon 7(2)
26. Uddin MJ, Parveen Z, Hossain MF (2016) Status of heavy metals in water and sediments of canals and rivers around the Dhaka City of Bangladesh and their subsequent transfer to crops. Adv Plants Agric Res 5(4):593–601

27. Whitehead PG, Bussi G, Hossain MA, Dolk M, Das P, Comber S, Peters R, Charles KJ, Hope R, Hossain S (2018) Restoring water quality in the Polluted Turag-Tongi-Balu river system, Dhaka: modelling nutrient and total coliform intervention strategies. Sci Total Environ 631(2):223–232
28. Whitehead PG, Bussi G, Peters R, Hossain MA, Softley L, Shawal S, Jin L, Rampley CPN, Holdship P, Hope R, Alabaster G (2019) Modelling heavy metals in the Buriganga river system, Dhaka. In: Bangladesh: impacts of tannery pollution control. science of the total environment, vol 697
29. Whitehead PG, Mimouni Z, Butterfield D, Bussi G, Hossain MA, Peters R, Shawal S, Holdship P, Rampley CPN, Jin L, Ager D (2021) A new multibranch model for metals in river systems: impacts and control of Tannery wastes in Bangladesh. Sustainability 13(6):3556

Chapter 13
Water, Sanitation-Hygiene Security of Urban Slum Dwellers During Covid-19 Pandemic: An Insight from Rajshahi City Corporation

Shehan Tawsif, Shitangsu Kumar Paul, and Md. Shohel Khan

Abstract Covid-19 pandemic has influenced the water, sanitation-hygiene security around the world. The current study aims to discover water, sanitation-hygiene (WASH) security of urban slum dwellers during covid-19. Rajshahi City Corporation slums were categorized into inner, middle and outer slum zones to conduct the study. Total of 361 slum household were determined as a sample population at 95% confidence level. A pre-tested semi-structure questionnaire was formulated to conduct household data collection. Data was collected from June–September 2022 and were analysed through SPSS and map was produced by ArcGIS 10.4 software. Study revealed that overall 37.5% respondent used mask to remain safe in pandemic. About 61.5% of the slum dwellers share one tube well and outer slum zone dweller were found higher (79.5%). About 53, 43.3 and 36.7% of outer, middle and inner zone dwellers use common toilet. Major defecation facility of inner slum dwellers was water sealed latrine (47.8%), followed by 24.5 (outer) and 10% (middle). The study also revealed that inner slum zone dwellers were highly (0.74) secured to WASH security and middle and outer zone dwellers were moderately secured. This study reflects that WASH practice needs to be more enhanced in slum communities to combat pandemic situation.

Keywords Pandemic · Sanitation-hygiene · Slum zone · Bangladesh

S. Tawsif (✉)
Department of Geography and Environment, Shahjalal University of Science and Technology, Sylhet, Bangladesh
e-mail: shehantawsif@gmail.com; stawsif-gee@sust.edu

S. Tawsif · S. K. Paul
Department of Geography and Environmental Studies, University of Rajshahi, Rajshahi, Bangladesh
e-mail: spaulrajbd@gmail.com; skp@ru.ac.bd

Md. S. Khan
Department of Environmental Science and Disaster Management, Noakhali Science and Technology University, Noakhali, Bangladesh
e-mail: sajibicb@gmail.com; sajibicb.esdm@nstu.edu.bd

G. M. Tarekul Islam et al. (eds.), *Sustainable Water Management and Environmental Resilience*, Springer Water, https://doi.org/10.1007/978-3-031-93350-9_13

13.1 Introduction

Bangladesh is extremely vulnerable to a variety of health and other external threats, such as, flood, cyclone, storm surges and several types of pandemic [35]. Higher population density and fast and chaotic urbanization, along with a large proportion of urban population living in slums render it vulnerable to the rising rates of morbidity from infectious disease epidemics, such as, cholera and dengue fever [49]. In 2019, World Health Organization (WHO) was informed of the cases of pneumonia of unknown etiology detected in Wuhan City, China [54]. Similarly, covid-19 pandemic trend has also affected Bangladesh which is renowned as a lower-middle-income country (LMIC) in Southeast Asia with a population of about 160 million [43]. This pandemic has wreaked havoc on communities all across the world, however, the elderly are particularly vulnerable to this deadly disease [40]. The major health issue of the current worldwide covid-19 pandemic has made water, sanitation-hygiene (WASH) behaviours for crucial contributors to the prevention and treatment of covid-19 infection. As a result, the role of WASH in the covid-19 response has been regarded as good hygiene, by ensuring frequent and appropriate hand washing technique [15, 50, 53]. Public health during infectious disease outbreaks such as the covid-19 pandemic, adequate WASH practices are essential to protect human lives [8, 22]. WASH is still inaccessible to one-third of the world's population, who are primarily marginalized and vulnerable to live in urban slums in many poor countries [30].

Several researches was investigated on various objectives around the globe i.e. present and future relevance on WASH practice in the case of covid-19 pandemic [20]. A online based cross-sectional study reported the knowledge, attitude and WASH practice during covid-19 pandemic [14]. As stated in a recent report, there are lots of challenges, impacts and mitigation strategies on covid-19 WASH practices [11]. Another study revealed the critical control on WASH measures in low-income countries during covid-19 pandemic [13]. A study explored the hand washing practice with soap and the overuse of water in Bangladesh perspective during covid-19 pandemic [37]. The waste disposal and WASH practice was reflected in Bangladesh during the covid-19 pandemic [22]. Additionally, machine learning applications can be a significant strategy to build up wash awareness during covid-19 [32]. One of the study critically reviewed that WASH can be an important measure that can control ongoing covid-19 [25]. Another study examined the WASH opportunity and accessibility in the urban slum dwellers of Nairobi, Kenya during covid-19 pandemic [26]. Furthermore, the WASH behaviour during covid-19 lockdown in rural Bangladesh was explored recently [48]. The studied about the progress on WASH practice on Brazil's school in pre and during stage of covid-19 pandemic was explored [34]. In addition, few studies were done around the globe by focusing on the slum dwellers WASH practice and security, such as, Nepal [38], India [3, 47], Sub Sharan Africa [2], Ethiopia [4]. However, as per our concern, we did not find research on WASH security of urban slum dwellers during covid-19 pandemic in Bangladesh. Again, WASH security in inner, middle and outer zone were not investigated earlier in the research. For this reason, the researchers explored these slum areas to implement this study. The objective of the present study was to explore the WASH security of urban slum dwellers during covid-19 pandemic.

13.2 Methodology

13.2.1 Selection of the Study Area

With a population of over 165 million, Bangladesh is among the most densely inhabited nations which consists of twelve city corporations and Rajshahi City Corporation (RCC) are listed among them [16]. Rajshahi is regarded as a peaceful city in north-west Bangladesh, located adjacent to the Padma River [17, 18]. It is renowned as third largest city of Bangladesh which is located between 24°21′ and 24°26′ North latitude and 88°28′ to 88°38′ East longitude, with a population of 0.85 million and an area of 96.72 km^2 [42, 44]. RCC were purposively selected for formulating the study (Fig. 13.1). First, 12 out of 39 slums of RCC were identified based on their location from the central business district (CBD). Second, the study area were categorized into three geographical zones, such as, inner (<1.5 km), middle (1.5–3.5 km) and outer zone (>3.5 km) to compare the WASH security of the slum dwellers [37, 39, 45].

13.2.2 Sampling and Collection of Data

Total of 4010 households were identified from three slum zones by the Community Development Committee (CDC) of RCC. The study used Kothari's sampling formula to calculate the sample size of the present study [27]. With 95% precision level and 5% sampling error, the sample size of this study was 361 for the known population which were consistently disseminated among three slum zones. Mainly the slum household heads (HH's) were surveyed and in terms of absence of the HH's, their spouses were surveyed for collecting the information [23]. The respondents were selected by simple random sampling procedure to conduct the household survey. A semi-structured and self-developed questionnaire was prepared based on the slum dwellers water, sanitation-hygiene practice during the covid-19 pandemic and the survey was conducted from the month of May to September 2022. After collecting the pre-tested questionnaire data from the slum HH's, it was carefully inserted in SPSS (version 25.0) and it was prudently checked before starting the coding or analysis. In addition, the study area map was formed by Arc GIS 10.4 software. However, a schematic overview of the present study is shown in Fig. 13.2 for better understanding of this research.

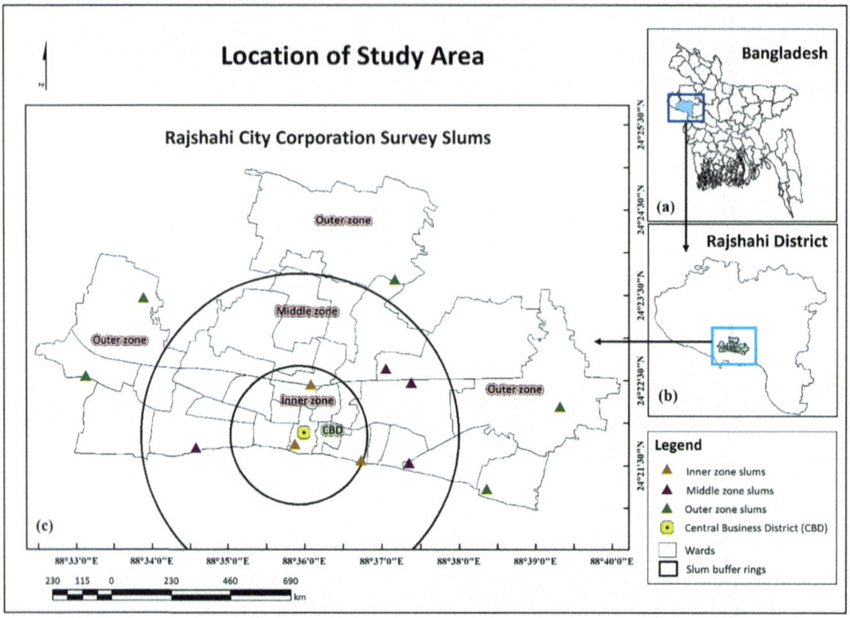

Fig. 13.1 Location of study area; **a** Bangladesh, **b** Rajshahi District, **c** Survey Slums

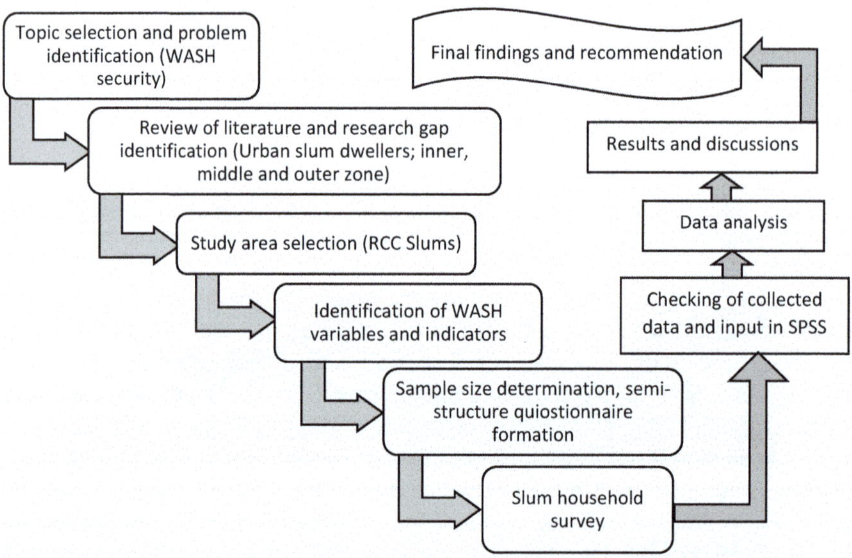

Fig. 13.2 Schematic overview of the study

Table 13.1 Assessment criteria for WASHSi and Likert scale

Practice criteria				
Never	In some extent	Sometimes	Often	Always
0.20	0.40	0.60	0.80	1.00
WASH security criteria				
Poor (< 0.34)	Moderate (0.34–0.66)		High (>0.66)	

13.2.3 Analysis Procedure

The current study used weighted average index (WAI) [19, 24, 33] to calculate WASH security. A five point Likert scale [24, 33] is used to calculate WASH security index (WASHSi) in the current study. Equation 13.1 is used to calculate the WASHSi of the slum dwellers. Available water for personal hygiene, washing hands, cleanliness after coming back from outside, use of antiseptic or disinfectant solutions, properly dispose of used mask, hand gloves and tissue in separate covered bins or bags, use of specific/ community waste disposal point are the variables to calculate WASHSi. However, the assessment criteria used in the present study has been shown in Table 13.1. In addition, the study used descriptive statistics, such as, chi square, ANOVA test etc. for evaluating the status of water, sanitation facilities of three different slum zones.

$$\mathrm{WASHSi} = [\{f\mathrm{N}(0.20) + f\mathrm{ISE}(0.40) + f\mathrm{S}(0.60) + f\mathrm{O}(0.80) + f\mathrm{A}(1.0)\}/\mathrm{N}] \tag{13.1}$$

where $f(\mathrm{N})$ = Frequency of never; $f(\mathrm{ISE})$ = Frequency of in some extent; $f(\mathrm{S})$ = Frequency of sometimes; $f(\mathrm{O})$ = Frequency of often; $f(\mathrm{A})$ = Frequency of always; N = Total number of observations.

13.3 Results

13.3.1 WASH Security

An important measure of livelihood security is how the condition of an individual communities water, sanitation-hygiene which is known as WASH. In order to safeguard public health during infectious disease outbreaks, such as, the covid-19 pandemic, adequate WASH practices are essential [8, 10, 22, 29]. According to WHO [50, 53], one of the most crucial steps that can be availed to avoid contracting the covid-19 virus is practicing regular and good hand hygiene. However, WASH professionals should improve facilities and employ tried-and-true behavior-change strategies to promote more frequent and regular hand washing [51]. In terms of interventions to improve water security, it should concentrate on four areas such as adequate water availability, acceptable water quality, water resource management and affordable access to WASH [9]. On the contrary, the current study reveals the WASH security of the slum dwellers by different indicators. The following sections have discussed the WASH status and security by giving the indicators weight.

13.3.2 Hygiene and Sanitation Status

The current study revealed the condition of the hygiene and sanitation of the slum dwellers in the different slum zones which are two important indicator to assess the status of the slum dwellers sanitation-hygiene security. It found that during the pandemic overall 91.5% of the slum dwellers did not maintain minimum hygiene or sanitation practice. Yet, after the pandemic most of the slum dwellers (93.1%) practice hygiene in all three slum zones. Individually, 93.3%, 91.7%, and 94% of inner, middle and outer slum zone dwellers respectively practice hygiene to combat the covid-19 pandemic. The study also revealed the types of hygiene practice by the three slum zone dwellers during the pandemic. Washing hands for remaining safe was one of the most common hygiene practice which is also expressed by World Health Organization (WHO). In this particular type of hygiene about 31% of inner, 40% of middle, and 38% of outer slum zone dwellers have practice it frequently but it was maintain without using any kind of soap. About 6, 2.7 and 3.5% of inner, middle and outer zone respectively used soap/sanitizer for remain safe during the pandemic. On the other hand, the use of a mask alone is not sufficient to offer an acceptable degree of protection against covid-19 pandemic; masks should be used as part of a complete strategy of actions to suppress transmission and save lives [52]. In addition, this study identified that about 41.7, 32.7% and 38.7% of inner, middle and outer slum zone dwellers respectively have practice wearing mask to remain safe from pandemic. Moreover, the slum dwellers were seen less in term of using soap/ sanitizer (3.9%) because most of them were economically poor to afford this safety measure. Similarly, 21.7% of total slum dwellers of the slum zones have availed all three hygiene practice (i.e. washings hands, using sanitizer and wearing face musk) in the time of covid-19 pandemic. In practicing all of these three hygiene and sanitation strategies the middle (24.5%) zone dwellers were most aware than inner (21.3%) or outer (19.7%) zone dwellers (Table 13.2).

13.3.2.1 Drinking Water Facilities

Water is also an important phenomena for WASH practice. Without safe water no one can live a secure life. The study revealed that overall majority percent (73.1%) of the slum dwellers depends on tube-well water. Individually, 54.4 (inner), 65 (middle) and 90.7% (outer) of slum dwellers respectively rely on tube well water facilities. On the contrary, majority percent (45.6%) of inner slum zone dwellers rely on pipe

Table 13.2 Slum dwellers hygiene and sanitation status

Types (during covid-19)	Survey slum zones							
	Inner		Middle		Outer		All	
	f	%	f	%	f	%	f	%
Washing hands without soap	26	31.0	44	40.0	54	38.0	124	36.9
Using soap/sanitizer	5	6.0	3	2.7	5	3.5	13	3.9
Wearing face mask	35	41.7	36	32.7	55	38.7	126	37.5
All of these	18	21.3	27	24.5	28	19.7	73	21.7
Total	84	100.0	110	100.0	142	100.0	336	100.0
Chi square	Value = 4.282, df = 6, p = 0.639							

line water comparative to middle (35%) and outer (9.3%) zone dwellers (Table 13.3). However, statistically significant difference exist in the three slum zones are found in terms of drinking water sources of the slum dwellers ($\chi^2(2) = 43.828$, $p < 0.05$). The study found less number of arsenic tested water sources as the arsenic is a big problem which was first observed in 1993 in Bangladesh [1]. It revealed that majority percent (47.1%) of the slum dwellers reported that their drinking water source is not arsenic tested. Similar percentage of all three slum zone dwellers expressed that their water source was not tested any kind of arsenic test. On the other hand, 31.1% of inner, 15% of middle and 15.9% of outer slum zone dwellers have reported that their drinking water source were tested and marked as safe drinking water. Moreover, due to lack of health and water security awareness 20, 40, and 36.4% of inner, middle and outer slum zone dwellers respectively have no idea about any arsenic tests (Table 13.3).

The current study founds the perception of the slum dwellers about their drinking water safety. It revealed that 96.7, 87.5, and 87.4% of inner, middle and outer slum zone dwellers respectively had mentioned that they are taking safe drinking water. Whereas, few number (10.2%) of the slum dwellers in all three slum zone have reported that they were taking unsafe water (Table 13.3). However, it found that majority percent of the slum dwellers (61.5%) are sharing one tube well for household water purpose, which indicates that insecure water security as they are gathering one place for water purpose (Table 13.3). Significantly outer slum zone dwellers share one tube well in higher percent (79.5%) comparative to inner (52.2%) and middle (45.8%). About 26.7, 32.5, and 8.6% of the slum dwellers of inner, middle and outer slum zone respectively share one tap for water purpose. Similarly, 16.7% of inner, 19.2% of middle and 7.3% of outer slum zone dwellers rely on portable water from far places of their dwellings. However, few slum dwellers have their own water sources, such as, own tube well for fulfil their household water need (Table 13.3). Likewise, statistically significant difference exists among the three slum zones in terms of water sharing by the slum dwellers ($\chi^2(6) = 42.586$, $p < 0.05$).

Table 13.3 Water facilities in slum zones by slum dwellers

Drinking water source	Survey slum zones							
	Inner		Middle		Outer		All	
	f	%	f	%	f	%	f	%
Tube well	49	54.4	78	65.0	137	90.7	264	73.1
Pipe line	51	45.6	42	35.0	14	9.3	97	26.9
Total	90	100.0	120	100.0	151	100.0	361	100.0
Chi square	Value = 43.828, df = 2, p = 0.000							
Arsenic tested drinking water facilities								
Yes	28	31.1	18	15.0	24	15.9	70	19.4
No	44	48.9	54	45.0	72	47.7	170	47.1
Do not know	18	20.0	48	40.0	55	36.4	121	33.5
Total	90	100.0	120	100.0	151	100.0	361	100.0
Chi square	Value = 15.497, df = 4, p = 0.004							
Drinking water safety by slum dwellers perception								
Safe	87	96.7	105	87.5	132	87.4	324	89.8
Unsafe	3	3.3	15	12.5	19	12.6	37	10.2
Total	90	100.0	120	100.0	151	100.0	361	100.0
Chi square	Value = 6.234, df = 2, p = 0.044							
Water sharing in the slum zones								
Portable water	15	16.7	23	19.2	11	7.3	49	13.6
Share one tube well	47	52.2	55	45.8	120	79.5	222	61.5
Share one tap	24	26.7	39	32.5	13	8.6	76	21.1
Others	4	4.4	3	2.5	7	4.6	14	3.9
Total	90	100.0	120	100.0	151	100.0	361	100.0
Chi square	Value = 42.586, df = 6, p = 0.000							

13.3.2.2 Sanitation Security

The current study revealed that about 45.7% of the total slum dwellers in all three slum zones have use common toilet for defecation. Which is a big matter of concern for sanitation security. Though, the study found statistically significant difference among the three slum zones in terms of using common toilet for defecation. It revealed that 36.7% of inner, 43.3% of middle, and 53% of outer slum zone dwellers are using the common toilet (Table 13.4). This findings indicated that inner slum zone dwellers have the ability of their own defecation facilities where middle and outer slum zone dwellers were in behind. Individually, more than half of the outer slum zone dwellers are using the common toilet for household defecation. The study also revealed that less than 10 members were using the common toilet in most of the cases in all three slum zones (81.8% in inner, 75% of middle, and 81.3% of outer slum zone). Similarly, higher percent (21.2%) of middle zone slum dwellers are using common toilet (more than 20 members) than outer (15%) or inner (12.1%) zone.

Average number of using common toilets by the slum dwellers were 5.67, 8.10, and 5.85 in inner, middle and outer zone respectively. It reflected that middle slum zone dwellers are more vulnerable in terms of using common toilet and increasing the sanitation insecurity among the three slum zones (Table 13.4). Slum dwellers households' defecation places significant vary among the three slum zones. The current study found that slum dwellers most common and preferable defecation place is ring slab (55.4%) in all three slum zone (Fig. 13.3).

Figure 13.3 shows the comparative use of defecation places in three separate slum zones. It revealed that, inner slum dwellers were more sanitation secure than other two slum zones as because majority (47.8%) of the slum dwellers use pucca (water sealed) defecation system. On the other hand, majority percent of using ring slab defecation system was observed in middle (57.5%) and outer (60.3%) slum zones respectively. However, open defecation was higher (22.5%) in middle slum zone than inner (2.2%) and outer (5.3%), which indicated that middle slum zone dwellers are more sanitation insecure than other two slum zone dwellers. Hanging defecation system also observed in all three slum zone (2.2% in inner, 1.7% in middle and 3.3% in outer slum zone). Similarly, pucca but un-hygienic defecation system was also observed significantly lower in inner (3.3%) than outer (6.6%) and middle (8.3%) slum zones. However, comparatively pucca water sealed defecation is the best defecation place for remaining secure in terms of sanitation security and it is poor in number in middle (10%) and outer (24.5%) slum zone than inner slum zone (47.8%). The current study reflects that, middle slum zone dwellers are more insecure than other two respective slum zone. Therefore, the inner zone slum dwellers were

Table 13.4 Defecation facilities in slum zones by slum dwellers

Use common toilet for defecation	Survey slum zones							
	Inner		Middle		Outer		All	
	N	%	N	%	N	%	N	%
Yes	33	36.7	52	43.3	80	53.0	165	45.7
No	57	63.3	68	56.7	71	47.0	196	54.3
Total	90	100.0	120	100.0	151	100.0	361	100.0
Chi square	Value = 6.455, df = 2, p = 0.040							
Number of members use common toilet by the slum dwellers in three zones								
< 10	27	81.8	39	75.0	65	81.3	131	79.4
11–20	2	6.1	2	3.8	3	3.8	7	4.2
> 20	4	12.1	11	21.2	12	15.0	27	16.4
Total	33	100.0	52	100.0	80	100.0	165	100.0
Average number of use common toilet	5.67		8.10		5.85		6.52	
ANOVA	F = 0.590, df = 2, p = 0.556							

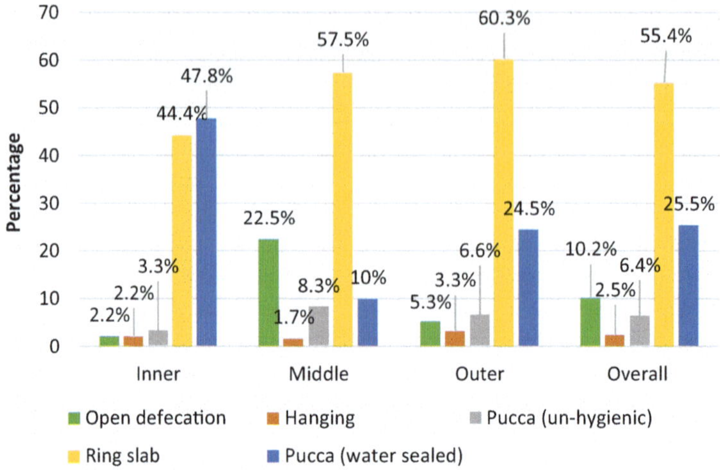

Fig. 13.3 Type of defecation by slum dwellers

relatively in better condition than the middle and outer zone slum dwellers. The main reason was that the inner slum zone dwellers were financially better to avail the sanitary latrine facilities in the study area.

13.3.3 WASH Practice

Water, sanitation-hygiene security which is representing as WASH, which is measured by descriptive statistics in this study in total slum dwellers perception. It found that available water for daily needs and personal hygiene with a mean value of 4.25 in scale of 1–5 (here, 1 denotes never and 5 denotes always). In Table 13.3 it revealed that the slum dwellers mostly avail water from tube well (73.1%) and pipe line (26.9%). In this case pipe line water was not available for whole day. Rather it remains from certain time of a day. Though, it reflected that most of the respondents have available water for the purpose of household sanitation and need. Frequently washing hands by soaps and water (after toilet and before meal) mean value is 3.79 which indicates that the slum dwellers perception of washing hand was practiced sometimes and some cases often (Table 13.5). As transmission is typically facilitated by physical contact between humans, hand washing is a widespread and crucial hygiene technique to protect against infectious microorganisms [6, 22, 29, 31]. Mean value of wash cloth, shoes, and other wearing after coming back home from outside is 3.17 (Table 13.5). Which indicates that slum dwellers were less in washing usable clothing and other materials. In term of cleaning room, house and toilets by using antiseptic or disinfectant solutions were less in mean value (2.50. It reflected that the slum dwellers are not aware about the disinfection or they could not able to afford the antiseptic materials for cleaning household. According to earlier studies, using antiseptics and disinfectants to clean your home is a good way to avoid viral infections as covid-19 [5, 7]. Another important WASH practice was that does they properly depose used mask, hand gloves and tissue in separate covered bins or bags. This indicator revealed average value of 2.42, which indicated that it was practiced rarely by the total slum dwellers. But in order to stop the spread of the virus and safeguard the environment, it's equally crucial to properly dispose of old protective equipment like masks and gloves [21, 22, 36, 41]. It is advised that old or used masks, gloves, and other infectious debris be disposed of separately to stop the spread of viruses and other harmful organisms [12, 22, 28]. Therefore, majority percent of the slum dwellers reported in the study that they through their used mask, hand globes and used tissues here and there beside their dwelling places. In addition, dispose of all types of waste together in a specific/community waste disposal point was also revealed and the mean score was 2.72 (Table 13.5). This finding indicated rarely disposal practice of household waste materials in specific point in the slum dwellers community. Overall, WASH practice by the slum dwellers was found poor in nature. Because unaware of severity of affected rate and unawareness of covid-19 pandemic the slum dwellers are not practicing WASH to remain safe.

However, the present study revealed WASH security of the slum dwellers by a five point Likert scale. It found that overall performance value of available water to maintain hygiene was 0.85, which indicates the slum dwellers have available water for daily use and maintain hygiene. Individually, inner (0.87) and outer (0.88) zone slum dwellers always get the water they need for household or hygiene purpose. Whereas, middle slum zone dwellers level of security was found often (0.80). Slum dwellers perception and the security of washing hand via soap in different time was measured. All three slum zone revealed that level of security was found often (0.76), which reflected that the slum dwellers often use to wash hands before meal or after toilet to remain safe during covid-19 pandemic. The study also found the level of security in case of personal cleanliness, such as, washing clothes, shoes, and others wear after returning home from outside. It indicated that inner (0.68) and outer (0.64) zone slum dwellers had often wash their clothes after returning home. On the contrary, the middle slum zone security value was found less (0.60) among the three slum zone. The covid-19 pandemic have manage to taught people some lesson about hygiene but the slum dwellers were very much behind in this particular case. In terms of house cleaning (i.e. cleaning rooms, house and toilets by antiseptic or disinfectant solutions) was practiced by few households which indicated low WASH security of the slum dwellers. The result revealed that all three slum zones had used the antiseptic or disinfectant solutions in sometime (0.50). Individually, the security values were 0.55 in inner, 0.41 in middle and 0.54 in outer slum zone respectively (Table 13.6). Disposal of all types of waste in a specific point is observed in the survey zones. The level security in this case was measured as sometimes (0.54). Though, highly secured value was observed in inner (0.82) zone. While middle (0.50) and outer (0.42) slum zones security value revealed that it was practiced in sometimes. It indicated that the inner slum zone dwellers were more secure in this case of WASH practice.

Table 13.5 Descriptive statistics of WASH practice during covid-19

WASH measures	Minimum	Maximum	Mean	Std. deviation	Skewness	Kurtosis
Available water for personal hygiene	1	5	4.25	0.856	− 0.963	0.288
Soap using practice (especially after toilet and before taking meal)	1	5	3.79	0.973	− 0.509	− 0.373
Wash wearable materials after coming back from outside	1	5	3.17	0.961	− 0.174	− 0.250
Use of disinfectant solutions in household	1	5	2.50	1.346	0.352	− 1.182
Properly dispose of used mask, hand gloves and tissue in separate covered bins or bags	1	5	2.42	1.607	0.630	− 1.222
Dispose waste in disposal point	1	5	2.72	1.766	0.288	− 1.702
Assessment scale						
Never (1)	Rarely (2)	Sometimes (3)	Often (4)		Always (5)	

WASHSi revealed that the inner slum zone dwellers security level was high (0.74) than middle and outer slum zone. The high level of WASHSi of inner slum zone dwellers was calculated because they maintain five types of level of security, such as, water availability for maintain hygiene, practice of soap using, personal cleanliness, house cleaning practice, and practice of using disposal points in slums were higher. But the middle and outer slum zone dwellers WASHSi were found moderate due to moderate practice of WASH during covid-19 pandemic. Overall, the slum dwellers WASHSi were found moderate (0.66) during covid-19 pandemic (Table 13.6). Which indicated that WASH security was hampered during the covid-19 in slum community. However, a radar diagram has been shown in Fig. 13.4 by using the WAI value revealed in the current study. This diagram symbolized the zero value in the core and increases access to WASH to a value of 1.00 [46]. The radar diagram indicated that inner zone slum dwellers were highly secured to WASH and all five WASH security variables resulted in a good pentagon structure for better calculated value than middle and outer zone. Inner zone slum dwellers practice of using disposal points were much better than the middle an outer zone slum dwellers (Fig. 13.4).

Table 13.6 Performance of WASH practice in slum zones

	Inner	Middle	Outer	Overall performance
Water availability to maintain hygiene				
WAI	0.87	0.80	0.88	0.85
Level of security	Always	Often	Always	Always
Practice of soap using				
WAI	0.78	0.71	0.79	0.76
Level of security	Often	Often	Often	Often
Persons cleanliness				
WAI	0.68	0.60	0.64	0.63
Level of security	Often	Sometimes	Often	Often
House cleaning practice				
WAI	0.55	0.41	0.54	0.50
Level of security	Sometimes	Sometimes	Sometimes	Sometimes
Practice of using disposal point in slums				
WAI	0.82	0.50	0.42	0.54
Level of security	Always	Sometimes	Sometimes	Sometimes
WASHSi				
WAI	0.74	0.60	0.65	0.66
Level of security	High	Moderate	Moderate	Moderate

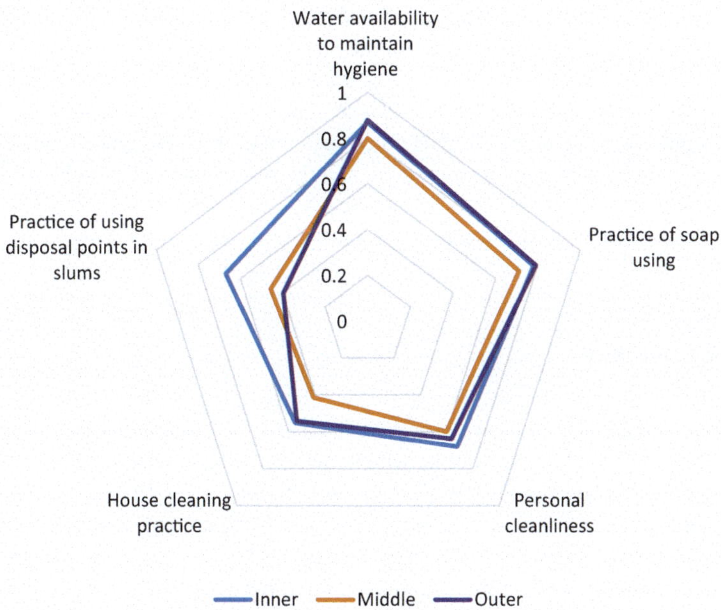

Fig. 13.4 WASH practice based on WAI

13.4 Discussion

The present study have revealed the WASH security of the urban slum dwellers of the RCC. It has also investigated the water, sanitation-hygiene sanitation status of the slum dwellers. During the covid-19 pandemic majority percent of the inner and outer zone slum dwellers used to wear face mask for remain safe. Whereas majority percent of the middle zone slum dwellers used the washing hands without soap. In terms of drinking water facilities majority percent of outer zone slum dwellers used to drink tube well water and their water source was not arsenic tested. In addition, the perception of the slum dwellers was that their drinking water sources are safe. Majority of the respondents of the all three slum zone reported that they share one tube well which indicated that the slum dwellers were mostly insecure. In terms of sanitation security, majority of the outer zone dwellers use the common toilet defecation which revealed that outer zone slum dwellers were insecure than the inner and middle zone slum dwellers. In addition, average number of common toilet user were revealed as higher in middle zone slum dwellers. Pucca (water sealed) defecation system were mostly observed in the inner zone than middle and outer zone. Though, ring slab defecation system were found in most cases in the middle and outer zone slum dwellers. Furthermore, the WASH practice were calculated in the present study which reflected that the available water for personal hygiene, use of soap, wash wearable materials after coming back from outside, use of disinfectant solutions in household, properly dispose of used mask, hand gloves and tissue in separate covered bins or bags, and dispose waste in disposal point were practiced poor by the slum dwellers. In addition, the WASH security revealed that inner zone slum dwellers were highly secured during the covid-19 pandemic because of perfect practice of WASH. Though, middle and outer zone slum dwellers were moderately secured. This moderate WASH security can be improved by the middle and outer slum zones by implementing a basic WASH practice in the future pandemic situation, such as, covid-19 can be prevented to spread out and the risk of illness through contagious germs and viruses which could be controlled by proper practice of hand washing with soap and clean water [13] which is found consistent with the present study.

13.5 Conclusion

The study revealed that slum dwellers WASH security varied among the slum zones. It reflected the status of the sanitation, water and hygiene of the slum dwellers before concluding the security status of WASH. Overall, majority (37.5%) of slum dwellers used to wear face mask which is most common hygiene and sanitation practice in the slum zones. It also reflected that overall 36.9% of the slum dwellers did not practice washing hands with soap during the pandemic which mainly introduce the insecurity of WASH. Water sharing was observed higher in all slum zones that indicated water insecurity. In addition, findings revealed moderate WASH security to combat pandemic by the middle and outer zone slum dwellers. Though, highly secure zone was revealed as the inner zone due to their minimum WASH practice during the covid-19 pandemic. Furthermore, slum dwellers are not aware about the spreading process of covid-19 because of low literacy and training that decreases the security of WASH practice. The major limitation of the present study is that it revealed WASH security status of the urban slum dwellers during phase of covid-19 pandemic. Another important limitation is that this study focused on the RCC slum areas, which need to be expanded to rural areas to reveal a broader scale of WASH security status of a region. Though, this study would be a preliminary guideline for the policy makers to generate new thinking of WASH as it revealed the WASH security and practice capability of the urban slum dwellers. Further research should be done by incorporating other city corporations of Bangladesh to investigate the status of the WASH practice and security of the poor urban slum dwellers.

Acknowledgements We acknowledge all the respondents who participated in our questionnaire survey. We also acknowledge the Institute of Water and Flood Management (IWFM), BUET, Bangladesh where this research was verbally presented on the 9th International Conference on Water and Flood Management (October 14–16, 2023).

References

1. Ahmad SA, Khan MH (2023) Groundwater arsenic contamination and its health effects in Bangladesh. In: Handbook of arsenic toxicology. Academic, pp 51–77
2. Amuakwa-Mensah F, Klege RA, Adom PK, Köhlin G (2021) COVID-19 and handwashing: implications for water use in Sub-Saharan Africa. Water Resour Econ 36:100189
3. Bera A, Das S, Pani A, Bera B, Shit PK (2022) Assessment of household water consumption during COVID-19 pandemic: a cross-sectional web-based study in India. Sustain Water Resour Manag 8(3):78
4. Berihun G, Adane M, Walle Z, Abebe M, Alemnew Y, Natnael T, Andualem A, Ademe S, Tegegne B, Teshome D, Berhanu L (2022). Access to and challenges in water, sanitation, and hygiene in healthcare facilities during the early phase of the COVID-19 pandemic in Ethiopia: a mixed-methods evaluation. PLoS One 17(5):e0268272
5. Biswal M, Kanaujia R, Angrup A, Ray P, Singh SM (2020) Disinfection tunnels: potentially counterproductive in the context of a prolonged pandemic of COVID-19. Public Health 183:48
6. Brauer M, Zhao JT, Bennitt FB, Stanaway JD (2020) Global access to handwashing: implications for COVID-19 control in low-income countries. Environ Health Perspect 128(5):057005
7. Centers for Disease Control and Prevention (CDC) (2020) Cleaning and disinfecting your home. Page Last Reviewed: 27 May 2020. Retrieved from: https://www.cdc.gov/coronavirus/2019ncov/index.html?CDC_AA_refVal=https%3A%2F%2Fwww.cdc.gov%2Fcoronavirus%2F2019-ncov%2Fprevent-getting-sick%2Fdisinfecting-your-home.html. Accessed 29 Nov 2022
8. Chand S, Shastry CS, Hiremath S, Joel JJ, Bhat CK, Mateti UV (2020) Water, sanitation, hygiene and biomedical waste disposal in the healthcare system: a review. Biomedicine 40(1):14–19
9. Cooper R (2020) Water security beyond Covid-19. K4D Helpdesk Report 803. Institute of Development Studies, Brighton, UK. Retrieved from: https://opendocs.ids.ac.uk/opendocs/handle/20.500.12413/15240. Accessed 29 Nov 2022

10. Das A, Ghosh S, Das K, Dutta I, Basu T, Das M (2020) Re:(In) visible impact of inadequate WaSH Provision on COVID-19 incidences can be not be ignored in large and megacities of India. Public Health 185:34
11. Desye B (2021) COVID-19 pandemic and water, sanitation, and hygiene: impacts, challenges, and mitigation strategies. Environ Health Insights 15:11786302211029448
12. Directorate General of Health Services (DGHS) (2020) National guideline on infection prevention and control in healthcare settings with additional measures for covid-19. Directorate General of Health Services, Ministry of Health and family welfare, Government of the people's Republic of Bangladesh, Dhaka. https://amr.cdc.gov.bd/wp-content/uploads/2018/10/02-IPC-guideline-v6.pdf. Accessed 29 Nov 2022
13. Donde OO, Atoni E, Muia AW, Yillia PT (2021) COVID-19 pandemic: water, sanitation and hygiene (WASH) as a critical control measure remains a major challenge in low-income countries. Water Res 191:116793
14. Ferdous MZ, Islam MS, Sikder MT, Mosaddek ASM, Zegarra-Valdivia JA, Gozal D (2020) Knowledge, attitude, and practice regarding COVID-19 outbreak in Bangladesh: an online-based cross-sectional study. PLoS ONE 15(10):e0239254
15. Giné-Garriga R, Delepiere A, Ward R, Alvarez-Sala J, Alvarez-Murillo I, Mariezcurrena V, Sandberg HG, Saikia P, Avello P, Thakar K, Ibrahim E (2021) COVID-19 water, sanitation, and hygiene response: review of measures and initiatives adopted by governments, regulators, utilities, and other stakeholders in 84 countries. Sci Total Environ 795:148789
16. Habib MA, Ahmed MM, Aziz M, Beg MRA, Hoque ME (2021) Municipal solid waste management and waste-to-energy potential from Rajshahi City Corporation in Bangladesh. Appl Sci 11(9):3744
17. Halder PK, Paul N, Hoque ME, Hoque AM, Parvez MS, Rahman MH, Ali M (2014) Municipal solid waste and its management in Rajshahi City, Bangladesh: a source of energy. Int J Renew Energy Res 4(1):168–175
18. Hasan S, Ullah SA (2022) State of E-governance at urban local government in Bangladesh: a paradigm of Rajshahi City corporation
19. Hossain MN, Paul SK (2018) Vulnerability factors and effectiveness of disaster mitigation measures in the Bangladesh coast. Earth Syst Environ 2(1):55–65
20. Howard G, Bartram J, Brocklehurst C, Colford Jr JM, Costa F, Cunliffe D, Dreibelbis R, Eisenberg JN, Evans B, Girones R, Hrudey S (2020) COVID-19: urgent actions, critical reflections and future relevance of 'WaSH': lessons for the current and future pandemics. J Water Health 18(5):613–630
21. Islam SDU, Safiq MB, Bodrud-Doza M, Mamun MA (2020) Perception and attitudes toward PPE-related waste disposal amid COVID-19 in Bangladesh: an exploratory study. Front Public Health 8:592345
22. Islam SDU, Mondal PK, Ojong N, Bodrud-Doza M, Siddique MAB, Hossain M, Mamun MA (2021) Water, sanitation, hygiene and waste disposal practices as COVID-19 response strategy: insights from Bangladesh. Environ Dev Sustain 23(8):11953–11974
23. Khan MS, Paul SK (2023) Fresh water management in coastal Bangladesh: preparedness and adaptation. Discover Water 3(1):27
24. Khan MS, Paul SK (2023) Sanitation-hygiene knowledge, practices and human health impacts: insights from coastal Bangladesh. Geosfera Indonesia 8(2)
25. Khatib MN, Sinha A, Mishra G, Quazi SZ, Gaidhane S, Saxena D, Zahiruddin QS (2022) WASH to control COVID-19: a rapid review. Front Public Health 10:976423
26. Kim J, Hagen E, Muindi Z, Mbonglou G, Laituri M (2022) An examination of water, sanitation, and hygiene (WASH) accessibility and opportunity in urban informal settlements during the COVID-19 pandemic: evidence from Nairobi, Kenya. Sci Total Environ 823:153398
27. Kothari CR (2004) Research methodology: methods and techniques. New Age International
28. Ma Y, Lin X, Wu A, Huang Q, Li X, Yan J (2020) Suggested guidelines for emergency treatment of medical waste during COVID-19: Chinese experience. Waste Disposal Sustain Energy 2:81–84

29. Mushi V, Shao M (2020) Tailoring of the ongoing water, sanitation and hygiene interventions for prevention and control of COVID-19. Trop Med Health 48(1):47
30. Nguyen TPL, Pattanarsi S (2022) WASH, vulnerability, severity, and the response of urban slum dwellers to the COVID-19 pandemic. J Water Sanit Hygiene Dev 12(8):600–611
31. Nicolaides C, Avraam D, Cueto-Felgueroso L, González MC, Juanes R (2020) Hand-hygiene mitigation strategies against global disease spreading through the air transportation network. Risk Anal 40(4):723–740
32. Pandey R, Gautam V, Pal R, Bandhey H, Dhingra LS, Misra V, Sethi T (2022) A machine learning application for raising wash awareness in the times of covid-19 pandemic. Sci Rep 12(1):810
33. Paul SK (2019) Post cyclone household food security in coastal Bangladesh. People at risk: disaster and despair. Disaster Research Training and Management Centre, University of Dhaka, Bangladesh, pp 185–209
34. Poague KI, Blanford JI, Martínez JA, Anthonj C (2023) Water, sanitation and hygiene (WASH) in schools in Brazil pre-and peri-COVID-19 pandemic: are schools making any progress? Int J Hyg Environ Health 247:114069
35. Rahman MM, Tasnim F, Uddin A, Chayan MSI, Arif MSI, Hossain MT (2023) Assessing vulnerability in ethnic Munda community: a study on a cyclone-prone area of Bangladesh. Int J Disaster Risk Reduct 95:103884
36. Rume T, Islam SDU (2020) Environmental effects of COVID-19 pandemic and potential strategies of sustainability. Heliyon 6(9)
37. Sayeed A, Rahman MH, Bundschuh J, Herath I, Ahmed F, Bhattacharya P, Hasan MT (2021) Handwashing with soap: a concern for overuse of water amidst the COVID-19 pandemic in Bangladesh. Groundw Sustain Dev 13:100561
38. Shrestha A, Kunwar BM, Meierhofer R (2022) Water, sanitation, hygiene practices, health and nutritional status among children before and during the COVID-19 pandemic: longitudinal evidence from remote areas of Dailekh and Achham districts in Nepal. BMC Public Health 22(1):2035
39. Shuaib ASM, Rana MMP (2020) Assessing water supply for the urban poor in Rajshahi City, Bangladesh. Manag Environ Qual: Int J 31(1):75–88
40. Shuja KH, Aqeel M, Khan EA, Abbas J (2020) Letter to highlight the effects of isolation on elderly during COVID-19 outbreak. Int J Geriatric Psychiatry 35(12)
41. Singh N, Tang Y, Ogunseitan OA (2020) Environmentally sustainable management of used personal protective equipment. Environ Sci Technol 54(14):8500–8502
42. Sultana R, Alam MS (2023) Access to ecosystem services: riverside informal settlement dwellers' perception in Rajshahi City, Bangladesh. Curr Res Environ Sustain 5:100216
43. Tabassum T, Farzana M, Nahar AN, Araf Y, Ullah MA, Rahaman TI, Faruqui NA, Prottoy MN, Anwar S, Ali N, Hosen MJ (2023) COVID-19 in Bangladesh: wave-centric assessments and mitigation measures for future pandemics. Heliyon 9(10)
44. Tawsif S, Alam MS, Al-Maruf A (2022) How households adapt to heat wave for livable habitat? A case of medium-sized city in Bangladesh. Curr Res Environ Sustain 4:100159
45. Tawsif S, Paul SK, Khan MS (2024) Livelihood vulnerability assessment of slum dwellers in Rajshahi Bangladesh: capital indices-based approach. J Environ Stud Sci pp 1–18. https://doi.org/10.1007/s13412-024-00988-0
46. Tawsif S, Paul SK, Khan MS (2024) Changing pattern of livelihood capitals of urban slum dwellers during COVID-19 pandemic. Int J Hum Capital Urban Manag 9(1):101–118
47. Tripathy S, Sahu L (2021) COVID 19 pandemic and behavioural change toward water hygiene and sanitation (WASH): a study in rural Odisha, India. Changes 3:5
48. van Gurp M, Riad IM, Islam KA, Islam MS, Geervliet RM, Bakker MI (2022) WASH, nutrition and health-seeking behavior during COVID-19 lockdowns: evidence from rural Bangladesh. PLoS ONE 17(12):e0278525
49. World Bank (WB) (2020) COVID-19 has worsened the woes of South Asia's informal sector. Retrieved from: https://blogs.worldbank.org/endpovertyinsouthasia/covid-19-has-worsened-woes-south-asias-informal-sector. Accessed 29 Oct 2022

50. World Health Organization (WHO) (2020) Timeline: WHO's COVID-19 response. Retrieved from: https://www.who.int/emergencies/diseases/novel-coronavirus-2019/interactive-timeline. Accessed 1 Sept 2023
51. World Health Organization (WHO), United Nations Children's Fund (UNICEF) (2020) Water, sanitation, hygiene, and waste management for the COVID-19 virus: interim guidance, 19 March 2020. World Health Organization. https://iris.who.int/handle/10665/331499. License: CC BY-NC-SA 3.0 IGO. Retrieved from: https://apps.who.int/iris/bitstream/handle/10665/331499/WHO-2019-nCoV-IPC_WASH-2020.2-eng.pdf?sequence=1&isAllowed=y. Accessed on 29 November, 2022
52. World Health Organization (WHO) (2020) Water, sanitation, hygiene, and waste management for SARS-CoV-2, the virus that causes COVID-19: interim guidance, 29 Jul 2020. Retrieved from: https://apps.who.int/iris/bitstream/handle/10665/333560/WHO-2019-nCoV-IPC_WASH-2020.4-eng.pdf. Accessed 1 Sept 2023
53. World Health Organization (WHO) (2021) Coronavirus disease (COVID-19) advice for the public: When and how to use masks. Retrieved from: https://www.who.int/emergencies/diseases/novel-coronavirus-2019/advice-for-public/when-and-how-to-use-masks. Accessed on 29 November, 2022
54. Zhu N, Zhang D, Wang W, Li X, Yang B, Song J, Zhao X, Huang B, Shi W, Lu R, Niu P (2020) A novel coronavirus from patients with pneumonia in China, 2019. New Engl J Med 382(8):727–733

Chapter 14
Solarine: A Solar Panel-Aided Efficient and Automatic Rooftop Rainwater Harvesting System Using Arduino

K. M. Sadman Sakib and Nazmush Shahadot Safin

Abstract Bangladesh endures recurrent water scarcity because of its geographical and climatic qualities. The traditional water supply infrastructure cannot handle the increased demand for potable water. Rainwater collection is a sustainable and ecologically aware alternative to traditional water procurement methods. The combination of this technology and rooftop solar panels is a fresh and inventive approach to solving Bangladesh's concurrent water and energy problems. In Bangladesh, there is little literature on rainwater collection, and scholarly research on the integration of this technology with rooftop solar panels is rare. Rainwater collection can provide a sustainable and viable water source in both urban and rural Bangladesh, according to empirical findings. This is where our hypothesis comes from, which is, combining the traditional solar panels in a v-shaped pattern can turn it into an effective rainwater catchment area, which can increase the harvest amount of rainwater greatly. Performing small scale experiments and analysing the captured data thoroughly, we have found our hypothesis is true, and combining solar panels as a v-shaped catchment area actually increases the harvest of rainwater. This study gives us the empirical proof that using this innovative idea helps significantly in both urban and rural areas for RRWHS along with solar energy conversion.

Keywords Environmental management · Rainwater harvesting · Water conservation · RRWHS · Solar energy

K. M. S. Sakib (✉)
Department of Electrical and Electronic Engineering, Ahsanullah University of Science and Technology, Dhaka, Bangladesh
e-mail: sadmansakibp2854@gmail.com

N. S. Safin
Department of Geography and Environment, University of Dhaka, Dhaka, Bangladesh

G. M. Tarekul Islam et al. (eds.), *Sustainable Water Management and Environmental Resilience*, Springer Water, https://doi.org/10.1007/978-3-031-93350-9_14

14.1 Introduction

A significant segment of the global population lacks access to secure water resources. According to estimates by WHO/UNICEF, approximately 1.1 billion individuals lack access to what is termed as "improved drinking-water sources" [1]. Despite substantial initiatives aimed at providing secure, piped community water to the worldwide populace, the actuality remains that the provision of safe water will not be universally attainable in the foreseeable future. The scenario in Bangladesh is no better, despite being a low-lying delta region, having a monsoon climate. Scarcity of freshwater prevails throughout the country, due to arsenic, pollution, poor management, In Bangladesh, a nation with a population of 165 million, the conditions are notably dire. The country is contending with a serious water scarcity predicament that is detrimentally impacting its economic stability and overall welfare. The situation is further aggravated by factors such as climate change, rapid urban expansion, and industrial development. Merely 62% of the population in Bangladesh has access to upgraded water sources, underscoring the gravity of the water accessibility challenges faced by a significant portion of the population [2]. Thus, finding an alternative solution for drinking water is in a dire need for the people of both rural and urban locality. Sole dependency on groundwater sources is plummeting the water reservoirs rapidly and could soon reach a vulnerable limit. On the other hand, being a monsoon country, Bangladesh receives an annual rainfall of 2200 mm, with at least 1500 mm in most regions of the country [3], which clearly hints us to use it as a pivotal source of freshwater. Implementing innovative modes of harvesting rainwater is becoming a necessary, rather than a fancy adjunction to a household. Harvesting rainwater from the rooftop is a rather clever way of water harvesting method, because the rooftop gets less dirty due to its even or almost even surface. This method, combined with the solar panel, which can be used as a catchment area for rainwater, is a marvellous approach to kill two birds with one stone—generate solar energy on a regular day, and capture the rainwater during a rainy day. This paper seeks the efficient way of using solar panels as a rooftop rainwater harvesting catchment area which can alternate between solar capture, as well as generating solar energy.

The small-scale experiment was conducted in Dhaka city. Although our study doesn't have a particular study area, due to having a small-scale hypothetical idea, we can safely say that the study area is Dhaka (Fig. 14.1).

14.2 Methodology

The proposed model shown below (Fig. 14.2) has four segments of workflow method. (a) Solar panel catchment area, (b) runoff channel, (c) waste water storage for cleaning purpose of the solar panels, (d) clean waterflow switch along with water flow metre.

The model will work with two modes, (a) Solar Energy Mode, and (b) Rainwater Harvesting Mode. While in the solar energy mode, the solar panels will work as a

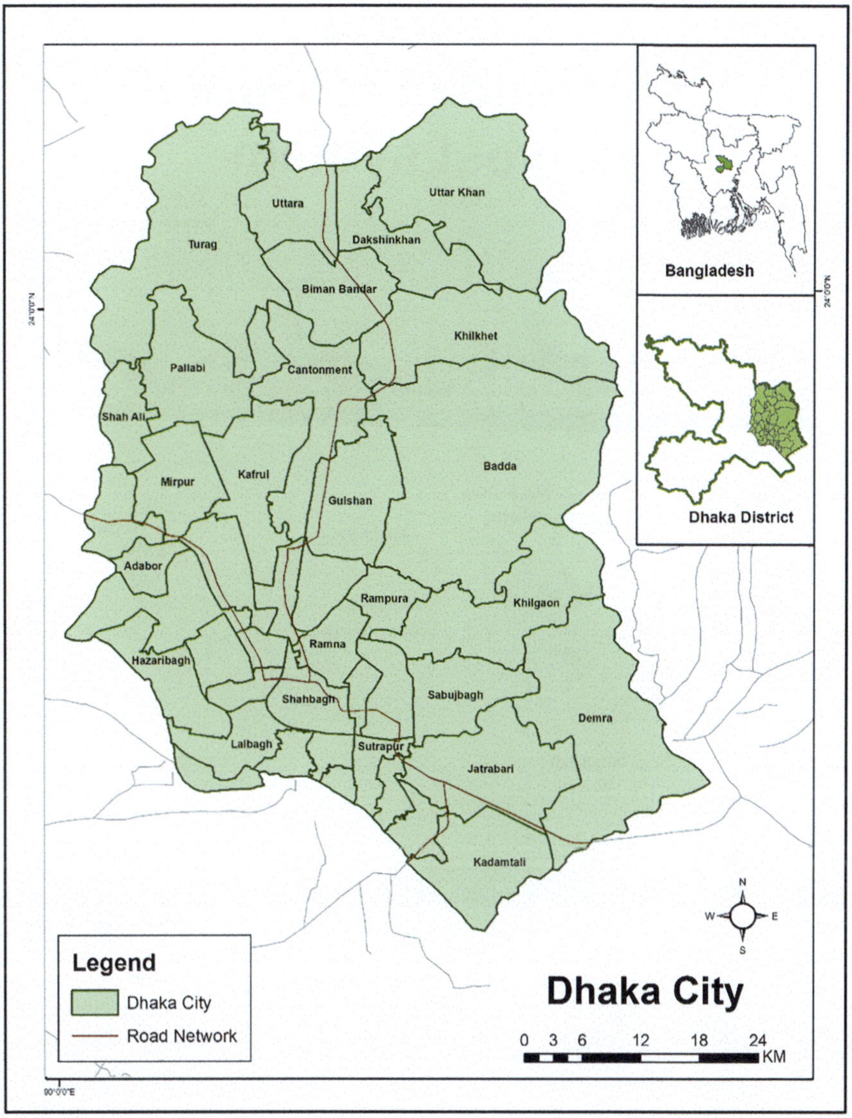

Fig. 14.1 Study area—Dhaka City

traditional solar energy capturing device, providing safe and renewable electricity to the household (Fig. 14.3).

And while triggering the device by sensing the presence of rain droplets through a rain sensor, the alternating solar panels will shift towards the opposite direction, creating a V-shaped area, which will function as rainwater harvesting catchment area (Fig. 14.4).

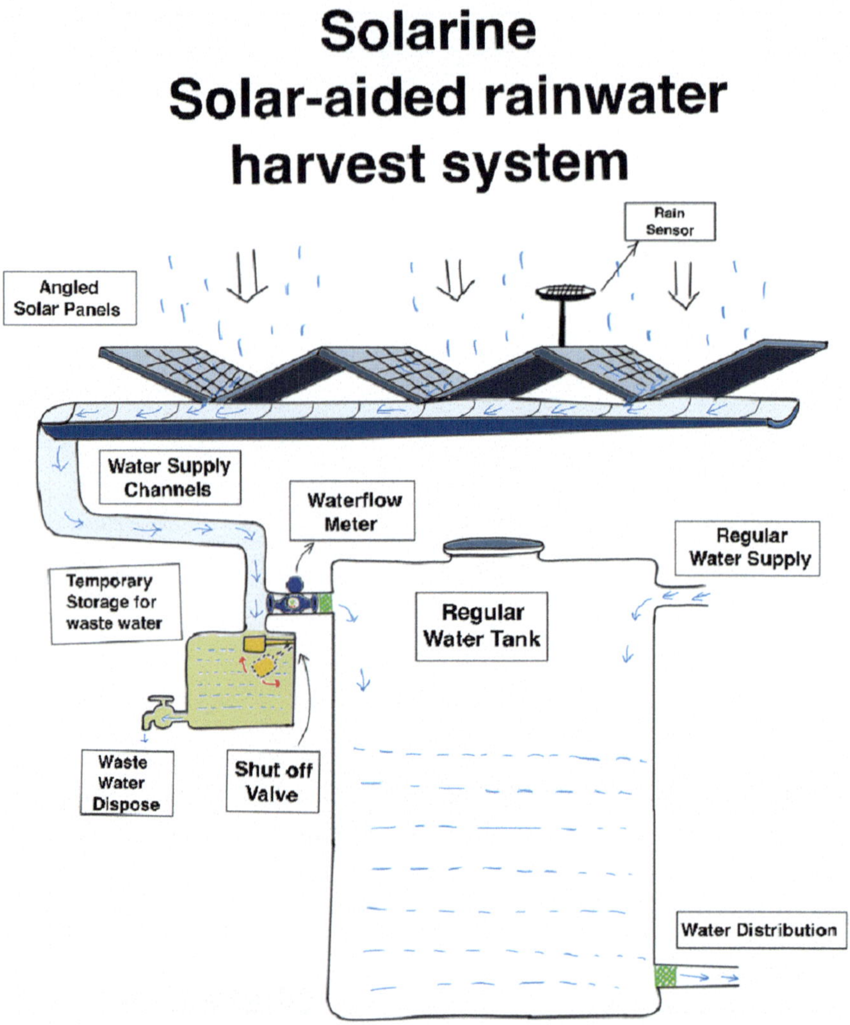

Fig. 14.2 Schematic overview: the Solarine—solar aided rainwater harvest system

Our primary goal was to prove that a "double-panel solar panel (v-shaped)" captures more water than a "single-panel solar panel" in a rainwater harvesting system. The optimal solar radiation collecting angle for Dhaka is 23° [4], and we maintained the catchment time constant to simplify the small-scale experiment that imitated real-life rainfall. Servo motors and Arduino are used to adjust the angle. In accordance with customs, the solar panel will be installed in a regular manner. During the rainy season, however, the rain sensor will detect rain, send a signal to our Arduino Uno [5], which will activate the servo motor to set our panels in a V-shape, allowing us to use the solar panels as a rainwater collection area. We assumed an

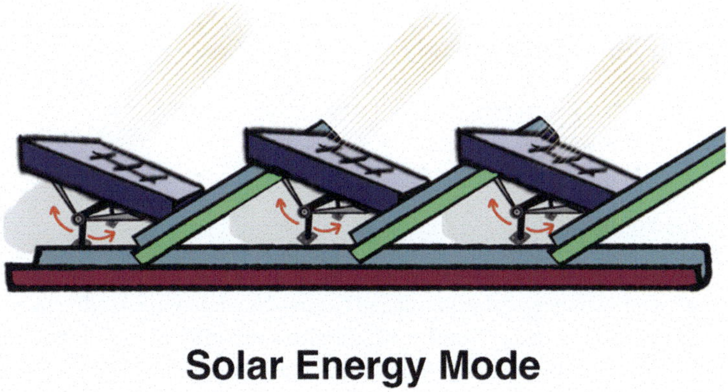

Fig. 14.3 Regular solar energy mode

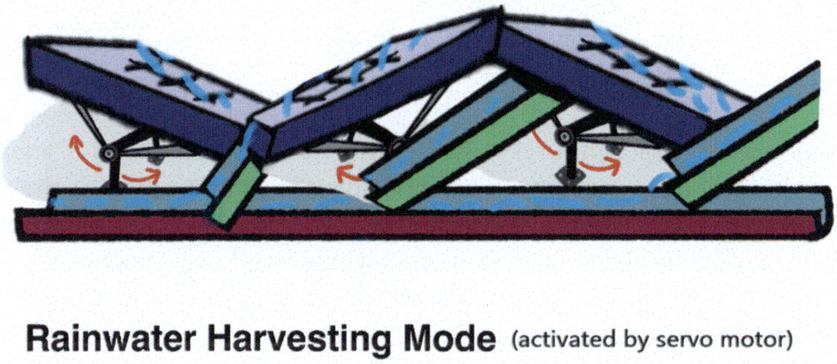

Fig. 14.4 Rainwater harvesting mode (activated by servo motor)

aggregate of low, medium, and high rainfall intensity. We found substantial differences in rainwater gathering between "double panel" and "single panel." To test both panels' catchments, we collected 20 samples per category. Analysing the data showed that "double-panel v-shaped solar panels" outperformed "single-panel solar panels" according to Formula (14.1).

$$p = \frac{(d - s)}{d} \times 100\% \tag{14.1}$$

here

p = percentage change
d = double panel average water catchment
s = single panel average water catchment.

The percentage change for each broad category also improved along with the increase of intensity for our artificial rainfall. By plotting the data, we found out that independent variable percentage change follows quadratic equations where the dependent variable is rainfall intensity.

$$p = 0.00176f2 + 0.03398f + 17.95637 \tag{14.2}$$

here

p = percentage change
f = rain intensity.

We have used Levenberg–Marquardt and Gauss–Newton optimization techniques, in python programming using NumPy, Matplotlib and SciPy library to find the optimised parameter of coefficient. Our hypothesis to use a Quadratic function was right. It actually goes along with the data points. But we don't actually know for absurd or unseen data sets how we can use this to evaluate all the unseen data by using the same a, b, c of the earlier equations founded by the known data. So now our goal is to find the optimum value of a, b, c where we used two popular optimization techniques, Levenberg–Marquardt and Gauss–Newton algorithm optimization techniques.

Gauss–Newton Algorithm

The Gauss–Newton algorithm serves as a computational approach for addressing non-linear least squares problems, encompassing the minimization of the sum of squared function values. This algorithm extends the principles of Newton's method, specifically tailored for the pursuit of minima within non-linear functions. Instances of non-linear least squares problems manifest in contexts such as non-linear regression, wherein the objective is to determine parameters within a model to optimise alignment with observed data. The nomenclature of this method derives from the renowned mathematicians Carl Friedrich Gauss and Isaac Newton.

Levenberg–Marquardt Algorithm

The Levenberg–Marquardt algorithm, recognized as a Damped Least Square Method (DLS), is employed for the resolution of non-linear least square problems, where the minimization task emerges in the context of least square curve fitting. This method integrates elements of the Gauss–Newton Algorithm (GNA) and employs the technique of Gradient Descent. Originally introduced in 1944 by Kenneth Levenberg during his tenure at the Frankford Army Arsenal, the algorithm experienced a subsequent re-emergence in 1963, independently by Donald Marquardt, a statistician at DuPont, and Girard, Wynne, and Morrison.

After using these two techniques we got the table.

As our dataset is very small both the optimizations show the same results for a, b, c. Now we can come back to our proposed function with the value of a, b, c.

14.3 Results

14.3.1 Dataset Explanation

Our experimental data are presented in Table 14.1, where we compare the quantity of precipitation collected during our experiment to the percentage change from a single to a double panel. There is also an increase in percentage change, which indicates that the greater the rainfall intensity, the more efficiently our model will function as the V-shaped catchment area will reduce water droplet ricochet (Table 14.2).

After Collecting the dataset, our main goal was to observe how our data is actually doing. So, we use the NumPy Library with Matplotlib to plot a graph with Python. NumPy Library takes the input as an array while Matplotlib Library helps to visualise the data (Fig. 14.5).

The data points shows Nonlinear properties that's why, we can compare it with a algebraic, quadratic functions which carries non-linear property. A general Quadratic functions

$$y = ax^2 + bx + c$$

After Implementing the data points as an input of Quadratic Functions we got the value of different y and then plot it below by using Matplotlib (Fig. 14.6).

Table 14.1 Algorithm analysis values

Gauss–Newton algorithm	Value	Levenberg–Marquardt algorithm	Value
a	0.001768	a	0.001768
b	0.033985	b	0.033985
c	17.956375	c	17.956375

Table 14.2 Precipitation droplets collection analysis

Type	Flow	Rain intensity (mm)	Average rain catchment in ml (according to experiment)	Percentage change
Single panel	Low	26.42	897	20.089
Double panel			1122.5	
Single panel	Medium	48.57	2147.5	23.780
Double panel			2817.5	
Single panel	High	88.57	3013.5	34.843
Double panel			4625	

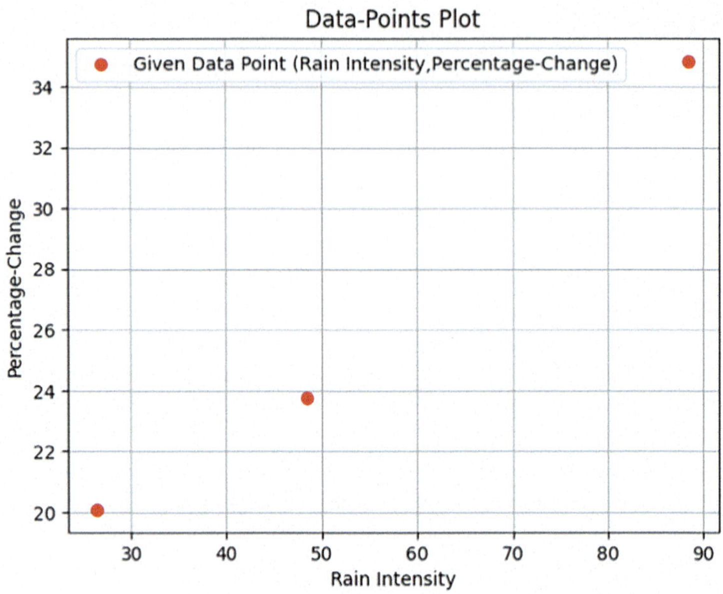

Fig. 14.5 Data points plot of known three types of flow

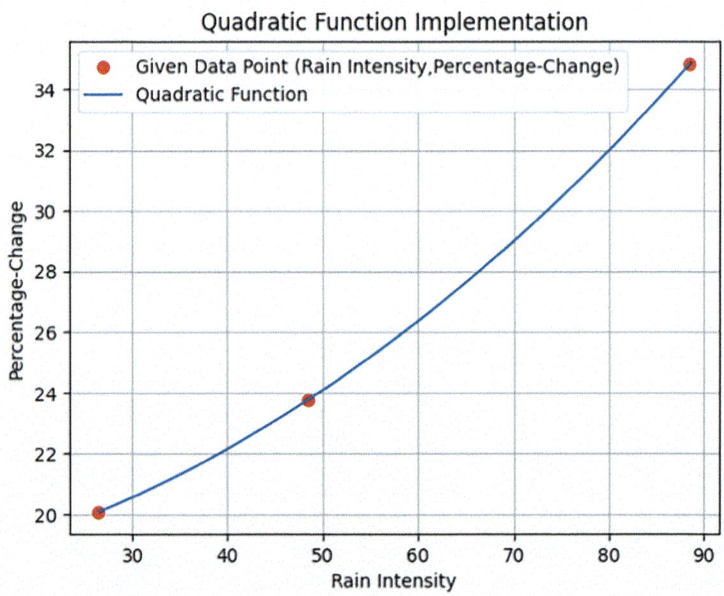

Fig. 14.6 Quadratic function implementation on known three types of flow

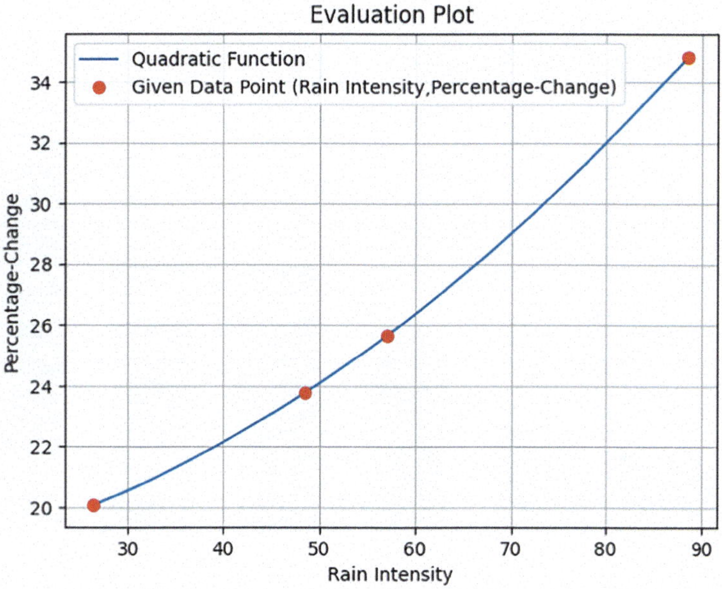

Fig. 14.7 Evaluation plot including unknown flow type

14.3.2 Dataset Evaluation

Using Eq. (14.1), we determined the percentage changes of 20.089, 23.780, and 34.843 for the variable flow and precipitation intensity, which follows the quadratic Eq. (14.2), with the tuned coefficient values of 0.00176, 0.03398 and 17.95637. By validating with a random sample, we gained the accuracy of our experimental data sample to approximately 98.23%. Due to a lack of access to the necessary equipment for simulating actual rainfall and the inability of our water discharge valve to precisely modify the discharge rate was a bottleneck for our three modes of precipitation intensity. Consequently, we were unable to validate our model with more than one validation sample, which was also difficult to obtain.

After that we used our proposed model to plot it with the help of Matplotlib (Fig. 14.7).

14.3.3 Dataset Analysis

Rainfall intensity is demonstrated here, along with three simplified modes, low rainfall, medium rainfall and high rainfall. For the sake of simplicity and easy to analyse, the lowest rainfall amount was 897 mm on average in the experiment, ranging to about 4625 mm (Fig. 14.8).

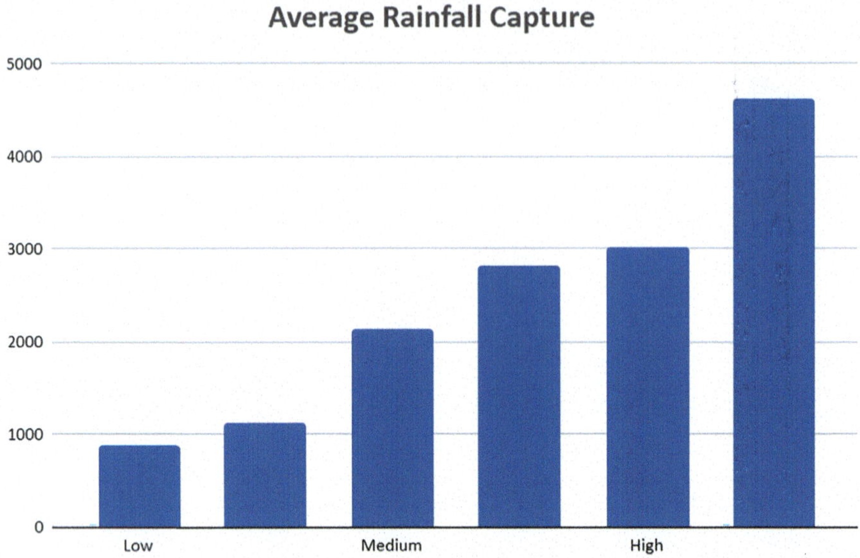

Fig. 14.8 Average rainfall capture without and with model for three types of flow

And this is where the model is visible to work. Percentage change of capturing increases along with the intensity of the rainfall increase. To be specific, the greater the rainfall happens, the more efficiently the model will work and it will capture more rain water due to the nature of water droplets being spread around on the surface and bouncing off (Fig. 14.9).

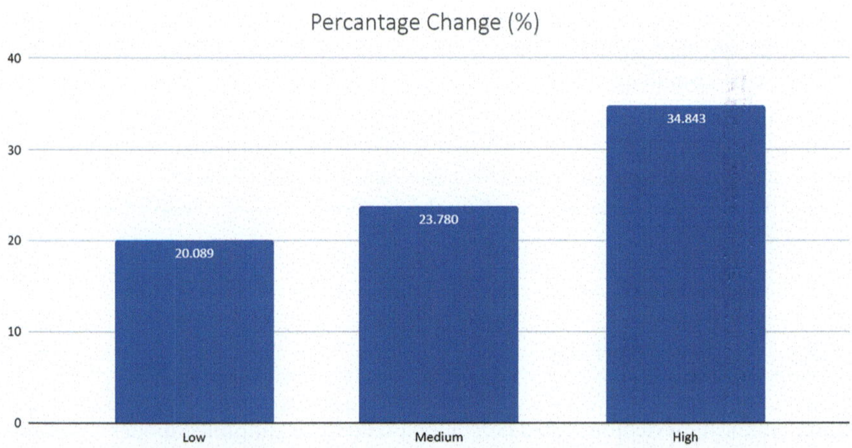

Fig. 14.9 Percentage change due to application of the proposed model

14.4 Discussion

The model works significantly greater along with the increase in the rainfall intensity, so as proving our primary hypothesis correct. These charts show the performance of v-shaped double panel solar panels is increasing as the rainfall increases (Fig. 14.10).

14.5 Conclusion

Through the execution of this investigation, valuable insights have been gained regarding the effectiveness of a V-shaped double-panel solar panel in capturing increased amounts of precipitation during heavier rainfall. The results suggest that this particular design is proficient in enhancing rainwater collection when faced with intensified rainfall conditions.

The adoption of such a model holds promising prospects, not only for bolstering our utilisation of environmentally friendly energy sources but also for serving as a significant reservoir of clean rainwater. This dual functionality has the potential to contribute substantially to advancing water conservation efforts, amplifying the efficiency of water utilisation, and consequently enhancing overall productivity.

In summary, the outcomes of this experiment advocate for the implementation of V-shaped double-panel solar collectors as a means to not only augment green energy utilisation but also to act as a noteworthy facilitator in the pursuit of sustainable water management practices.

Fig. 14.10 Significant increase in performance of the double panel in every condition

Acknowledgements First, we would like to thanks the almighty Allah, for providing enough courage, health and determination to pull this innovative idea off from our thoughts to this paper. We would like to express our sincere gratitude to our friends and family who volunteered us financially for the experiment to carry on. We have completed our mini-scaled experiment through hurdle as it was very tough to mimic the Rain Simulation. Our family's continuous motivation helps us to go through this rigorous process. Special thanks to our dearest friend, Mr. Sakib Hasan Plabon, who have motivated us throughout our whole journey, showered us with his criticisms and advices, with which we have successfully completed our vision.

References

1. WHO/UNICEF (2000) Global water supply and sanitation assessment 2000 report. World Health Organization. https://www.who.int/publications-detail-redirect/9241562021
2. World Bank (2018) Promising progress: a diagnostic of water supply, sanitation, hygiene, and poverty in Bangladesh. World Bank. https://openknowledge.worldbank.org/handle/10986/29450
3. World Bank (2020) World bank climate change knowledge portal: climatology | climate change knowledge portal. https://climateknowledgeportal.worldbank.org/country/bangladesh/climate-data-historical
4. Nazmul RB (2017) Calculating optimum angle for solar panels of Dhaka, Bangladesh for capturing maximum irradiation. In: 2017 IEEE international WIE conference on electrical and computer engineering (WIECON-ECE). IEEE, pp 25–28
5. Rani KP, Srija K, Jyothianvitha A, Ashasri M, Mamatha I, Rajesh G (2021) Rain water harvesting for smart water management using IoT. In: 2021 5th international conference on intelligent computing and control systems (ICICCS). IEEE, pp 475–478

Chapter 15
Effectiveness of Using Water Hyacinth and Submerged Aerator to Accelerate the Ammonia-Nitrate Conversion in Trimohoni Area

Anannya Ghosh Tusti, Rizwanur Rahman, and Md. Delwar Hossain

Abstract The rapid population growth and uncontrolled pollutant discharge through various point and non-point sources have endangered the water body of Trimohoni, Dhaka. The water restoration here is necessary to support economic activity, sustainability, and ecological balance. The industrial and domestic waste mixed with Trimohoni's water has caused eutrophication, resulting in algal bloom, loss of DO and aquatic species, and high turbidity. The case-controlled study assessed the efficiency of integrated use of aeration and phytoremediation for higher efficiency in improving several water quality parameters within 5 days in the laboratory setup. The reactor with water hyacinth and submerged aerators had greater ammonia–nitrogen (AN) conversion efficiency (93.9%), which was 2.3 times better than using only submerged aeration and 3 times better than natural aeration. The water quality also got better in terms of BOD, COD, TDS, and turbidity. The reactor with water hyacinth and submerged aerators was consistently seen to work very well. For better performance and further successful implications, a real-life, large-scale assessment is recommended.

Keywords Aeration · Ammonia-nitrate (AN) · Removal efficiency · Trimohoni · Water hyacinth (WH)

A. G. Tusti (✉) · R. Rahman · Md. D. Hossain
Department of Civil Engineering, Bangladesh University of Engineering and Technology, Dhaka, Bangladesh
e-mail: 1704043@ce.buet.ac.bd; anannyatusti@gmail.com

R. Rahman
e-mail: 1704178@ce.buet.ac.bd; rz12938@gmail.com

Md. D. Hossain
e-mail: delwar@ce.buet.ac.bd

G. M. Tarekul Islam et al. (eds.), *Sustainable Water Management and Environmental Resilience*, Springer Water, https://doi.org/10.1007/978-3-031-93350-9_15

15.1 Introduction

Water is an essential resource required to support economic activity, sustain life, and maintain the environment. Water is contaminated by different point and non-point sources that make the waterbody unsuitable for drinking, cooking, fish farming, irrigation, and recreational purposes. It has been necessary to treat the water bodies effectively and sustainably. According to Bashar and Fung [4], five industrial clusters in Dhaka discharge 31,685 m^3 of wastewater into waterways daily, with 31,765 kg of BOD. Trimohoni, a crucial water body in southeast Dhaka, receives intake from Jirani Canal, Norai River, Balu River, and Rampura Canal. Once a main water source for residents there, it is now polluted by domestic and industrial pollutants. Trimohoni's degraded water quality is evident from its eutrophication, offensive odor, turbid water, and poor physical and chemical parameters. As the wastewater treatment system poses a high cost, it is not feasible to treat the water in Trimohoni using a water treatment plant. This study aims to propose a cost-effective solution to the existing water pollution problem in Trimohoni, which includes the improvement of water quality parameters and the removal of ammonia–nitrogen in a significant ratio.

Nitrogen is a fundamental element present in water in different forms, namely total nitrogen (TN), total Kjeldahl nitrogen (TKN), ammonia nitrogen (NH_3–N), organic nitrogen (org-N), nitrate ($-NO_3$), and nitrite ($-NO_2$) [15]. Among them, the higher presence of ammonia–nitrogen (AN) represents the mixture of industrial effluent in the water body. As studied by Culp et al. [5], AN poses a threat to the water quality by (i) depleting the oxygen level, (ii) increasing the toxicity in water bodies (impacts biological life) and (iii) forming chloramine by reacting with chlorine (disrupts disinfection). Anderson et al. [3] reported in his study that AN also promotes eutrophication, which is a major environmental issue due to its participation in excessive toxic algae blooms. Thus, removing AN from wastewater to an acceptable limit has become necessary, which can be done using the phytoremediation technique. According to Fox et al. [7], Phytoremediation is the process of using plants to remediate polluted soil or water. As reported by El-Gendy et al. [6] and Moorhead et al. [11], aquatic microphytes and floating water hyacinths are proven to remove nutrients from wastewater at low cost and are adaptable in a wide range of pH, electric conductivity (EC), and temperature, which makes them an easy and desirable remediation technique. Effective phytoremediation relies on the choice of aquatic microphytes. Xu et al. [17] stated that *Eichhornia crassipes*, or water hyacinth (WH), is a free-floating, perennial aquatic plant with high growth rates that produces high-value products and biofuels and treats wastewater, which makes this weed the most suitable aquatic plant for phytoremediation. WH is efficient in improving water quality parameters such as biochemical oxygen demand (BOD), chemical oxygen demand (COD), turbidity, total dissolved solids (TDS), total solids (TS), total suspended solids (TSS), and nutrients. However, the use of only phytoremediation may take a longer period to significantly reduce the AN and other nutrients. In one study by Anandha Varun and Kalpana [2], it took 20 days to reduce AN and P

Table 15.1 Acceptable surface water quality limits per WHO [16] and ECR’2023

No.	Parameter name	WHO	ECR’2023			
		Drinking	Drinking	Recreation	Fishing	Industrial
1	DO (mg/L)	≥ 6.5	≥ 6	≥ 5	≥ 5	≥ 1
2	BOD (mg/L)	≤ 4.5	≤ 2	≤ 3	≤ 6	12
3	COD (mg/L)	10	10	10	50	100
4	NH_3–N (mg/L)	0.2	0.1	0.3	0.3	2.7
5	NO_3-N (mg/L)	50	7	7	7	–
6	TDS (mg/L)	≤ 300	1000			

**WHO* world health organization, *ECR* Bangladesh environment conservation rules

nearly to an acceptable limit. Combining aeration with phytoremediation can accelerate treatment efficiency by reaching permissible limits (Table 15.1) and reduce remediation time.

In diffuser aeration (a type of Submerged Bubble Aeration), air bubbles are blown from the bottom of the water body, which transfers oxygen to the body of water across the boundary of the liquid film. Oxygen supplied by submerged aerators contributes to reducing BOD and COD and increasing the AN conversion rate. Al-Ahmady [1] reported that the efficiency of oxygen transfer depends upon many factors, such as the geometry of the diffuser and tank. Numerous studies have been conducted on the efficiency of water hyacinths in nutrient removal or on the efficiency of different submerged aerators. However, to date, few studies have been conducted on the integrated use of submerged aeration and water hyacinth for high AN conversion efficiency and rate assessment. Along with this, this study also gives a comparative picture of AN conversion for three different conditions applied. Hadiuzzaman et al. [8] examined water pollution in areas adjacent to Trimohoni, such as the Balu River and Shitalalakhya River. However, no comprehensive studies have been conducted on the deteriorating water body of Trimohoni, nor have any measures been implemented to address the issue. The initiative to use water hyacinth and aeration in an integrated way for the Trimohoni water body is observed to maintain the water quality parameters in an acceptable range, proving this method to be highly recommended for the restoration of the Trimohoni area.

15.2 Methodology

15.2.1 Plant and Wastewater Sampling

Water hyacinth, selected for phytoremediation, was sampled from Trimohoni’s middle part using 15 × 18-in. polyethylene bags and visually inspected for young plants with fresh leaves and healthy roots. Algae in the root system were removed using water flow and lab forceps after careful cleaning. The plants were then adapted in a lab container with fresh water and sunlight for a week before being introduced into the experimental setup with 3–4 leaves and weighing 120 gm each. Wastewater was collected from Trimohoni Middle, after being assessed as highly polluted based on water quality parameters DO, BOD, COD, EC, turbidity, NH_3-N, TDS, and H_2S, compared to other water collection zones (Rampura Canal, Norai River, Jirani Canal, Trimohoni Middle). This water receives effluent from various industries like paints, glass, textiles, and the Dasherkandi Sewage Treatment Plant (DSTP). A 300 L capacity PVC container tank transported the 180 L wastewater sample to the lab.

15.2.2 Experimental Design

The experiment conducted was a case-controlled experiment with three treatments. The treatments consisted of (i) Natural Aeration (T1), (ii) Submerged Aeration (T2), and (iii) Submerged Aeration with water hyacinth (T3). For each experimental observation (T1, T2, and T3), three separate identical plastic containers with an 80-L capacity, 3168 cm^2 surface area, and 37.5 cm depth each were filled with wastewater. Each tub contained 50 L of wastewater, placed under natural sunlight with a water temperature of between 25 and 30 °C during a 7-day observation period to ensure the proper sunlight necessary for hyacinths and simulate the natural environmental conditions. Submerged aeration in the T2 and T3 setups used three 5W aerators per setup, each connected to two 15 cm diameter, 90 gm air stone bubble diffusers producing coarse bubbles. Each tub had six diffusers, totalling 12 diffusers and six aerators, operating 24/7 for five days. The physical water qualities such as temperature, turbidity, EC, TSS, TDS, and chemical water quality parameters such as pH, DO, BOD, COD, NO_3–N and NH_3–N were experimented using standard methods for each tub at 0, 3, 6, 24, 48, 72, and 120 h until the water quality parameters improved to a significant value (Table 15.2).

15.2.3 Statistical Analysis

All statistical analysis were performed using OriginPro 2024 for Windows. Data are presented as mean ± 1 SD from repeated experiments. Polynomial regression fit lines were plotted to show the relationship between mean concentrations and time. Removal efficiency for each treatment unit was calculated using the formula: Removal Efficiency = $(1 - C_f/C_i) \times 100\%$, where C_f and C_i are the final and initial concentrations, respectively.

Table 15.2 The initial and final concentrations of water quality parameters in three treatment tubs as mean ± 1 SD

Parameters	Permissible limit	Initial	Final concentration			Method used
			T1	T2	T3	
BOD (mg/L)	≤ 4.5	37 ± 1.96	2.83 ± 0.74	10.9 ± 1.43	1.17 ± 0.12	SM 5210 D
COD (mg/L)	10	82 ± 2.82	11.97 ± 1.38	13.94 ± 1.5	6.23 ± 1.07	SM 5220 D
TDS (mg/L)	≤ 300	315 ± 2.74	270 ± 1.87	275 ± 1.72	270 ± 1.86	SM 2540 B-D
EC (μS)	1500	445 ± 2.1	432.7 ± 1.98	430.6 ± 0.63	404 ± 1.29	SM 2510 B
NH_3-N (mg/L)	0.20	1.5 ± 0.14	1.00 ± 0.07	0.87 ± 0.10	0.085 ± 0.10	SM 4500-NH_3B
NO_3-N (mg/L)	50	0.27 ± 0.06	0.55 ± 0.08	2.25 ± 0.11	0.75 ± 0.08	SM 4500-NO_3B
Turbidity (NTU)	≤ 5	22 ± 1.89	5.08 ± 1.17	9.44 ± 1.20	1.85 ± 0.93	SM 2130B

T1 natural aeration, *T2* submerged aeration, *T3* submerged aeration with WH

15.3 Results and Discussions

15.3.1 *Water Quality Parameters*

The wastewater under observation in the tubs was tested for the water quality parameters after 0, 3, 6, 24, 48, 72, and 120 h, and a significant amount of change in the value of parameters was observed. The initial BOD was as high as 37 ± 1.96 mg/L, which did not meet the permissible limit of 4.5 mg/L [16]. T3 performed best in reducing the BOD by increasing DO. For each of the parameters BOD, COD, EC, and ammonia nitrogen; T3 containing water hyacinth performed better reaching a final concentration of 1.17 ± 0.12 mg/L, 6.23 ± 1.07 mg/L, 404 ± 1.29 μS, and 0.085 ± 0.1 mg/L, respectively, which meets the limit for drinking purposes (Table 15.1).

The initial concentration of nitrate was significantly lower than the acceptable limit of 50 mg/L. A study found that water hyacinth can directly consume nitrate from wastewater, converting it primarily to N_2 [15]. Thus, nitrate within permissible limits will not impact water quality for various purposes. Water hyacinth's extensive root system and surface area effectively remove water contaminants. It is also observed by a study that the negatively charged roots adsorb cations and accumulate solids, reducing water turbidity [14]. The T2 tub's turbidity wasn't significantly reduced due to continuous mixing from bubble flow, whereas the T1 tub with natural aeration allowed smaller particles to settle. T2 had higher TDS levels. However, Water hyacinth's sorption, sedimentation, and chemical transformation were key in reducing TDS to 270 < 300 NTU and turbidity to 1.85 ± 0.93 < 5 NTU.

15.3.2 *Ammonia Nitrogen Conversion*

Among the many studies conducted for ammonia–nitrogen removal, the AN removal in T3 for this study was as high as 93.9% within only 2 days. Only 45.6 ± 1.3% TKN removal and 23.3 ± 1.4% nitrate removal were observed in one study by Kumari and Tripathi [9]. In the T3 setup, aeration increased DO, boosting the nitrification rate. High ammonia–nitrogen levels are toxic to aquatic life as they can cross epithelial membranes. The nitrification process requires microorganisms (mainly Nitrosomonas and Nitrobacter bacteria) to convert the AN to nitrate. T1 and T2 setups lack the conditions for high-rate AN to nitrate conversion. T2 provides oxygen through aeration, but T3 performs best, as water hyacinth roots host beneficial microorganisms and ample oxygen supports their activity. Plants also directly assimilate ammonium. In phytoremediation, AN is converted in several steps. Water hyacinth and root microorganisms, aided by aerator-supplied oxygen, convert AN to nitrate. As stated by Mayo et al. [10], the plant consumes much nitrate, while microorganisms convert some to organic N, which cycles back to AN. Excess nitrate undergoes denitrification. So, the primary focus is converting AN to nitrate, as nitrate levels are already well below toxicity limits and gradually transform into stable compounds over time.

In this study, initial ammonia nitrogen concentration exceeded the acceptable limit at 1.5 mg/L, rising to 3.2 mg/L within hours across all setups, likely due to organic N mineralization [10]. In T3, AN rapidly converted to nitrate, reaching near 0.2 mg/L at 1.8 mg/L/day within 2 days, maintaining this level for 3 days (Fig. 15.3). Nitrate gradually increased from 0.27 to 2.75 mg/L over 2 days, indicating effective nitrification. However, subsequent nitrate decline suggests denitrification and microorganism uptake. The 2nd order polynomial lines plotted for AN and nitrate concentrations having R^2 values 0.65 and 0.81 represent 'moderate' and 'good' fits accordingly, indicating the acceptability of the AN conversion observation method and data collection. More repetition of the method followed is recommended for getting more statistically significant outcomes. In T1 and T2, gradual AN reduction occurred the 5 days but failed to reach permissible limits. A longer observation period can solve this problem. T3 successfully met the study's goal for rapid water quality recovery, especially AN conversion to nitrate. There is also scope for this study to accelerate the AN conversion using fine bubble aerators, though cost efficiency will still be a concern. Nitrate concentration slightly increased to 0.75 mg/L, suggesting incomplete nitrification, possibly due to lacking plant microorganisms and nitrification bacteria. These limitations can be addressed in further studies. AN reduction implies microorganism uptake and conversion to organic N, with conversion rates in T1 and T2 at 0.46 mg/L/day and 0.5 mg/L/day, respectively (Figs. 15.1 and 15.2).

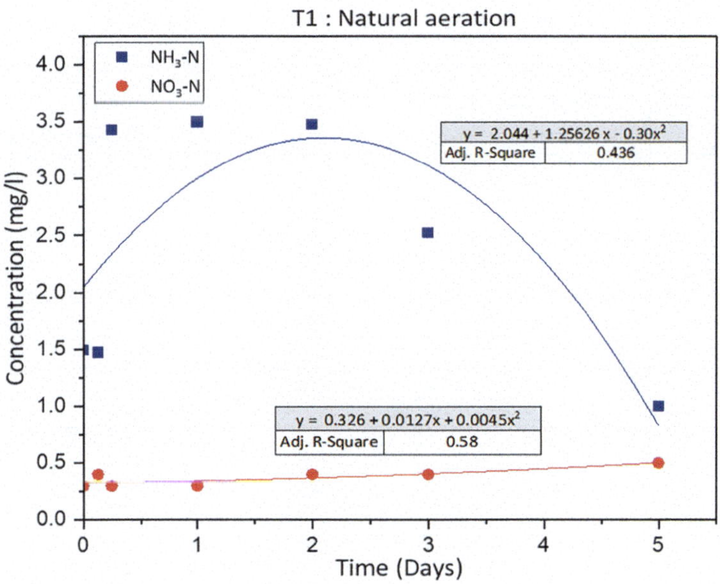

Fig. 15.1 Conversion of ammonia nitrogen in T1: natural aeration

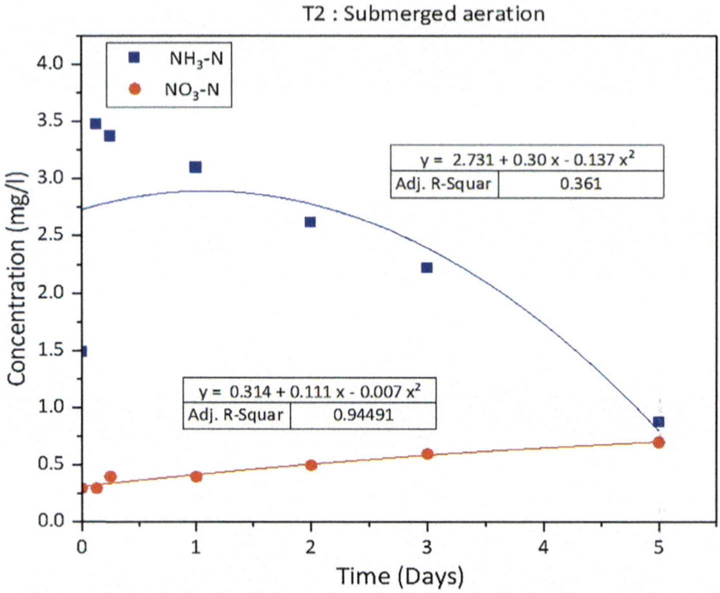

Fig. 15.2 Conversion of ammonia nitrogen in T2: submerged aeration

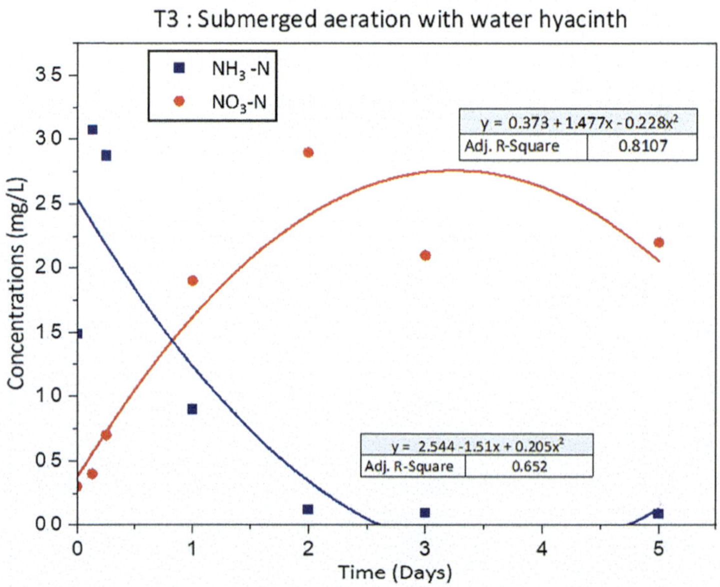

Fig. 15.3 Conversion of ammonia nitrogen in T3: submerged aeration with water hyacinth

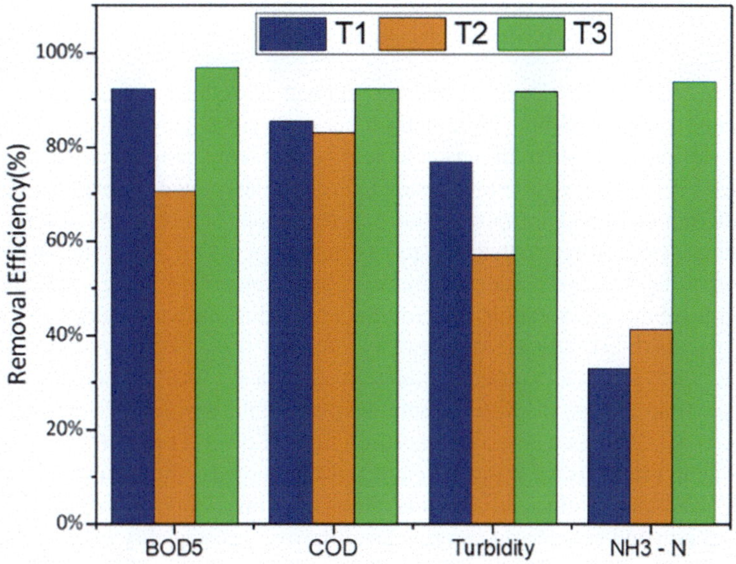

Fig. 15.4 Removal Efficiency (%) in T1, T2, and T3 setup

15.3.3 Removal Efficiency

T3 consistently outperformed T1 and T2 in removing BOD, COD, turbidity, and NH_3-N. BOD removal rates were 92%, 70.52%, and 96.84% for T1, T2, and T3, respectively (Fig. 15.4). Natural aeration surpassed submerged aeration in increasing DO due to the mixing effect in T2. This increase in DO was further supported by the photosynthetic activity of water hyacinth, which, as noted by Reddy [13], reduces dissolved CO_2 and promotes aerobic bacterial activity, ultimately leading to a greater reduction in BOD and COD.

Qin et al. [12] demonstrated that water hyacinth has a root surface area of 0.97–1.10 m^2/g fresh weight, an active absorption area, an active leaf area, higher root activity, and a higher net photosynthetic rate. These factors contribute to its high efficiency in removing turbidity (93.9%) and ammonia nitrogen (NH_3–N) (91.6%) in T3. Thus, WH in T3 proves highly efficient in rapidly improving water quality, including BOD and COD within 5 days, and ammonia nitrogen (NH_3–N), TDS, and turbidity within just 2 days. Water hyacinth plants should be monitored and removed when they begin to degenerate. Rapidly growing bushy leaves can block sunlight, and degraded plants can complicate the process. Cultivating WH within defined boundaries is recommended for real-world applications. Thus, this study underscores the significant potential of integrating water hyacinth and submerged aerators to swiftly enhance water quality and achieve rapid ammonia conversion within 2 days.

15.4 Conclusion

This study successfully demonstrated the removal of organic and inorganic pollutants from Trimohoni. While the primary focus was on AN conversion and removal efficiency using an eco-friendly, sustainable, low-cost remediation process, improvements in other physical and chemical parameters were also noted. Managing pollution from different point and non-point sources across water branches poses challenges. Initial nutrient levels highlighted the cause of high algal bloom, necessitating nutrient reduction in the water body. The purposes were accomplished by using water hyacinth along with submerged aeration. The following conclusions can be drawn from the study conducted: (i) only aeration cannot improve the AN conversion, whereas adding water hyacinth to a particular amount can convert 93.9% ammonia nitrogen to nitrate within 2 days; (ii) water hyacinth enhanced the removal of organic (BOD: 96.84%, COD: 92.4%), and inorganic (turbidity: 91.6%, TDS: 15.6%) pollutants; (iii) the incorporated use of water hyacinth and submerged aeration outperformed by showing an ammonia–nitrogen conversion rate of 1.8 mg/L/day; (iv) incorporation of water hyacinth and submerged aeration can improve water quality to an acceptable limit for water pollution load similar to Trimohoni within just 5 days. This study addresses Trimohoni's low-to-medium pollutant concentration. Similar setups can treat wastewater with comparable pollution levels. As population grows, increased ammonia–nitrogen production may lead to further study on conversion efficiency for complex nitrogen pollutants. Exploring various aerators and additional phytoremediation plants can offer cost-effective, sustainable solutions for water quality issues in Bangladesh and globally.

Acknowledgements The authors thank BUET for the research grant and the faculty and lab assistants for their support.

References

1. Al-Ahmady KK (2006) Analysis of oxygen transfer performance on sub-surface aeration systems. Int J Environ Res Publ Health 3(3), Article 3
2. Anandha Varun R, Kalpana S (2015) Performance analysis of nutrient removal in pond water using water Hyacinth and Azolla with Papaya Stem. Int Res J Eng Technol 2(1):444–448
3. Anderson DM, Burkholder JM, Cochlan WP, Glibert PM, Gobler CJ, Heil CA, Kudela RM, Parsons ML, Rensel JEJ, Townsend DW, Trainer VL, Vargo GA (2008) Harmful Algal blooms and eutrophication: examining linkages from selected coastal regions of the United States. Harmful Algae 8(1):39–53
4. Bashar T, Fung IWH (2020) Water pollution in a densely populated Megapolis, Dhaka. Water 12(8), Article 8
5. Culp RL, Wesner GM, Culp GL (1978) Handbook of advanced wastewater treatment, 2nd edn. Van Nostrand Reinhold Co. Ltd
6. El-Gendy AS, Biswas N, Bewtra JK (2004) Growth of water hyacinth in municipal landfill leachate with different pH. Environ Technol 25(7):833–840

7. Fox LJ, Struik PC, Appleton BL, Rule JH (2008) Nitrogen phytoremediation by water hyacinth (*Eichhornia crassipes* (Mart.) Solms). Water Air Soil Pollut 194(1):199–207
8. Hadiuzzaman M, Basar Baki A, Mostafa Khan S (2006) Pollution status and trends in water quality of the Shitalakhya and Balu Rivers. J NOAMI 23:1–22
9. Kumari M, Tripathi BD (2014) Effect of aeration and mixed culture of *Eichhornia crassipes* and *Salvinia natans* on removal of wastewater pollutants. Ecol Eng 62:48–53
10. Mayo AW, Hanai EE, Kibazohi O (2014) Nitrification-denitrification in a coupled high rate—water hyacinth ponds. Phys Chem Earth Parts A/B/C 72–75:88–95
11. Moorhead KK, Reddy KR, Graetz DA (1988) Nitrogen transformations in a Waterhyacinth-based water treatment system. J Environ Qual 17(1):71–76
12. Qin H, Zhang Z, Liu M, Liu H, Wang Y, Wen X, Zhang Y, Yan S (2016) Site test of phytoremediation of an open pond contaminated with domestic sewage using water Hyacinth and water lettuce. Ecol Eng 95:753–762
13. Reddy KR (1981) Diel variations of certain physico-chemical parameters of water in selected aquatic systems. Hydrobiologia 85(3):201–207
14. Rezania S, Taib SM, Md Din MF, Dahalan FA, Kamyab H (2016) Comprehensive review on phytotechnology: heavy metals removal by diverse aquatic plants species from wastewater. J Hazard Mater 318:587–599
15. Ting WHT, Tan IAW, Salleh SF, Wahab NA (2018) Application of water Hyacinth (*Eichhornia crassipes*) for phytoremediation of ammoniacal nitrogen: a review. J Water Process Eng 22:239–249
16. WHO (2002) Guidelines for drinking-water quality. World Health Organization
17. Xu J, Li X, Gao T (2022) The multifaceted function of water hyacinth in maintaining environmental sustainability and the underlying mechanisms: a mini review. Int J Environ Res Public Health 19(24):16725

Chapter 16
A Sustainable Waste Management Model for Passenger Ships in Bangladesh: A Step Towards Climate Change Adaptation

Md. Mahmudul Hasan Akib, Zobair Ibn Awal, and Mohammad Tanvir Hossain

Abstract Bangladesh's river network of almost 230 rivers, serves as crucial transportation routes for millions of passengers annually. Despite ratifying international conventions on marine pollution, Bangladesh lacks a National Marine-Environmental Protection Policy. Therefore, dumping untreated waste into water bodies has become habitual due to the absence of monitoring of effluent-releasing standards. This research aligns with the urgent need for a sustainable waste management model in the maritime sector, considering the inadequate waste management practices and the country's vulnerability to climate change. The proposed model integrates a biogas-composting plant within a passenger ship and a plastic segregation chamber. This innovative solution addresses organic waste from ships, generating biogas for cooking, a sustainable alternative to LPG. The plastic segregation chamber facilitates the proper storage and sale of plastic waste for recycling, fostering a circular economy and reducing the demand for new plastic production. Aligned with multiple Sustainable Development Goals (SDGs), the research promotes environmental protection, energy independence through biogas, and a more resilient infrastructure. It aims to reduce greenhouse gas (GHG) emissions and improve air and water quality by solving the longstanding waste management challenges in the maritime sector of Bangladesh.

Md. M. H. Akib (✉)
Department of Naval Architecture and Marine Engineering, Sonargaon University, Dhaka, Bangladesh
e-mail: akib.name06@gmail.com

Z. I. Awal
Department of Naval Architecture and Marine Engineering, Bangladesh University of Engineering and Technology, Dhaka, Bangladesh
e-mail: zobair@name.buet.ac.bd

M. T. Hossain
Reactron Bangladesh Private Limited, Motijheel, Dhaka, Bangladesh

G. M. Tarekul Islam et al. (eds.), *Sustainable Water Management and Environmental Resilience*, Springer Water, https://doi.org/10.1007/978-3-031-93350-9_16

Keywords Sustainable waste management · Marine pollution · Water management · GHG reduction · Biogas

16.1 Introduction

According to the recent report [1], climate change manifests in rising global temperatures, sea level rise, and increased extreme climatic conditions such as drought, floods, and rain patterns. The Intergovernmental Panel on Climate Change (IPCC) identifies Bangladesh as highly vulnerable to climate change, with tropical cyclones costing about $1 billion annually [2]. Climate variability will likely reduce agricultural GDP by one-third by 2050 and displace 13.3 million people. Bangladesh's river network, vital for transportation, faces rising marine pollution, threatening human health and biodiversity. Annually, ferries and coastal cargo services transport 225 million passengers and 25 million vehicles through these waterways, contributing to marine pollution [3]. The study found that [4] in Dhaka, only 20% of the city has a sewerage network, with 1250 million liters of sewage entering rivers daily. Rivers carry millions of tons of plastic waste into the Bay of Bengal [5], with the Padma, Meghna, and Jamuna rivers carrying 73,000 tons daily [6]. According to the report [7], the World Bank highlights the need for 29 projects to combat river pollution, and the government has committed $20 billion over seven years to the "Umbrella Investment Plan—Dhaka Rivers".

Figure 16.1 illustrates waste management practices and types of waste generated in passenger ships, highlighting the deficiency in monitoring and enforcement of effluent-releasing standards, leading to untreated waste being dumped directly into water bodies, significantly harming the environment and public health. Despite ratifying international conventions on marine pollution, Bangladesh lacks a National Marine-Environmental Protection Policy [8].

Organic waste, including kitchen and fecal matter, depletes dissolved oxygen in water bodies, potentially causing fish deaths and ecosystem disruption. One kilogram of untreated sewage is estimated to contaminate 10 kg of groundwater and 5 kg of

Fig. 16.1 Passenger density and waste management practices, including waste types generated on passenger ships (location: Sadarghat)

surface water [9]. Besides, according to the [10], organic waste produces biogas, with methane being 28 times more potent than carbon dioxide, contributing significantly to climate change. Passenger ships also generate inorganic waste like plastic and aluminium. As stated in a report [11], the "Plastic Pandemic" in Bangladesh is concerning as plastic waste takes hundreds of years to degrade, harming aquatic life. This research fills a critical gap in Bangladesh's maritime waste management, paving the way for a greener industry, cleaner waterways and providing a strategic pathway towards resilient development and sustainable water management.

16.2 Methodology

A blend of methods was adapted for this study due to the lack of local research in Bangladesh. The unique environmental, economic, and cultural context necessitated a tailored approach to accurately assess waste generation on Bangladeshi passenger ships. This research focuses on implementing a biogas composting plant with a plastic segregation chamber, featuring an anaerobic digester and a water treatment facility for safe river discharge (Fig. 16.2).

16.2.1 Waste Collection and Digestion Tank

The ship will have two storage and digestion tanks positioned transversely below the main deck along the centreline, alternating operation every six months. These insulated tanks will hold sludge, capturing solids with a geotextile fabric lining while allowing water to pass. Anaerobic digestion of sewage and kitchen waste will produce biogas, with an agitator to maximize production and prevent foam formation. The tanks will feature a pressure indicator and relief valve [12]. The tanks will feature a pressure indicator and relief valve.

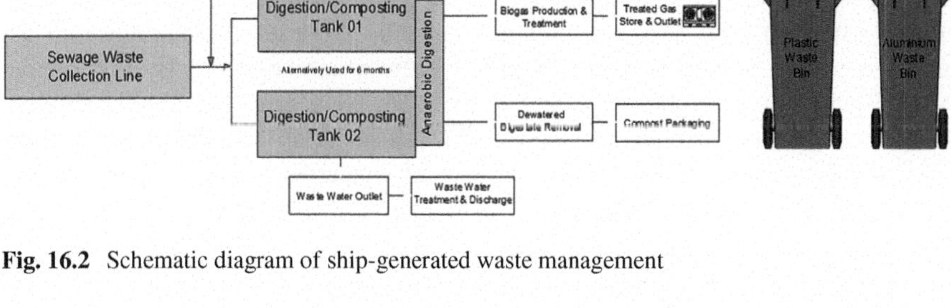

Fig. 16.2 Schematic diagram of ship-generated waste management

16.2.2 Biogas Production, Treatment, and Storage

Biogas will be released from the digestion tank in batches, and treated via water scrubbing to remove impurities, followed by Iron (III) Oxide treatment to eliminate hydrogen sulphide [13]. Iron (III) oxide, known for its simplicity and cost-effectiveness, removes hydrogen sulphide with an efficiency of up to 99.98%, reducing levels from thousands of ppm to as low as one ppm. However, the treated gas will be stored in a storage bag for end users. This process involves gas entering from the bottom while water enters from the top of a tower filled with packing material. The treated gas exits from the top, and the water, discharged from the bottom, is reused after spraying to separate gaseous elements [12].

16.2.3 Wastewater Treatment and Discharge to the River

Wastewater treatment involves storing water from the digestion tank in a reserve tank below the main deck and pumping it to a treatment plant on the main deck. It will pass through a trickling filter with a grated diffuser plate and a Slow Sand Filter (SSF) [14]. This filter removes 40–65% of BOD_5 [15]. Next, the water will pass through a ventilated Slow Sand Filter (SSF) with a Schmutzdecke layer, reducing *E. Coli* by 3.23 logarithmic units, total coliforms by 2.98 logarithmic units, turbidity by 60–95%, and apparent colour by 50–90% in 30–40 days [16]. An SSF consists of layers of fine sand, coarse sand, fine gravel, coarse gravel, and large pebbles. Multiple SSFs in series can remove 79–92% of BOD_5, 50–67% of COD [17], and 45–67.5% of nitrogen [18]. They can also remove over 90% of heavy metals. Biochar can enhance SSF efficiency [19]. SSFs will have a backwashing mechanism to prevent clogging. Chlorination may be added for disinfection, ensuring the water meets freshwater standards.

16.2.4 Digestate Management Through Composting

Each digestion tank will function as a dry composting pit for six months, producing biogas. Afterward, the tank will undergo anaerobic digestion, converting into an aerobic digestion chamber with thermophilic aerobic microbes decomposing organic matter. Temperatures inside the tank may rise to 50–70 °C due to biological activity, gradually decreasing to ambient temperature. A curing phase of at least four months will follow, producing homogeneous, biologically inactive compost, ready for use or sale [12]. The details of the proposed model are illustrated in Fig. 16.3.

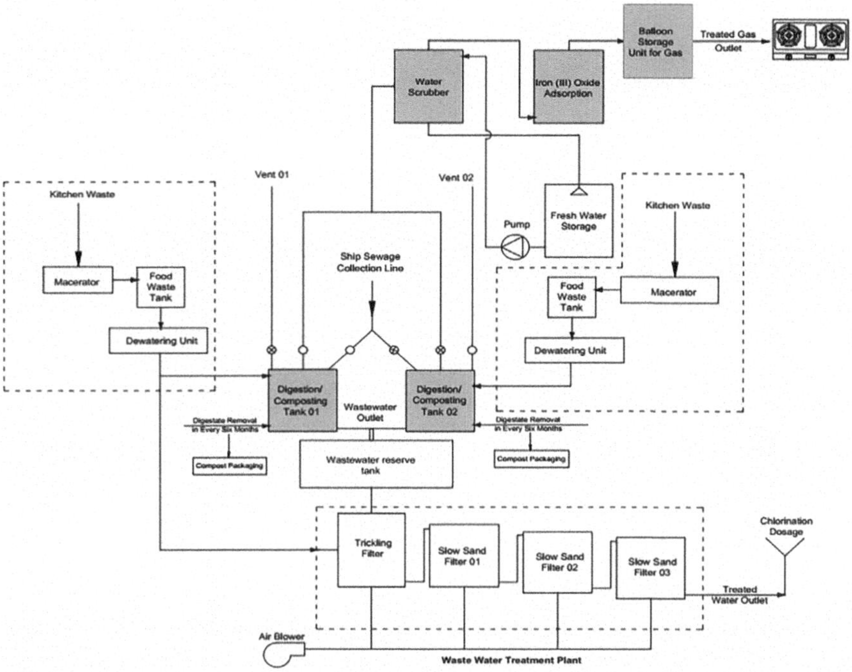

Fig. 16.3 Detailed diagram of the proposed model

16.3 Development of Mathematical Model: Results and Analysis

The model depends on parameters such as Passenger Number (P), Average Fecal Waste per capita (AF_eW_{PC}), Percentage of fecal waste discharging people (P_{F_e}), Food Waste Generated per capita (AF_dW_{PC}), Percentage of food waste-producing people (P_{Fd}), and others. These parameters predict CO_{2eq} (equivalent) emission reduction and the corresponding biogas production. Therefore, the mathematical expression of the model can be formulated as a function of the following variables (unit = kg):

Amount of CO_{2eq} reduction (ton/year) from Organic Waste,

$$CO_{2\,Red-OW} = f\left(P, AF_eW_{PC}, P_{F_e}, AF_dW_{PC}, P_{F_d}\right)$$

$$\text{Biogas Production}\left(\text{m}^3\right),\quad BP = f\left(P_{CH_4}, P_{CO_2}, GWP_{CH_4,CO_2}, BPC\right)$$

where, P_{CH_4}, P_{CO_2} = Component Ratio of Methene and Carbon dioxide, GWP = Global Warming Potential, BPC = Biogas Production Capacity (m^3/kg).

Likewise, in the context of plastic recycling, the process is contingent upon specific input parameters and factors, including Recycled Plastic's GHG emissions reduction capacity ($P_{GHG\text{-}Red}$), CO_2 emission from Producing and Burning 1 kg of New Plastic

(NP_{CO2}), Percentage of Plastic Waste Producing People (P_{PW}), Average weight of Empty Plastic Bottle (AWP). The mathematical model can thus be articulated as a function of the following variables (unit = kg):

Amount of CO_{2eq} reduction from Plastic Waste Recycling (ton/year),

$$CO_{2\,Red-PR} = f\left(P_{GHG-Red}, NP_{CO_2}, P_{PW}, AWP\right)$$

Essential data, such as passenger numbers and voyage durations, are collected from the Bangladesh Inland Water Transport Authority (BIWTA). Dhaka is connected through 47 river routes and has the highest traffic and passenger density, contributing significantly to river pollution due to the existing conventional waste management practices. The Dhaka–Barisal route is chosen as a standard due to its highest daily passenger count and voyage frequency.

16.3.1 Estimation of the Potential Reduction of CO_{2eq} Emissions from Organic Waste

Firstly, the passenger ships that run between Dhaka and Barisal are given in Table 16.1. The table shows the names of the ships and corresponding passenger numbers for Day (D) and Night (N) travel.

Assuming a ship travels daily from Dhaka to Barisal with an average of 800 passengers. The relevant factors and parameters for this calculation are detailed in Table 16.2.

Using the data and factors from Table 16.2, the proposed model calculates the potential reduction in carbon emissions. The results show an annual reduction of

Table 16.1 List of passenger ships on the Dhaka–Barisal route

No.	Ship's name	Passenger number (D/N)		No.	Ship's name	Passenger number (D/N)	
1	MV Surbhi-7	742	999	11	MV Parabat-12	944	1350
2	MV Surbhi-8	890		12	MV Subhraj-9	481	807
3	MV Surbhi-9	1274		13	MV Parabat-10	925	
4	MV Sundarban-10	921	1309	14	MV Kirtankhola-2	1120	1410
5	MV Sundarban-11	727	1091	15	MV Parabat-11	751	1025
6	MV Parabat-18	774	1255	16	MV P. Awlad-10	782	1019
7	MV Parabat-9	534	910	17	MV Ketankhola-10	1020	1550
8	MV Adventure-1	736	863	18	MV Kuakata-2	705	921
9	MV Adventure-9	972	1350	19	MV Sundarban-16	890	1106
10	MV Manami	914					

Source BIWTA

Table 16.2 Important factors and parameters to estimate CO_{2eq} emissions from organic waste

Avg. number of passenger travel per ship (day)	800	
Voyage duration (h)	08	
Avg. fecal waste per capita (kg)	0.20 [20]	
Percentage of faecal waste discharging people (factor)	25%	
Food waste generated per capita (kg)	0.22 [21]	
Percentage of food waste-producing people (factor)	75%	
Proportions of major components	CH_4 = (60%)	CO_2 = (35%)
Density (kg/m^3)	0.656	1.98
Global warming potential (GWP)	28 [10]	1
Biogas production capacity (m^3/kg)	0.05	
1 m^3 biogas equivalent to	0.46 kg LPG [22]	

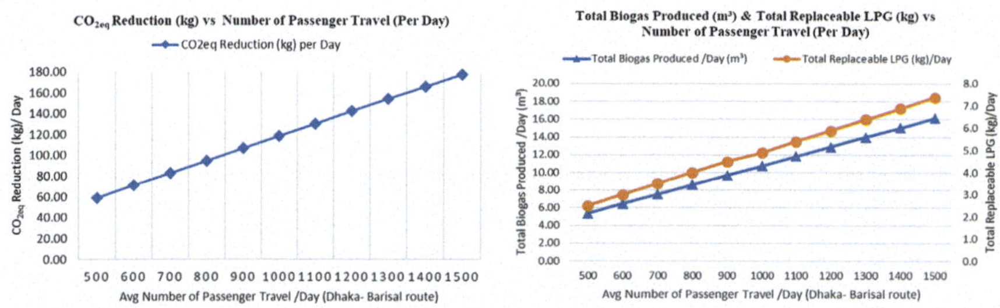

Fig. 16.4 CO_{2eq} reduction (kg), total biogas produced (m^3) and total replaceable LPG (kg) versus number of passenger travels (per day)

34.59 tons of CO_{2eq} emissions per ship, equating to a daily reduction of approximately 94.78 kg of CO_{2eq}. Figure 16.4 illustrates a daily production of 8.6 m^3 of biogas, which can replace 4 kg of LPG for onboard cooking, saving energy costs.

Table 16.3 details the impact of varying the percentages of individuals contributing to fecal waste (25%) and food waste (75%) on daily and annual CO_{2eq} reduction.

The findings demonstrate the proposed waste management model's potential for CO_{2eq} emissions reduction, pollution mitigation, and biogas production, benefiting Bangladesh's marine sector. While specific to a particular route, Fig. 16.5 shows the model's applicability to other routes throughout Dhaka Port and beyond, detailing the total number of voyages and corresponding daily carbon emission reductions (kg).

Table 16.3 Sensitivity analysis by varying the percentage of fecal waste discharging people and percentage of food waste producing people

Considering 800 passengers onboard per day (Dhaka-Barisal route)					
Fecal waste			Food waste (kitchen waste and food residue)		
Percentage of population discharging fecal waste (%)	CO_{2eq} reduction (kg)/day	CO_{2eq} reduction/ year (ton)	Percentage of population producing food waste (%)	CO_{2eq} reduction (kg)/day	CO_{2eq} reduction/ year (ton)
20	17.63	6.44	65	63.04	23.01
25	**22.04**	**8.05**	70	67.89	24.78
30	26.45	9.65	**75**	**72.74**	**26.55**

Bold values represent the reference scenario selected for calculating the final carbon emission reduction outcomes, based on 25% of passengers discharging fecal waste and 75% producing food waste

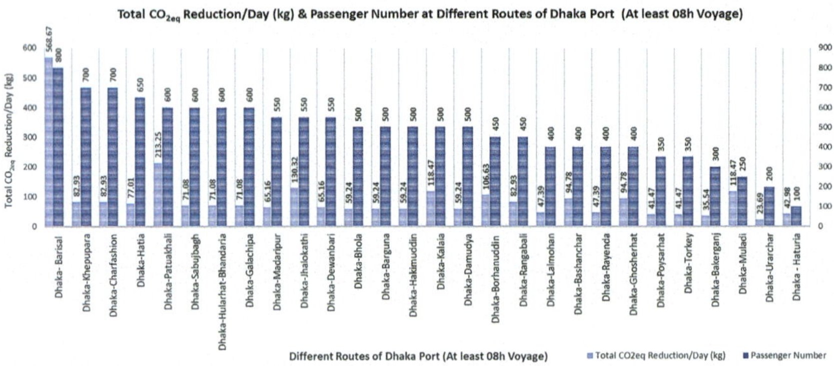

Fig. 16.5 Total CO_{2eq} reduction/day (kg) and passenger number at different routes of Dhaka Port (at least 08-h Voyage)

16.3.2 Estimation of the Potential Reduction of CO_{2eq} Emissions from Inorganic Waste

In Bangladesh, passenger ships generate plastic waste from PET and HDPE bottles. Producing a 1-L PET bottle requires 2 L of water, 4 million joules of energy, and emits 3 tons of CO_2. According to a statistical report [23], producing and burning 1 kg of PET plastic releases 6 kg of CO_2. And recycling PET, however, reduces greenhouse gas emissions by 30–70% [24]. Recycling aluminium uses only 5% of the energy needed to produce new aluminium, significantly cutting emissions and costs [25]. The average plastic bottle weighs 30 g (Fig. 16.6).

Based on these references, the Estimation of the Potential Reduction of CO_{2eq} Emissions from Plastic Recycling is given in Table 16.4. Hence, the proposed

Fig. 16.6 Weight of different plastic bottles. *Source* Author Md. Mahmudul Hasan Akib

Table 16.4 Important factors and estimation of CO_{2eq} emissions from plastic recycling

Recycled PET can reduce GHG emissions	30%
CO_2 emission from producing and burning 1 kg of new PET	6 kg
Avg. number of passenger travel per ship/day	800
Percentage of plastic waste producing people (factor)	30%
Avg. weight of empty plastic bottle	0.03 kg
CO_2 emission from producing new PET	43 kg
Potential reduction of CO_2 emissions from plastic recycling	13 kg
Potential reduction of CO_2 emissions per year	**1.8 tons**

Bold value highlights the final estimated potential reduction of CO_2eq emissions per year through plastic recycling, derived from the preceding input parameters

waste management model offers significant annual CO_{2eq} emission reductions per ship. Implementing this model across all navigational routes in Bangladesh could potentially reduce thousands of tons of greenhouse gas emissions annually.

16.4 Discussion

This research proposes a waste management model for passenger ships on Bangladesh's rivers. The model uses anaerobic digestion to reduce carbon emissions, promote biogas for cooking, and lower costs compared to LPG, which has safety advantages. It can be applied across Bangladesh's waterways, supporting environmental protection, climate adaptation, and generating revenue from compost and recycled plastic, aligning with UN sustainability goals for 2030. However, there are some limitations to this research like the assumption of optimal operating conditions, which may not reflect real-world variations that could impact the efficiency of the model. Additionally, significant initial investment, regular monitoring, and variable waste composition could affect the results.

16.5 Conclusion

This research proposes a comprehensive waste management model for passenger ships navigating Bangladesh's river network. This model leverages anaerobic digestion and recycling to address both organic and inorganic waste streams. It offers significant potential for CO_{2eq} emission reduction, with estimates indicating an annual decrease of 34.59 tons per ship solely in a specific route from organic waste processing. Furthermore, plastic recycling provides additional reductions. This dual approach not only mitigates pollution but also generates valuable resources: biogas for fuel and compost for soil enhancement. While initial investment costs and variations in waste composition present implementation challenges, widespread adoption across Bangladesh's waterways could lead to substantial greenhouse gas reductions, significantly contributing to the nation's sustainability objectives.

Acknowledgements The authors thank the Marine Safety and Traffic Management Department of BIWTA for their invaluable support in data collection. We extend our gratitude to the anonymous reviewers of the 9th ICWFM 2023, organized by IWFM, BUET. Special thanks to Professor Dr. Goutom Kumar Saha of the NAME Department, BUET, for his pivotal guidance in the data collection process.

References

1. Ministry of Environment, Forest and Climate Change (2022) National adaptation plan of Bangladesh (2023–2050). Government of the People's Republic of Bangladesh
2. The World Bank Group (2022) Country and climate development report for Bangladesh. Retrieved from https://www.worldbank.org/en/news/feature/2022/10/31/key-highlights-country-climate-and-development-report-for-bangladesh
3. Biswas JC, Haque MM, Maniruzzaman M, Kalra N (2021) Coastal and marine pollution in Bangladesh: pathways, hotspots and adaptation strategies. Eur J Environ Earth Sci 2(4):26–34
4. Brouwer R, Sharmin DF, Elliott S, Liu J, Khan MR (2023) Costs and benefits of improving water and sanitation in slums and non-slum neighborhoods in Dhaka, a fast-growing mega-city. Ecol Econ 207:107763
5. The Business Standard (2021) Bangladeshi rivers receive 24 times more plastic waste from neighbours. Retrieved from https://www.tbsnews.net/environment/plastic-waste-rivers-neighbours-dump-24-times-higher-bangladesh-218230
6. Yoshijima N, Yi K, Montoya H (2021) Towards a multisectoral action plan for sustainable plastic management in Bangladesh. The World Bank
7. The Business Standard (2023) Can Dhaka (rivers) be saved, still? Retrieved from https://www.tbsnews.net/bangladesh/environment/can-dhaka-rivers-be-saved-still-686338
8. Alam MW, Bhuyan MS, Xiangmin X (2021) Protecting the environment from marine pollution in Bangladesh: a brief in legal aspects with response to national and international cooperation's. Int J Mar Sci 37(2):871–881
9. Tymczyna L, Chmielowiec-Korzeniowska A, Saba L (2000) Effect of a pig farm on the physical and chemical properties of a river and groundwater. Polish J Environ Stud 9(2):97–102
10. Intergovernmental Panel on Climate Change (IPCC), C.8 (2022) Climate change 2022: impacts, adaptation, and vulnerability. Cambridge University Press. https://doi.org/10.1017/9781009325844

11. Fatema K, Rahman T, Islam MJ, Sumon KA, Uddin MH, Hasan SJ, Rashid H et al (2023) Microplastics pollution in the river Karnaphuli: a preliminary study on a tidal confluence river in the southeast coast of Bangladesh. Environ Sci Pollut Res 30(13):38853–38868
12. Hossain MT, Zaman M, Dey D (2023) Sustainable solution for household organic and faecal waste management: a case study of Dhaka City. In: 8th international conference on integrated solid waste and faecal sludge management. In: Proceedings of the waste safe, Khulna, Bangladesh, p 143
13. Wilken D (2017) Biogas to biomethane: flexible energy supply from biomass. Fachverband Biogas eV
14. Ranjan P, Prem M (2018) Schmutzdecke—a filtration layer of slow sand filter. Int J Curr Microbiol Appl Sci 7(07):637–645
15. Liu DH (1997) Process integration. In: Environmental engineers' handbook, vol 142
16. Lubarsky H, Fava NDMN, Souza Freitas BL, Terin UC, Oliveira M, Lamon AW, Fernandez-Ibañez P et al (2022) Biological layer in household slow sand filters: characterization and evaluation of the impact on systems efficiency. Water 14(7):1078
17. Farooq S, Al-Yousef AK (1993) Slow sand filtration of secondary effluent. J Environ Eng 119(4):615–630
18. Nakhla G, Farooq S (2003) Simultaneous nitrification–denitrification in slow sand filters. J Hazard Mater 96(2–3):291–303
19. Kaetzl K, Lübken M, Nettmann E, Krimmler S, Wichern M (2020) Slow sand filtration of raw wastewater using biochar as an alternative filtration media. Sci Rep 10(1):12–29
20. Barbosa MR (2012) Chemical composition and formation of human feces–problems and solutions of large mergers demographics in developing countries. In: Tenth international symposium on recent advances in environmental health research, vol 39
21. Dagiliūtė R, Musteikytė A (2019) Food waste generation: restaurant data and consumer attitudes. Environ Res Eng Manag 75(2):7–14
22. Wahyuni S, Sutjahjo SH, Purwanto YA, Fuah AM, Kurniawan R (2018) Application of small digester biogas for energy supply in rural areas. IOP Conf Ser Earth Environ Sci 141(1):012035
23. APR AO (2019) Association of plastic recyclers (APR): recycled plastics reduce energy consumption, GHG emissions. Retrieved from Waste Dive: https://www.wastedive.com/news/apr-recycled-plastics-reduce-energy-consumption-ghg-emissions/547027/
24. Shen L, Nieuwlaar E, Worrell E, Patel MK (2011) Life cycle energy and GHG emissions of PET recycling: change-oriented effects. Int J Life Cycle Assess 16:522–536
25. Aluminium TA (nd) The aluminium association. Retrieved from Infinitely Recyclable: https://www.aluminum.org/Recycling

The manufacturer's authorised representative in the EU is Springer Nature Customer Service Centre GmbH, Europaplatz 3, 69115 Heidelberg, Germany. If you have any concerns regarding our products, please contact ProductSafety@springernature.com

Printed and bound by CPI Group (UK) Ltd, Croydon, CR0 4YY
05/10/2026
02228127-0006